This is a detailed account of the British and German Steel industries' performance during three decades which were marked by radical changes in technology, in sources of raw materials, and in product markets. Relying on governmental and corporate archives as well as on the contemporary trade literature, Professor Wengenroth has drawn a meticulous picture of how managements in the two countries met the strategic problems raised by these changes. The author does not however merely trace technological developments; rather, he uses them as a backdrop for a contribution to the long-running debate on Britain's relative industrial decline in the late nineteenth century. Was this the result of massive entrepreneurial failure, or was it merely the by-product of evolutionary changes that bestowed competitive advantage on latecomers such as the Germans? The author argues a detailed case for the latter scenario, and in doing so makes a major contribution to the debate on the 'Great Depression'.

Enterprise and Technology

Enterprise and Technology

The German and British Steel Industries, 1865–1895

Ulrich Wengenroth

Translated by Sarah Hanbury Tenison

CAMBRIDGE
UNIVERSITY PRESS

Published by the Press Syndicate of the University of Cambridge
The Pitt Building, Trumpington Street, Cambridge CB2 1RP
40 West 20th Street, New York, NY 10011–4211, USA
10 Stamford Road, Oakleigh, Victoria 3166, Australia

First published 1994

Printed in Great Britain at the University Press, Cambridge

A catalogue record for this book is available from the British Library

Library of Congress cataloguing in publication data

Wengenroth, Ulrich.
Enterprise and technology: the German and British steel industries, 1865–1895 / Ulrich Wengenroth: translated by Sarah Hanbury Tenison.
p. cm.
Translated from German.
Includes index.
ISBN 0 521 38425 7
1. Steel industry and trade – Germany – History. 2. Steel industry and trade – Great Britain – History. I. Title.
HD9523.5.W427 1992
338.4'7669142'0941–dc20 92–10572 CIP

ISBN 0 521 38425 7

WD

Contents

Figures

Tables

Preface

Comparative research into one branch of industry in two countries is not only peculiarly fascinating and, I hope, enlightening; it is also very expensive and I was only able to complete this project thanks to much generous support. My thanks are due in the first place to the German Historical Institute in London and the Deutscher Akademischer Austauschdienst; between autumn 1976 and spring 1978 they provided financial support which allowed me to engage in uninterrupted research in British archives and libraries. I am obliged too, to my former colleagues in the German Historical Institute for their help and ingenuity, which enabled me to employ this time usefully right from the start. The British Steel Corporation and its archivists, Mrs Hampson, Mr Charman, Mr Emmerson, Mr Hassall and Mr Newman were invariably obliging in providing excellent working conditions in the firm's regional archives. The same applies to their colleagues in the Public Record Office and the archive of the National Library of Wales.

In Germany too, it was primarily firms' archivists and their colleagues who gave me practical support in securing source material. I owe sincere thanks to Frau Dr Köhne-Lindenlaub in the Historisches Archiv of Fried, Krupp GmbH, Frau Kühlborn in Werksarchiv Bochum of Fried. Krupp Hüttenwerke AG, Herr Dr Baumann, the former head archivist, in the archive of August-Thyssen-Hütte, Herr Herzog in the Historisches Archiv of Gutehoffnungshütte AV and to Herr Dr Hatzfeld, the former head archivist in the archives of Mannesmann AG. I am equally grateful to the archivists of the Bundesarchiv in Koblenz and the Archives Nationales in Paris.

The results of this work are critically dependent on evaluating the contemporary technical literature of many countries. That this was possible was largely due to the excellent holdings and unparalleled hospitality of the Stiftung Eisenbibliothek of Georg Fischer AG, in Schaffhausen, whose former librarian, Frau Anne-Marie Kappeler, supported my research with great patience and helpfulness.

The fact that an academic monograph has finally emerged from all this financial and practical support is thanks above all to my academic teacher Professor Dr Akos Paulinyi, who not only gave me the opportunity of working with him after my time in London but also took great personal interest in my work, encouraging me to continue and complete it. It acquired its shape during my countless conversations with him. Professor Dr Dirk Ipsen gave friendly advice which helped me sharpen my economic questions and made them operational.

Following my primary research in Great Britain the book took shape during my employment at the Institut für Geschichte in the faculty for Gesellschafts- und Geschichtswissenschaften in the Technische Hochschule Darmstadt, where it was accepted as a thesis in the summer of 1982. That same year it was awarded the Rudolf-Kellermann prize for history of technology by the Verein Deutscher Ingenieure. The thesis has been abridged and re-worked for publication. In this I received valuable advice from Professor Dr Knut Borchard and Dr Peter Alter. All errors are mine of course.

Since the first German version of this book was published in 1986, further works about the German and English steel industries at the end of the nineteenth century have appeared. In spite of this, I decided to leave my text unchanged for translation, since its basic approach, that of researching the business history of this branch from the perspective of production technology, has not been attempted again in this intensive form. As before, I am convinced that this approach produces insights which cannot be achieved by other means. In business history we must penetrate as deeply into the technology of the enterprises as into their finances and marketing etc., if we want to understand the motives and conditions that determined entrepreneurial business.

Moreover, several changes have occurred with regard to the location of archival material. The locations and document numbers given in this book refer to conditions between 1976 and 1980, when I viewed the material.

Finally, I want to thank the Press Syndicate of the University of Cambridge for having made it possible to produce an English version of this book. My thanks too to Sarah Hanbury Tenison for supplying the translation, for which the Stiftverband für die Deutsche Wissenschaft has generously provided financial assistance.

Ulrich Wengenroth
Munich, February 1993

Abbreviations

BITA	British Iron Trade Association
BITA (year)	British Iron Trade Association, annual report for the year reported on
BT	Board of Trade
BSC	British Steel Corporation
CIA	Cleveland Iron Masters' Association
CRO	County Records Office
GHH	Gutehoffnungshütte
HA	Historisches Archiv
ICTR	*Iron and Coal Trades Review*
IRMA	International Railmakers' Association
JISI	*Journal of the Iron and Steel Institute*
PEE	Protokolle über die Vernehmung der Sachverständigen durch die Eisen-Enquête-Kommission (minutes of the examination of the experts by the Iron Inquiry Commission)
RC	Royal Commission
Reichsstatistik (year)	The production of the mines, saltworks and ironworks in the German Reich and in Luxemburg for the year reported on
RRC	Regional Records Centre
RSW	Rheinische Stahlwerke
SC	Select Committee
VDESI	Verein Deutscher Eisen- und Stahlindustrieller
VDI	Verein Deutscher Ingenieure
ZBHSW	*Zeitschrift für das Berg-, Hütten- und Salinenwesen in dem Preussischen Staate* (Journal of the mines, ironworks and saltworks in the Prussian state)

Introduction

'The age of steel' is what David Landes called the last thirty years of the nineteenth century in his standard work on the industrialisation of Western Europe.[1] He felt that this epoch was wholly dominated by the rise of modern mass production of steel. This was also a period when Great Britain was forced to surrender her previously unchallenged position as the 'workshop of the world' and as the most powerful industrial nation. The USA and Germany proved better adjusted to the demands of the new steel age. Not only did they both overtake their British tutor's initial lead, but they soon established themselves at the forefront of world steel production. By the end of the eighties American steel production, and by the beginning of the nineties German steel production too, had exceeded the output of British works, which in the seventies had still dominated the world market.[2] Although the American steel industry presented scarcely any direct competition to British steel, having unique opportunities for expansion within its enormous and protected internal market, when the production and export success of the German steelworks was compared with the relatively backward British performance, it was not long before the British industry was charged with neglecting to follow up its technological advances.

Given that all the standard innovations of the new steel technology, the Bessemer, Thomas and Siemens-Martin processes, had been developed during the sixties and seventies in Great Britain[3] – the last with French assistance – British entrepreneurs were reproached for failing to exploit to optimum advantage the opportunities which had been offered first to them. Appearances seemed to indicate that in Germany, well-developed business strategies were promoting technical progress more rapidly and applying it more efficiently in

[1] David S. Landes, *Der entfesselte Prometheus. Technologischer Wandel und industrielle Entwicklung in Westeuropa von 1750 bis zur Gegenwart*, Cologne 1973, pp. 236–53.

[2] Re the individual stages and forms of this process see above and Peter L. Payne, Iron and Steel Manufactures, in: Derek H. Aldcroft (ed.), *The Development of British Industry and Foreign Competition 1875–1914*, pp. 71–99, esp. pp. 71–9.

[3] The Bessemer process had already been invented in 1856 but was applied industrially only from the 1860s onwards. See too Gerhard Mensch, *Das technologische Patt. Innovationen überwinden die Depression*, Frankfurt am Main 1975, p. 163.

the key industry of the day. Apparently, not even the especially severe and protracted depression in the seventies and eighties in Germany had damaged her steel industry's growth potential. Rather, it seems that it achieved a particularly high degree of competitivity on the world markets behind the protection of cartels and tariffs, i.e. in an environment which was basically hostile to competition.

The present work will examine whether there was a superior German way of developing mass steel production in the last thirty years of the nineteenth century and how this differed from the British way. We are therefore not concerned with giving a rounded business history of steelmaking in the two countries with all its characteristics and problems, such as Wilfried Feldenkirchen has undertaken for the Ruhr.[4] Our special line of inquiry leads us into a comparative examination of the relationships between business strategies and technical progress, with a view to delivering substantiated observations in the final chapter about the relative efficiency and capacity for innovation of the German and British steel industries, and of the entrepreneurs who led them. Our inquiry concludes with an analysis of the debate about 'entrepreneurial decline' in late-Victorian Britain, which centres basically on the controversy about the relative performances of the German and British steel industries. Unlike previous contributions to this debate which have compared the industries on the basis of static definitions of efficiency and which have relied, for the bulk of their argument, on treating technical progress and its practical application as a 'black box', to be inferred from its effects on price and quantity change, in this study we will place the strategies which businessmen employed to confront their commercial problems at the centre. The historical process which, since the 1860s, led to the formation of the different structures of the German and British steel industries, both in the technical and in the institutional spheres, will be made clear.

In order to make it easier to understand the technological dimension of this process, this analysis will be preceded by a chapter about the technology of steelmaking in the late nineteenth century. The characteristics of the various processes set objective limits to the management's 'room for manoeuvre' in the technological field. We will refer constantly to this during our discussion and assessment of individual entrepreneurial decisions.

We are confronted by an apparently paradoxical situation, in that the period of the new steel industry's breathtaking expansion is presented both by contemporaries and by economic historians also as a period of crisis and slump, for the industry in particular. Although the degree of crisis is assessed very variously, no voice has described the years of especially rapid growth after

[4] Wilfried Feldenkirchen, *Die Eisen- und Stahlindustrie des Ruhrgebiets 1879–1914. Wachstum, Finanzierung und Struktur ihrer Grossunternehmen*, Wiesbaden 1982.

1873 as a period when it prospered. Expanding production and winning new markets, generally proven means of increasing profits, appear in this period not to have benefited the steel industry, or only very inadequately so. Nevertheless both courses of action were extensively pursued, especially in Germany. It also looked as if the most disastrous consequences were experienced there during the mid-seventies. We will examine this development in Chapter 2.

It must also be asked why the firms were set so definitely on the course of expansion when, to all appearances, this was not accompanied by an improvement in their profits. This soon led to recriminations about 'overproduction', which was supposed to have brought about 'falling prices' and, consequently, the economic crisis. All the same, it is not very credible that the firms should have headed straight for disaster with their eyes open and persisted year after year in mistaken overproduction, without learning anything from an error that was clear to all. Ulrich Troitzsch has stated that our understanding of the Great Depression, the first protracted world-wide crisis of industrial capitalism, still awaits thorough research into the 'origins, processes and results of technological changes and their connection with economic development'.[5] He anticipated that an important contribution to this would be made by research into the technical rationalisation of ironmaking which he qualified as a pressing research problem. We have responded to this initiative in Chapter 3.

In the case of steelmaking the connection postulated by Troitzsch seems all the more evident, given that immediately prior to the 'Great Depression' the industry experienced its decisive technological breakthrough in the two basic innovations of the Bessemer and the Siemens-Martin processes. This stage of production, with its new and still unperfected mass production processes, must have appeared as a prior candidate for comprehensive rationalisation, as opposed to established and long-proven processes for extracting pig iron in blast furnaces and shaping it in rolling mills. Indeed, it was still not clear which way efforts could be directed successfully. The works managers had no previous experience to go by. The firms were therefore confronted by wholly new tasks, and their continued existence depended precisely on solving them in the conditions set by the looming economic crisis. And although their first defensive reaction against the Slump did, as Troitzsch supposed, consist in rationalisation measures, as we know these did not lead to renewed prosperity. The question must be asked why the rationalisation measures missed their mark. Given that they related to the technological aspects of the firms' operations, an answer can be found only with the help of detailed technical-historical investigation.

[5] Ulrich Troitzsch, Technische Rationalisierungsmassnahmen im Eisenhüttenwesen während der Gründerkrise 1873–1879 als Forschungsproblem, in: *Hamburger Jahrbuch für Wirtschafts- und Gesellschaftspolitik* 24 (1979), pp. 283–96.

This innovatory interference in the production process must have entailed unwanted and even unforeseeable consequences for the enterprises, which in turn created new problems to which they had to adjust. This new adjustment could consist of further activities internal to the works, or assume political forms external to the works, which are the domain of traditional economic history. In our case, this was represented by the well-researched agitation about protective tariffs for German heavy industry, which ended in the 'rye and iron' alliance and contributed to the conservative re-formation of the German Reich in 1879. However, without research into the technical-historical aspect of internal measures undertaken by the enterprises, the significance of the steel industry's cartel and customs policy remains unclear. The research by Hardach and Lambi notwithstanding, it still cannot be stated whether this policy was the last refuge of an industry threatened with collapse, after every attempt within the industry to stave off the crisis had already been exhausted, or whether a favourable political constellation was merely used as a rallying point. The question closely associated with this, whether the new cartels within the steel industry really were the 'children of necessity', can only be answered when it has been established that all other attempts at stabilising profits had failed. These forms of collective crisis management and their assimilation into the strategies of individual works are investigated in Chapter 4. Our final review of the various strategies employed by German and British entrepreneurs will show that forms of reaction to economic crisis which developed in the 1870s continued to determine the collective behaviour of the steel industry of each nation until the turn of the century and beyond.

Whereas the structure of the German and British steel industries exhibited great similarities until the early eighties, their subsequent development was along very different lines. In both cases new opportunities for expansion were created for the steel industry through diversification of processes and products. However, this took different routes in Germany and in Great Britain. The reasons for these divergent developments and the forms they took are dealt with in Chapter 5. We thereby lay the ground for a comparative assessment of German and British entrepreneurial efficiency and capacity for innovation in the final Chapter 6. By then, we will be in a position to review the debate about 'entrepreneurial decline' in the light of a comparison between individual and collective business strategies and, thanks to our previous technical-historical analysis, we are not obliged to rest our discussion about the possible failings of British steel industrialists on mere conjecture. One important finding of this chapter is that various conjectures about technological relationships and opportunities, which were fundamental to the various arguments previously advanced in this debate, do not stand up when confronted with the sources.

The central thrust of this inquiry is directed at the relationship between firms' strategies and technological progress. What forms did the growth of this branch of industry and the development of its productive capacity take? To what extent can regional and national differences in the structure of the firms be attributed to their strategic decisions, and to what extent to external conditions in the wider sense – supplies of raw materials, market outlets and national economic policies? Did firms try to modify or compensate for these external conditions? Which means did they employ to this end? In what way was technical advance an object of strategic decision-making? Was it perceived as a dimension which could be directed or influenced? Were there different strategies in the technical field and how were these determined?

Our research has been guided by these and similar questions. Common to them all is the notion that firms not only developed explicit strategies in the administrative field, an aspect of business history which has recently been better researched, but that they did the same thing in the technical field. Technological progress was not viewed merely as the application of technology, which was mainly introduced to the firms from outside, but as a process determined by strategic decisions on the part of the firms and steered in its direction. If technical progress as a whole means raising the productivity of the factors of production,[6] it also means that the firms strove very purposefully to raise the productivity of specific parts of their production. Consequently, technical progress could be achieved not only by overall investment in new-fangled or improved production plant, but also precisely by developing the existing organisation of production.

Although it would be possible at this point to separate technological progress from rationalisation measures, as required by the theory of production cost,[7] we have not done so. Research into concrete rationalisation steps has shown that these have always been supported by a whole bundle of measures, and that they were generally carried out under varying conditions within individual works as and when valuable technical information was acquired. It would at the very least be extremely difficult to establish empirically whether a particular bundle of measures ultimately led to the establishment of a fundamentally new production function or only to a shift in an already familiar one. Since it was anyway a matter of indifference whether the firms drew their profits from a novel production function or one already in existence somewhere, this distinction is not relevant to our particular line of inquiry.

The same applies when we treat innovation theoretically and attempt to

6 This is the general definition provided in Kurt Elsner, Wachstums- und Konjunkturtheorie, in: *Kompendium der Volkswirtschaftslehre*, Werner Ehrlicher et al. eds., Göttingen 1975 (2), pp. 246–96, see here p. 281.

7 For this delimitation, which places special importance on the necessity of combining technological progress with the acquisition of new technical knowledge and the installation of a fundamentally new function of production, see J. Heinz Müller, Produktionstheorie, in: *Kompendium der Volkswirtschaftstheorie*, I, pp. 57–113, p. 105.

evaluate technological innovations from a macro-economic perspective. The innovations handled in the course of our inquiry would have to be roughly classified in this context as improvements; innovations which represented a further development of the previous basic innovations, in this case the Bessemer and Siemens-Martin processes.[8] We are convinced that the technological progress resulting from these efforts reflects the strategic decisions taken by the firms, which they came to in the pursuit of their predominant aim, that of achieving and stabilising satisfactory returns.

The strategy concept employed here belongs to the type developed by A. Chandler, E. Learned et al. and P. de Woot,[9] as opposed to the strategy concept of the long-range planning school of thought or of the corporate strategy approach, as propounded for instance by H. I. Ansoff.[10] Older strategy concepts from Thucydides to Clausewitz, which were motivated by political and military aims, form no part of our discussion. According to the schools of thought represented by the authors mentioned first, strategy means a problem-solving process, whose chief purpose is to adjust the firm to changes, as well as to define the firm's most important aims and the ways of achieving them. Thus it is a truly comprehensive concept, which includes all longer-term entrepreneurial decisions, as well as those of a social and political nature. According to an early proponent of this school, strategy is 'analysing the environment to determine how best to use existing resources and capabilities of the organisation'.[11] In this sense the internal dynamics of the decision-making process do not interest us, unlike the two other schools of thought mentioned above, but rather the content of the decision. The individual stages of the intellectual and social process which finally leads to a strategic decision are largely left out of our discussion.

The limitation of this strategy concept as developed in countless case studies is explained by the context in which it has been applied in books written mostly by members of the Harvard Business School circle. Their aim was to provide firms with criteria for a concrete strategy advisory service. This meant that the content of individual decisions, as manifested in investment and organisational reforms, was at the outset more important than the decision-making process itself. Their failure to subject the concept to a far-ranging

[8] Re the relationship between basic innovation and improvement innovation, see Mensch, pp. 54f.

[9] Alfred D. Chandler, *Strategy and Structure. Chapters in the History of the Industrial Enterprise*, Cambridge, Mass. 1962, esp. pp. 11–16. E. P. Learned, C. R. Christensen, K. R. Andrews and W. D. Guth, *Business Policy, Texts and Cases*, Homewood, Ill. 1965, esp. p. 16. Philippe de Woot, *Stratégie et management*, Paris 1970, esp. pp. 37–41.

[10] Ansoff is much more interested in the enterprise's external connections than in its internal development and he attempts to formulate a strategic method. H. Igor Ansoff, *Corporate Strategy*, New York 1965, esp. pp. 18–22.

[11] Paul Selznick, *Leadership in Administration*, Evanston 1957, p. 149.

abstract analysis was thus based expressly on their greater proximity to empirical data, which proximity was thereby maintained. This sometimes leads to sharp contrasts with a model economy, which are thoroughly tested by applying the necessary discipline of reality.[12]

The test for the existence and nature of a strategic decision is as follows: 'In the absence of explicit statement the student may deduce from operations what the goals and policies are.'[13] In this way a strategic decision may be taken on the edge of an unknown process or even by chance. This constitutes one of the basic differences between the Harvard strategy concept and the other strategy concepts, as well as being a reason why it should prove operational for our purposes. The content of strategic decisions is far more likely to trickle into historical sources than intellectual processes in the minds of people about whom not much has been written. Statements about enterprise strategy then also become possible in places where the source material about the actual decision-making processes no longer exists. This strategy concept can thereby be directly applied to research into business history since it helps in surmounting one of the weightiest problems with the source material; that of the documentation of decision-making processes. In fact, it has implicitly always played an important role in the compilation of business history.

In spite of the great practical advantages of understanding strategy concept in this way, its principal limitation should, however, not be overlooked. While it leads to the description of as it were 'objective' enterprise strategy, yet it tells us nothing about the actual aspirations of the management during the planning phase of 'the contents', which came into being later on. These aspirations could, under certain circumstances, as for instance in our case study of the Consett Iron Co. diverge substantially from the ultimate 'objective' strategy. This is why we have, where the sources permit, deviated from the Harvard Strategy Concept (which we have otherwise applied) and have tried, for a few case studies in particular, to reconstruct the 'subjective' enterprise strategies as well and to compare them with actual developments.

Unless one is aware of the commercial motives which gave rise to firms and industries, any statements about their structures must be unsatisfactory for the purposes of historical analysis. As it is, this situation is inevitable in many cases due to the patchy nature of the surviving source material. This is precisely why the limited Harvard Strategy Concept has of necessity been so well suited to our research, because it can be supplied from the available source material and so helps prevent the construction of empirically unreliable theses. Although it would be desirable for the purposes of historical analysis to formulate a broader strategy concept, which would include the internal

12 James L. Bower, *Managing the Resource Allocation Process*, Boston 1970, pp. 28ff.

13 E. P. Learned et al., p. 17.

structure of the decision-making process, our knowledge of the sources is such that this could not be seriously sustained when it came to making statements about *the* German and *the* British steel industries. We have therefore limited its application to a few case examples, where it can be sustained.

Our statements about the steel industrialists' commercial motivation therefore do not claim the same overall viability as our statements about the structure of the whole branch. Nevertheless, they are much more than merely illustrative in character, since these model inquiries into the decision-making process in individual firms have clearly exposed the unreliability of using the contents of strategic decisions to deduce retrospectively the underlying commercial motivations. The special value of the case studies within the context of this book, which primarily deals with the development of the whole industry, lies precisely in having shown up this possible difference. At this point, the importance of luck as an element in the development of individual firms must finally be conceded. Given that every entrepreneurial decision was necessarily based on limited information, the attempt to achieve the greatest possible rationality in the strategies adopted by the firms was still not sufficient to guarantee success, in the same way as inadequate decision-making processes could ultimately and fortuitously result in the optimal firm structure. However, it was not our intention to produce, by so to speak adding-up individual and 'subjective' firms' strategies, a rounded picture of the particular commercial motivations and the particular decision-making processes involved. We have not pursued organisational and sociological lines of inquiry, which could have provided information about this, nor, by the same count, have we attempted a cumulative business history of the German and British steel industries.

The structure of firms is viewed, by Chandler especially, as a result of their strategies, with the former in its turn establishing the data for further strategic decisions. Structure is a function of strategy, or, in an assessment of Chandler's work, 'strategy is a principle of internal coherence and in relation to the environment, a historical link between the past, the present and the future of the enterprise'.[14] The objective character of Chandler's strategy concept is all the more obvious in that it has been tied to the actual and not to the envisaged structure. He equates strategy with the structure-forming process; we too, in the course of our inquiry, can, by applying his concept, develop a historically enlightening analysis of the structure of the German and British steel industries at the end of the nineteenth century.

The source basis for researching the complex questions referred to above has consisted primarily of archival material from fourteen British and German

14 Hubert Heyyaert, *Stratégie et innovation dans l'entreprise* (Université catholique de Louvain, faculté des sciences économiques, sociales et politiques, nouvelle série no. 110), Louvain 1973, p. 225.

steelworks (Steel Company of Scotland, Consett, Bolckow, Vaughan and Co., Bell, S. Fox, Dowlais, Nantyglo & Blaina and Blaenavon on the one hand, and Krupp, Bochumer Verein, Gesellschaft für Stahlindustrie, Rheinische Stahlwerke, Gutehoffnungshütte and Phönix on the other). As was only to be expected, this primary material has proved very variable. On the whole, the holdings of German firms are more extensive than those of British firms. While the minutes of the Board of Directors' meetings and the shareholders' annual general meetings were in every case still available, records of cost accounting were not universally preserved. For most of the period researched, the cost accounts, sometimes in very detailed form, still exist only in Consett, Dowlais, Krupp and the Gutehoffnungshütte. The Steel Company of Scotland, Tredegar, Workington, Bochumer Verein, Rheinische Stahlwerke and Phönix could produce only sporadic notes of cost accounts.

With reference to the kinds of source material described above there are no significant differences between German and British holdings. Examples of the greatest prolixity as of the very briefest notices turned up to an almost equal extent in both countries. It was however remarkable that the often extensive correspondence accompanying the execution of central instructions in Germany was almost wholly absent from Great Britain. In the British archives, submissions to the Directors' meetings and reports from individual parts of the works and from firm representatives were generally only preserved if they had been extensively quoted in the minutes of the Directors' meetings. While legally binding business of a directly commercial nature, such as sales and finance, was generally well documented in the minutes, it became abundantly clear that information about production technology and the organisation of production was wholly lacking. Although the greater flow of source material in the German firms did not reveal a composite picture in their case either, it did at least produce a few important criteria. It was clear that, in matters of production organisation especially, only a very limited amount of written material was produced in the works. This is where the lectures, reports and above all the discussions which took place in the technical associations and institutions can provide valuable supplementary information. In these organisations the works managers and their assistants engaged in a brisk exchange of information; they also appear to have acted as a recruitment pool for filling vacant leading positions in technical management. In particular, the discussions and lectures which took place in the British Iron and Steel Institute and were published in its magazine, were strongly geared towards the direct exchange of practical experiences. It is thus not surprising that a whole series of German works managers were also members and took part in its discussions.

A comparatively academic style predominated in the relevant technical literature produced on the Continent – in Germany, Austria and to some extent in France and Belgium too. In it, reports dealing with direct experience appeared, especially at the start of our research period, often only as travellers' reports from more advanced ironmaking districts, primarily from Great Britain. They give the impression that people on the Continent tried to compensate for their lack of empirical knowledge by engaging in more intensive theoretical study of the technological processes in a variety of academic institutions, which had no parallel in Great Britain. The fact that about half-way through our research period, one of these academic teachers, the Austrian Peter Tunner, was known as the 'Father of scientific iron production', while the ironworks of Europe were still orientated towards the methods employed by their British colleagues, sheds some light on the situation. For our purposes, this means that it is precisely the trade literature in those more backward regions which contains vital information about the most modern practice in steelworks at the time, in a very differentiated form.

The reconstruction of activities external to the works also relies strongly on the holdings of firms' archives and on trade literature. In this context, the two weekly trade newspapers of British heavy industry, the *Iron and Coal Trades Review* and *Iron*, played the most important part. Next in importance were, on the part of the industrial associations in Germany, the archives of the Verein Deutscher Eisen- und Stahlindustrieller (Association of German Iron and Steel Industrialists) and, in Great Britain, the legacy of the Cleveland Ironmasters' Association, a regional cartel, and the reports of the British Iron Trade Association. In this way internal information could also be gleaned about those firms whose archives could not be included or viewed.

1 The technology of late nineteenth-century steelmaking

Nowadays we understand 'steel' to mean 'all iron or iron alloy produced by smelting which is capable of being wrought or rolled'.[1] Such a metal had indeed existed before the three steelmaking processes in which we are interested were developed by Bessemer, Thomas and Siemens-Martin.

The starting product for the steelmaking is and was, apart from insignificant exceptions, pig iron made in blast-furnaces. This metal is not suitable for most technical uses since it cannot be rolled or wrought, being brittle and unmalleable. In its pure form, it can only be used for making cast-iron wares. The reason pig iron is brittle lies in the smelting process in the blast furnaces, when both carbon and other trace elements in the iron ore are absorbed into the molten iron. Pig iron produced in this way is therefore a very complicated alloy, whose components can be influenced only to a strictly limited extent. According to the iron ore employed, its composition will vary between 92 to 95 per cent iron, 3 to 4 per cent carbon, and small quantities of silicon, manganese, phosphorus and sulphur. The brittleness of pig iron can be traced to the 'impurity' resulting from the 5 to 8 per cent of other elements, in particular to its high carbon content.

Now, in order to make the material called steel from pig iron, the greater part of these other elements must be removed. This procedure is called 'refining'. (The German term is *Frischen* which has no equivalent in English. While in English texts 'decarburizing' is normally used in this context, *Frischen* is more than just removing carbon. Hence the somewhat outdated term 'refining' is used in this text.) Before the mass production of steel was introduced, between two-thirds and four-fifths of all iron within the German Customs Union was refined.[2] Conditions in Great Britain must have been similar. Precise data are not available, given that it is only since 1881 that the production of the

[1] *Brockhaus Enzyklopädie in zwanzig Bänden*, 17th wholly new edn, 17, Wiesbaden 1973, 'Stahl', p. 837.

[2] Horst Wagenblass, *Der Eisenbahnbau und das Wachstum der deutschen Eisen- und Maschinenbauindustrie 1835 bis 1860. Ein Beitrag zur Geschichte der Industrialisierung Deutschlands*, Stuttgart 1973, p. 248.

iron-processing industry of Great Britain has been fully recorded; however, both Clapham's and Hyde's estimates tend in this direction.[3]

What is certain is that by far the largest part of the iron produced in both countries was refined and thereby processed into what, going by the definition stated above, we nowadays call 'steel'. In those days, however, the designation 'steel' was reserved only for a small quantity of refined iron with a carbon content of about 1 per cent. This was an iron which could be hardened, for which a carbon content of between 0.5 per cent and 1.5 per cent is required. The closest approximation to this is our present-day term 'tool steel'. However, the mass of refined iron contained only a very small amount of carbon – generally less than 0.2 per cent; not because there was an especially great need for this, but because the production process – puddling – led to this result. Iron made according to this process was called 'puddled iron' or 'wrought iron'; in German it was sometimes also called *Schweisseisen* or *Schmiedeeisen*, from the processing stage which followed immediately after.

As the old puddling process disappeared, so did these various terms for iron with a low carbon content, capable of being wrought and rolled, and for iron with a high carbon content, leaving only the comprehensive term 'steel'. Nevertheless, these differences are extremely important for understanding the steel industry and its markets in the later nineteenth century. For this reason the present study retains the old terms instead of adopting modern usage.

1 The puddling process

The puddling process was first used in 1784 by Henry Cort, with the aim of refining pig iron using the cheaper pit-coal.[4] Since pig iron acquired the impurities, which it should have been freed of, in the blast furnace, by being in constant contact with coke or coal, the idea was to ensure that during the

[3] John H. Clapham, *An Economic History of Modern Britain*, 2, Cambridge 1932, p. 52. Charles K. Hyde, *Technological Change and the British Iron Industry 1700–1870*, Princeton 1977, pp. 166ff.

[4] The development of the puddling process from its introduction by Cort to our research period is presented in Isaac L. Bell, *Principles of the Manufacture of Iron and Steel, with some notes on the Economic Conditions of their Production*, London 1884, section XIII, pp. 351–80. The metallurgical details in our description of the puddling process are based, unless otherwise stated, on Bernhard Osann, *Lehrbuch der Eisenhüttenkunde, verfasst für den Unterricht, den Betrieb und das Entwerfen von Eisenhüttenanlagen*, 2: *Erzeugung und Eigenschaften des schmiedbaren Eisens*, Leipzig 1926 (2), pp. 43–9. Our description of the working method is based on Carl Hartmann, *Der practische Puddel- und Walzmeister, oder Anleitung zum Verpuddeln des Roheisens mit Steinkohlen, Braunkohlen, Torf, Holz und brennbaren Gasen, sowie zur weiteren Verarbeitung des Puddeleisens und Puddelstahls zu Stabeisen aller Art, zu Eisenbahnschiene, Spurkranzeisen, zu Blech und Draht*, Weimar 1861 (2), 4th section. A summary of the main characteristics of puddling from a modern viewpoint is provided in: *Gemeinfassliche Darstellung des Eisenhüttenwesens*, ed. Verein Deutscher Eisenhüttenleute, Dusseldorf 1937 (14), pp. 90–3.

refining process the iron would be exposed only to the heat and not to the fuel itself. Cort solved this problem in his puddling furnace by separating the fuel from the iron by means of fire bricks. His furnace was heated by causing the flames to pass horizontally above the surface of the iron, which was laid flat on a hearth, thereby melting and finally oxidising the unwanted elements.

While this was happening, the puddler had to stir the bath of molten iron energetically with a paddle so as to expose all the pig iron to the oxidising effect of the flames. Given an initial charge of above five hundredweight, which, once it had turned into a glowing, doughy mass, had to be thoroughly stirred and turned over, this work was, in the unanimous opinion of many former furnace workers, the hardest and most physically deleterious job in ironmaking. There were not many old puddlers.[5] As the decarburizing process advanced, the iron curdled into a spongy mass, which the puddler then shaped into lumps or balls, each weighing about a hundredweight. These balls were then removed from the furnace and compressed into compact blocks by hammering, which removed as much of the remaining slag as possible. Since the size of individual balls was limited by the puddler's strength, large pieces, for instance the iron ingots required for rolling rails – the chief product – could only be produced by welding several individual pieces together.

The greatest problem about puddling was that it was very difficult to produce a homogeneous material of a permanently consistent quality. In any case, the product could never be wholly homogeneous on account of the slag which always remained in it. Puddled or rather wrought iron was never simply the same as other wrought iron – its quality could vary quite considerably, sometimes even between the individual balls produced in the same charge. If the furnace cooled too fast, or if the slag did not contain enough oxygen, a sort of half-steel would emerge because the decarburizing process had not been completely carried through. The level of carbon content was a matter of luck. Only highly qualified puddlers using specific raw materials could consciously control so intrinsically indeterminate a process and achieve high-carbon iron (i.e. steel). The quality of the iron or steel thus produced was subject to permanent variations, the range of which depended primarily on the individual puddler's experience.

A natural limit to the production capacity of a puddling furnace was set by the physical endurance of its workers. Productivity remained constant for a long time and depended primarily on the composition of the pig iron employed. Furnace capacity could not be expanded as long as manual labour could not be dispensed with. Attempts at improvement were therefore directed almost

[5] In the course of a discussion about the Thomas process an employee of a British ironworks stated that during warm weather it was 'no uncommon thing to see a puddler drop down dead'. Edward Riley in: *Iron* 26 May 1882, p. 411.

exclusively at conserving heat during the process. Fuel could be economised by building two puddling furnaces together. The gases escaping up the chimney could be used for producing steam. Nevertheless, none of this affected the daily rate of production, which remained in the order of between 27 and 38 hundredweight in the course of a 12-hour shift, depending on the consistency of the pig iron and the number of assistants.[6]

Among the sequential stages of ironmaking – blast furnaces, refining furnaces and rolling mills – the middle stage, the puddling furnaces, already represented something of an anachronism at the beginning of our research period. The capacity of a puddling furnace had remained practically unchanged since the early nineteenth century, because it was directly linked to human physical capacity. Under the most favourable conditions, the maximum annual production of a single puddling furnace was about 1,000 tons, but generally it lay at between 500 and 800 tons according to the sort of pig iron and the working methods employed.[7] In Great Britain, however, the average annual production of blast furnaces between 1815 and 1871 rose from 1,500 to 9,845 tons per annum, and in the case of the modern blast furnaces in Cleveland, reached even 14,700 tons.[8] In their case, unlike the puddling furnaces, technological progress could effect an increase in production capacity, since the role of human labour in the blast furnaces changed; human labour was limited to supervising the process, and no longer directly worked the materials.

The same applied to the rolling-mill stage that followed the refining process. Cylindrical rollers differed from hammering in that they took on both the shaping and the moving of the worked material. Thus the capacity of a rolling mill depended primarily on the quality of its rollers and on the power of the engines that drove them, and could be altered by changing the size of these. No meaningful estimate of average capacity per rolling mill can be given, since milled products exhibited a far greater variability than pig iron and wrought iron, and rolling mills of very different dimensions corresponding to every kind of specific demand were erected. For instance, the Britannia Ironworks, which were set up in 1870 in Cleveland specially for making rails, was equipped with 120 puddling furnaces but only a single rolling mill line for iron ingots and a single mill line for rails. Its annual production is said to have been in the region

6 Bell, p. 527.

7 Hyde, gives 500–600 tons for Great Britain. In the German Reich during the 1870s, between 570 and 827 tons were achieved. My own calculations are based on *Reichsstatistik*: 1871, pp. 44–50; 1872, pp. 156–66; 1873, pp. 80–3, 88–90; 1874, pp. 82–5, 90–3; 1875, pp. 82–5, 90–3; 1876, pp. 82–5, 90–3; 1877, pp. 102–5, 112–15; 1878, pp. 104–7, 114–17; 1879, pp. 108–11, 118–21.

8 Hyde, p. 174. *Mineral Statistics* 1871, pp. 88f.

of 60,000 tons of rails.[9] Its pig-iron requirements could be met by five blast furnaces. This works was exceptional in that it was intended solely for the optimal production of railway rails. The general rule in the construction of new puddling works in the 1860s was sixteen to twenty puddling furnaces together with one rolling mill, or double the number of both.[10] Such a works would process the product of one or two blast furnaces. Thus the size of an ironworks was determined by the number of blast furnaces and rolling mills and their common multiples, not by its puddling furnaces, the size and productivity of which was still unchanged.

Only in the case of puddling were the output and quality of the product still determined by the physical skill and strength of the workers. With blast furnaces and rolling mills, by contrast, production proceeded as chemical and mechanical processes, without direct human manipulation, which could be planned with absolute accuracy. Human labour was restricted to managing this process and to a few ancillary and maintenance jobs. The quality of blast-furnace and rolling-mill products depended almost exclusively on the raw materials employed. If these did not change, neither did the end product. The specific components of the process could be minutely quantified and measured. This meant that management could determine the composition, quality and quantity of the product within narrow limits. Conversely, puddled blooms were still very much individual products. This could be an advantage, in that an experienced puddler could produce a different sort of iron for a specific purpose with every charge. Indeed, the puddling procedure survived into our century for such special orders. However, the mass demand emerging at the time, especially in the case of the railways, required the cheapest possible consistently uniform product, primarily for rails. There was no shortage of complaints from the railway companies about the lack of uniformity and poor quality of rails. Mass rail demand also exposed the fact that the puddling process had long since become a bottle-neck in iron manufacture. The capacity of blast furnaces and rolling mills could easily be extended and their workforce could be trained very quickly, but there was no point in building new puddling furnaces if there were not enough qualified puddlers. Their training was extremely protracted. They generally learnt their skill by assisting a trained puddler for three or four years, before themselves acquiring the experience sufficient for economical (in terms of fuel consumption, metal wastage) and uniform working. The more uniform the desired product was to be, the more

[9] J. K. Harrison, The Production of Malleable Iron in North East England and the Rise and Collapse of the Puddling Process in the Cleveland District, in: *Cleveland Iron and Steel, Background and 19th Century History*, C. A. Hempstead (ed.), The British Steel Corporation 1979, pp. 117–55, here p. 141.

[10] Harrison, pp. 128–31. Cleveland has been selected to illustrate these ratios in the 1860s, since this was a very recent iron district, where there was no old plant to distort the overall picture.

reliable the puddler had to be. Any rapid expansion of production thus had to occur at the expense of quality.

Various attempts were made to overcome this bottle-neck by mechanising the process. Beck mentions a first attempt as early as 1836, after which came a veritable flood of patents for improving puddling furnaces until the sixties.[11] Ironmaking firms hoped thereby to be able to extend their managerial autonomy in two directions. On the one hand they wanted to reduce their dependence on the puddlers, who had been the first section of the English iron industry's force to demonstrate by means of trade-union organisation and strikes that they were fully aware of the power which their central position in the production process gave them.[12] In addition, management hoped that the mechanisation of puddling would at least limit the puddled product's unwanted and strong tendency to vary. Thus the process would become more 'transparent' to management and more capable of rational planning. In this way, the application of technological methods could have brought them closer to solving their organisational problem.[13] The puddlers were very aware of this dual strategy, as was demonstrated by their periodic energetic opposition to attempts at mechanisation, which would have represented more than just 'humanising' their work, even if promoted under this flag. However, the search for solutions so close to the old puddling procedure did not yield any viable alternatives to it. Until the introduction of the Bessemer process, puddling by hand remained the only possible way of refining pig iron, other than older and quantifiably insignificant processes. Around three-quarters of all iron had to pass through the puddling bottle-neck.

2 Crucible steel

At least one technical solution to the problem of the homogeneity of the material, steel, existed; iron which had been refined to steel could be smelted down again in a fire-proof crucible until it had turned completely liquid.[14] To do so it had to be heated to temperatures of *c.* 1,600 degrees centigrade, the highest ironworks could achieve in those days, which then had to be maintained for quite a long time. This allowed the carbon to be distributed evenly through the iron and thus achieve the homogeneity which did not occur with puddled steel. Meanwhile all the slag (which could never be entirely removed

[11] Ludwig Beck, *Die Geschichte des Eisens in technischer und kulturgeschichtlicher Beziehung*, 5: *Das 19. Jahrhundert von 1860 an bis zum Schluss*, Braunschweig 1903, pp. 112f.

[12] Beck, *Geschichte*, p. 213 and p. 914. Harrison, pp. 131ff. J. C. Carr and W. Taplin, *History of the British Steel Industry*, Oxford 1962, p. 56.

[13] This wish was expressed with great emphasis when mechanical puddlers were envisaged. See for instance the Report of the Belgian Commission on the Working of Danks's Rotary Puddling Furnace, in: *Journal of the Iron and Steel Institute*, 1872, 2, pp. 311–33, especially p. 332.

[14] Osann, pp. 521ff.

from the metal in the puddling process) collected on the surface of the steel, where it was easy to pour it off. Sheffield, in England, and in Germany, Krupp and the Bochumer Verein, were the centres for the fabrication of crucible steel, which was always surrounded by an aura of secrecy. The steel produced in this manner was in any case first-rate and for some special uses it has, like puddled steel, only been replaced this century.

In the nineteenth century crucible steel was used for heavy-duty pieces such as tools, for mechanical engineering and for armaments and railways. It was the railways with their huge demand for high-grade axles, steel tyres and suspension springs which gave the manufacture of crucible steel its boost.[15] Apart from the fact that there was no alternative refining process, the production of crucible steel had one decisive disadvantage: it was extraordinarily expensive. Although crucible steel could have been used in all sorts of fields on technical grounds, especially on account of its tensile strength, it was about five to seven times more expensive than puddled iron.[16] Between puddled iron and crucible steel there thus yawned not only a great gulf in quality but also a considerable price difference. Crucible steel could only penetrate those areas where material costs did not matter very much; as for the new mass products such as rails, metal plates etc., its use was prohibitive on grounds of cost alone.

While there certainly was a receptive market for a steel superior to that produced in puddling furnaces and cheaper than crucible steel, which would thus fill the gap, the product as yet did not exist. It was only after Bessemer's invention that it became possible to make homogeneous steel of this kind for mass consumption.

3 The Bessemer process

Bessemer came to metallurgy, which, according to his own testimony, he originally did not understand at all, through his interest in developing artillery shells. It emerged from this that the usual bronze or cast-iron guns of those days were not adequate for his newly developed shells. If these were to be used by the artillery, the gun barrels would first have to be strengthened. So Bessemer started to look for a metal which was tougher than cast iron but cheaper than

[15] Walther Däbritz, *Bochumer Verein für Bergbau und Gußstahlfabrikation in Bochum, neun Jahrzehnte seiner Geschichte im Rahmen der Wirtschaft des Ruhrbezirks*, Düsseldorf 1934, pp. 71–83. *Krupp 1812–1912, Zum 100jährigen Bestehen der Firma Krupp und der Gußstahlfabrik zu Essen/Ruhr*, Essen 1912, pp. 158–61. In 1926 Osann could still name the following as within the field of application of crucible steel: tool steel, armaments steel, and highly sought-after machine, locomotive, car and ship parts, Osann, p. 522.

[16] William Fairbairn, *Iron – its History, Properties, & Processes of Manufacture*, Edinburgh 1869 (3), p. 156. According to Fairbairn, closing this price and quality gap was the greatest advantage of the Bessemer process.

crucible steel. Around the same time, Krupp had begun making gun barrels out of crucible steel. Although superior to all others, as demonstrated by the Prussian army's tests, they were so exorbitantly expensive that few could afford them in the 1850s.[17] Bessemer envisaged a kind of wrought iron similar to puddled iron, being both tough and cheap.[18] But wrought iron could not be cast in moulds, since the very essence of the puddling process was that the iron thickened again in the course of the refining process. And even a Herculean puddler would not be able to produce a bloom large enough for a gun barrel, quite apart from the practical impossibility of hammering such a bloom into a tube.

The only way in which refined iron could be cast was still the expensive crucible casting process. The protracted and convoluted history of the development of the Bessemer process until ready for production is not part of our theme and has no bearing on the assessment of later enterprise strategies.[19] So we will limit ourselves to describing the process in its developed form, when it was first applied on a large industrial scale at the beginning of the 1860s.

The Bessemer process operated on the principle of refining molten pig iron in a pear-shaped converter by blowing air through it.[20] The impurities contained in pig iron were burnt off by the oxygen in the air as it was compressed through the molten iron, and were thus removed as slag or gas. As the air was blown at high pressure from the nozzles at the base it had a simultaneous chemical and mechanical effect: chemical, in that the unwanted trace elements were oxidised by the oxygen in it; mechanical, in that the molten iron was thoroughly stirred by it. Consequently, the strenuous work of puddling was now superfluous. Nor did the temperature of the iron fall during the process, as many sceptical contemporaries expected it to; on the contrary it rose considerably although no additional heat was introduced from outside. Indeed, it had to, given that the smelting point for refined iron is several hundred degrees higher than that for pig iron. The temperature of iron refined by the Bessemer process lay in fact way above 1,500 degrees centigrade, so that it could easily be cast into ingot moulds.

[17] *Krupp 1812–1912*, p. 142, Däbritz, p. 38.

[18] Henry Bessemer, *An Autobiography*, London 1905, pp. 136ff.

[19] Re the history of the invention, see W. K. V. Gale, The Bessemer Steelmaking Process, in: *Transactions of the Newcomen Society*, 46, 1973/4, pp. 17–26. See, too, the interesting dispute about Bessemer's metallurgical knowledge and its importance for his discovery, which was then conducted between Eberhard Schurmann, Der Metallurge Henry Bessemer, in: *Stahl und Eisen* 76 (1956), pp. 1013–20, and his critic Eduard Maurer, Empirie, Produktionszahlen und Wissenschaft bei der Stahlherstellung, in: *Abhandlungen der Deutschen Akademie der Wissenschaften zu Berlin, Klasse fur Mathematik, Physik und Technik*, Jahrgang 1961, no. 3.

[20] The description of the metallurgical changes during the Bessemer process follows that given by Osann, pp. 143–64 and *Gemeinfassliche Darstellung*, pp. 93–101.

The heat which kept the steel molten[21] was generated exclusively by the oxidation of the unwanted elements in the pig iron, which occurred with such energy that the whole refining process was over in less than half an hour. The actual heat levels generated by the individual elements during oxidation varied considerably. When iron is heated to a temperature of 1,500 degrees centigrade, the rise in temperature resulting from the oxidation of every 1 per cent of

carbon is 6 degrees centigrade;
silicon is 190 degrees centigrade;
manganese is 46 degrees centigrade;
phosphorus is 120 degrees centigrade.

Thus silicon or phosphorus can be considered primarily as heat generators,[22] since burning off the 4 per cent carbon contained in pig iron does not produce a significant rise in temperature.

To Bessemer's great dismay, the phosphorus did not burn, thus rendering the steel brittle and worthless. In puddling furnaces the phosphorus was oxidised to form phosphorus pentoxide (P_2O_5) and so immediately combined with the basic slag, which was rich in ferrous oxide ($_3FeO{\cdot}P_2O_5$). This basic slag was absent from the Bessemer converter, however, so that without anything to bind with, the phosphorus pentoxide immediately dispersed and the phosphorus remained within the iron solution. The reason a basic slag could not form was due to the converter's peculiar fireproof lining. The lining used in the puddling furnaces was rich in iron oxide and so could not withstand the high temperatures in the converter; Bessemer responded by using ganister, a clay which was already known to be rich in silicic acid. Although this solved the temperature problem, at the same time it led to the formation of an acid slag in the converter, which could not bind phosphorus. Adding lime, in other circumstances a well-known and proven means of binding phosphorus, simply resulted in destroying

[21] Converter-refined iron was also called steel in the common usage of the time if the carbon content lay below 0.5 per cent and the metal could consequently not be tempered. The early term of 'ingot iron' which was still used to describe it, was soon abandoned. From 1877 onwards, the official German term, which could also be found in all the statistical records, was *Flusseisen*. However, this expression, the equivalent of 'ingot iron', did not become part of everyday speech. Thus our present-day term 'steel' became prevalent at the same time as the Bessemer process was introduced, emphasising yet again its epoch-making significance for ironmaking in general.

[22] *Gemeinfassliche Darstellung*, p. 94. These values were altogether unknown to Bessemer and his contemporaries; they were discovered only much later on. In those days, ideas about proceedings in the converter were very diffuse, a fact which did not, however, inhibit successful steelmaking. Initial considerations still started from the premise that the heat in the converter was generated by setting the carbon alight and burning off 10 per cent of the iron. See Hermann Wedding, Die Resultate des Bessemer'schen Processes für die Darstellung von Stahl, und Aussichten desselben für die rheinische und westfälische Eisen- resp. Stahlindustrie, in: *Zeitschrift für das Berg-, Hütten- und Salinenwesen in dem preussischen Staate*, 2, Berlin 1863, pp. 232–70, see especially pp. 235f.

the Bessemer converter's fireproof lining. Since lime has a greater affinity to the silicic acid present in ganister than to phosphorus, it could only bind with the latter after the lining had been completely destroyed – which made it practically impossible to remove phosphorus inside the converter.[23] This meant that the Bessemer process was limited to processing the sorts of pig iron which were free of phosphorus but rich in silicon (for generating heat!).

The refining process ran as follows (see Figure 1):[24] molten pig iron is poured into a converter which has been previously heated to glowing point. The blowing machine is then turned on and the converter stood upright, allowing the air to stream through the molten iron from beneath. Silicon and manganese are the first to oxidise, leading to a rapid rise in temperature. They both turn into slag. The carbon begins to oxidise only a few minutes later, disappearing in a glittering bright flame as fuel gas (CO). This flame then burns out, indicating that the process is over. The converter is tipped over again and the blowing machine turned off. By now the iron is almost free of carbon. However, towards the end of the refining process a small amount of oxidised iron (FeO) dissolves back into the steel, and does not turn into slag as in the puddling process. This oxygen content has to be removed, since it would otherwise result in red brittleness (brittle in the red-hot state, thus incapable of being wrought or rolled). This means that the steel has to be de-oxidised, which is achieved by adding Spiegeleisen (manganese iron) after the refining process. Spiegeleisen was a pig iron with an 8 to 20 per cent manganese content, at which the crystals combine to form reflective surfaces, hence its name.[25] The decisive element at this stage was the manganese, which could only be obtained in combination with iron; the proportion of manganese determined the price of Spiegeleisen. The manganese extracted the oxygen from ferrous oxide (FeO) and transferred it to the slag. Since Spiegeleisen, like every other pig iron, contained a small percentage of carbon, the steel was inevitably re-carbonised in the course of this procedure and ended up harder than puddled iron, which did not require the addition of Spiegeleisen.

Bessemer steel produced in this way had a carbon content of between 0.2 and 0.6 per cent.[26] These base values were only seldom exceeded, since this involved extraordinary difficulties. Nor could the lower content be reduced any further on account of the laborious de-oxidation process using Spiegeleisen. Impediments also stood in the way of achieving a higher carbon content, which could initially only be done by adding pig iron. Achieving a carbon content of

[23] *Gemeinfassliche Darstellung*, p. 101.

[24] Illustration taken from *Gemeinfassliche Darstellung*, p. 104.

[25] Wilhelm Pothmann, *Zur Frage der Eisen- und Manganerzversorgung der deutschen Industrie*, Jena 1920, p. 261.

[26] Bessemer steel containing 0.13 per cent carbon was already being called 'subcarburised'. J. S. Jeans, *Steel: Its History, Manufacture, Properties and Uses*, London 1880, p. 795.

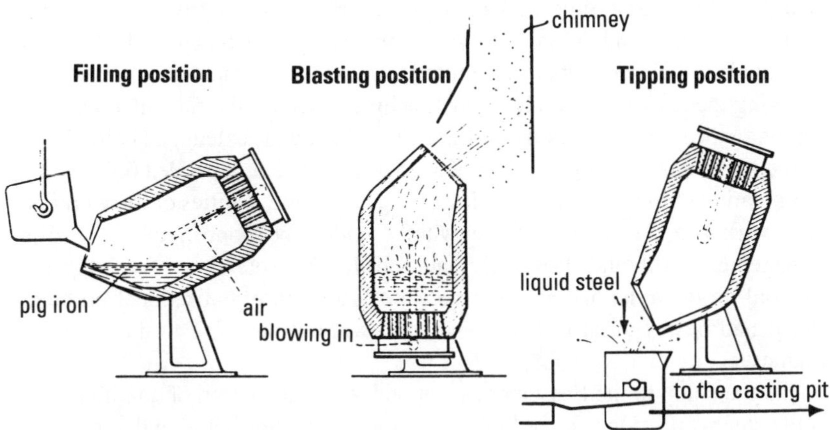

1 The stages of the Bessemer process

1.5 per cent in finished steel, for instance, would have involved mixing pig iron and converter-refined iron in a ratio of around 40:60, since the carbon content of pig iron seldom lay above 4 per cent. However, this would to a great extent have reversed the refining process, since such a mixture would have resulted in the steel re-acquiring not only carbon but also other elements in possibly undesirable concentrations.[27] Making a controlled and to a great extent pure iron and carbon combination with a carbon content of around or over 1 per cent was consequently still the privilege of the crucible process, although it could use Bessemer steel as a starting product instead of puddled steel, for instance.

Nevertheless, Bessemer had achieved his original aim. Pig iron could be refined in the Bessemer converter which, on account of the large quantity of surplus heat thereby generated, could then also be cast painlessly in ingot

[27] R. Mushet had proposed this method for producing Bessemer steel with a high carbon content, but had only gone as far as a mixture in a ratio of 2:1. He wanted Swedish charcoal pig iron for his carburizing process, which was especially pure but also expensive. In fact, Bessemer steel with a high carbon content was being produced in small quantities in Sweden in this way. Yet processing charcoal pig iron in a Bessemer converter was not only anachronistic but also much too expensive to realise large sales. Beck, *Geschichte*, p. 144. During the early period of the Bessemer process the attempt had also been made to achieve this aim by interrupting the blast at a moment when the iron still contained a lot of carbon. Jeans, *Steel*, pp. 569f., who in this connection states incorrectly that this practice was the rule in Germany. Compare the thorough presentation of the practice of the Bessemer process in Germany by Friedrich C. G. Müller, Untersuchungen über den deutschen Bessemerprocess, in: *Zeitschrift des Vereins Deutscher Ingenieure* 22 (1878), Sp. 385–404, 453–70. It would not have been possible to establish the practice of interrupting the blast process, since the refining process took so little time that there was great danger of selecting the wrong moment. Furthermore, as well as carbon, not all the other trace elements would have been burnt off, leaving unwanted impurities. In spite of every effort, steel of the quality of crucible steel could not be achieved in the Bessemer converter.

moulds. At the same time he had freed the iron industry from its puddling bottle-neck – although initially only for part of its products, but he had shown the way. Work in steelworks was limited from now on to managing the process – serving the converters and blowing machines, loading the raw materials and removing the steel. The quality of the steel could now be calculated beforehand by the works manager or the analytical chemist, since it depended only on the combination of raw materials employed and not on the abilities of the workers. The work process thereby came wholly under the control of the firm's management. Although Bessemer had fulfilled his intention of solving his material problem, he had at the same time solved an organisational problem which had been central to heavy industry, without meaning to do so in the slightest.

The transition to the Bessemer system led to a polarisation of qualifications in the steelworks. One single highly qualified man supervised and controlled the refining process from his managerial desk, while all the other ancillary tasks could be performed by semi-skilled workers.[28] Far greater demands were made of this plant manager than previously in the puddling works, since the quality and costs of the entire day's production now depended on him alone. He alone directed the converter and the blowing machines. Any initiative in the Bessemer plant had to come from him alone, which meant that he also bore entire responsibility for a successful product. The other workers were merely expected to carry out his instructions implicitly. Thus the plant manager's primary task was to organise the production process, which was why the introduction of the Bessemer converter to the old works in South Wales, the bastion of the puddling process, was greeted with such enthusiasm; at last they would be able to rid the refining process of its refractory and assertive puddlers, and replace them with cheap Irish immigrants.[29]

4 The Thomas process

Although the Thomas process constitutes technically only a minor modification to the Bessemer process, it was nevertheless one which increased its scope most decisively. In 1878 Sidney Thomas discovered a method which made it possible to refine phosphoric pig iron in a converter. In order to use the phosphorus as a heating agent and to bind it like silicon to the slag, a fireproof lining had first to be found for the converter, which was capable of combining the chemical properties of the puddling furnace's lining with the heat resistance of acid ganister.

[28] In many places this trend even led to setting up a podium for the works manager, right in the middle of the Bessemer plant. M. J. de Macar, Note sur les Aciéries allemandes et belges en 1882, in: *Revue universelle*, 2nd series, 12 (1882), pp. 142–78, here p. 155.

[29] *ICTR* 27 January 1882, p. 110.

This objective had long been recognised as such. In 1872 George James Snelus, director of the West Cumberland Steel Co., had already secured a patent for the basic converter lining as a prerequisite for removing phosphorus.[30] At the time, however, he was not in a position to produce a durable material of this kind, which meant that he was ultimately unable to hold on to the patent rights. Thomas finally solved the problem, which he had been working on since 1874, by using burnt dolomite, which he soon afterwards mixed with tar to make it easier to work with.[31] The dolomite not only withstood the high temperatures in the converter but, when lime was added, it also allowed a basic slag to form, which bound the phosphorus. This reaction could not occur in the Bessemer converter with its acid lining, since the lime would have dissolved the silicious lining instead of combining with the phosphorus. Thus, providing the converter with a basic dolomite lining was only the precondition which allowed the phosphorus to be bound by adding lime.

Although the stages in refining a charge in the Thomas converter or, as it was called in England, the basic converter, differed only in a few points from the stages in a Bessemer converter, these differences were essential. The similarities were limited to the first two phases, when as in the Bessemer process, first silicon and manganese, and then carbon were oxidised. Both phases were over after about ten to twelve minutes. They took less time than in the acid Bessemer process because Thomas pig iron had a lower silicon content. This meant that the silicon oxidized sooner, but also that it generated relatively less heat. In order to prevent the charge in the converter from 'freezing', the Thomas pig iron was previously heated to a higher temperature than the Bessemer pig.[32] However, the excess heat present after the carbon had been removed would still not have been sufficient to enable the steel to be cast safely. This was only achieved in the basic converter by means of 'continued blowing'. This term was derived from the practice of continuing the air flow after the carbon flame had burnt out, which in the acid Bessemer process indicated the end of the refining process. Only then did the phosphorus oxidise and heat the molten steel to the high temperatures which Bessemer knew about and wanted to achieve. By the end, the duration of the whole refining process differed hardly at all from Bessemer's acid process.

Altering the stages in the refining process did, however, have important effects on the finished steel, a fact which was only gradually appreciated by the early users of the Thomas or basic Bessemer process. It differed essentially from steel which was refined in the acid Bessemer converter in that basic steel was practically free of silicon and carbon. Since Thomas pig iron from the start

30 Jeans, *Steel*, p. 592.

31 Osann, p. 58.

32 This practice was already known in Germany from earlier Bessemer practice. Müller, *Untersuchungen*, pp. 396ff.

contained only slight quantities of silicon, almost all of it could be oxidised before the carbon started to burn: the process was further assisted by the addition of lime, which could not be introduced to the acid Bessemer converter precisely because of its affinity to silicon. Finally, the basic converter was by far and away the refining container best adapted to achieving the lowest possible carbon content.[33] Since de-phosphorisation began only after the carbon flame had burnt out, it was certain that all the carbon had combusted. Although continued blowing could also have been performed with the Bessemer process to burn off all the carbon, it would have led to heavy oxidation of the steel and so have rendered it useless. In the Thomas process, on the other hand, the steel was protected by the oxidation of the phosphorus during continued blowing. Since both silicon and carbon make steel extra hard, this meant that in the Thomas converter, where both these elements were removed as far as was possible, a particularly mild steel could be produced, one far easier to hammer and work than Bessemer steel and which opened to converter steel a whole new range of uses previously reserved for puddled steel.

In the eyes of contemporaries and technical-historical literature, however, the main advantage of Thomas's discovery was that it was now possible to refine high phosphoric pig iron to steel in the converter. At the same time, however, it was and is often overlooked that the refining process in the Thomas converter was not at all suited to every kind of phosphoric pig iron. To generate the necessary heat the phosphorus content had to be at least 1.7 per cent,[34] a high content which was not present in most sorts of pig iron. Consequently the Thomas process, like the Bessemer process, remained limited to particular sorts of pig iron. It was not a universally applicable refining process like puddling. Pig iron with a phosphorus content of more than 0.05 and less than 1.7 per cent could not be processed to usable steel in the Bessemer or the basic converter respectively. These sorts of pig iron remained, as before, the preserve of the puddling process, during which the heat was introduced from outside and so made no demands on the initial composition of the pig iron. As a result, puddling was for the time being still the most versatile refining process.

33 Karl Ernst Mayer and Helmut Knüppel, Entwicklungslinien des basischen Windfrischverfahrens in Deutschland, in: *Stahl und Eisen* 74 (1954), pp. 1267–75, see p. 1268 here.

34 When the basic Bessemer process was first applied in Cleveland, and during the first tests in Hörde, types of pig iron containing only 1.25 per cent phosphorus were also blasted, but they contained a further 1.5 per cent silicon to meet the heat requirements. The limit of 1.7 per cent phosphorus content applied only when the pig iron did not contain any significant quantities of silicon and when the heat necessary to the process was supplied solely by burning the phosphorus. Nevertheless, attempts were soon being made to achieve precisely this, since this very presence of silicon in pig iron interfered with the refining process. See too Chapter 5, s. 1, pp. 157f.

5 The open-hearth process

Although the open-hearth process was developed at about the same time as the Bessemer process and was later to surpass it in importance, it was completely overshadowed by it until the end of the seventies. Unlike the converter process, but similar to puddling, in the open-hearth process the charge was refined to steel on a hearth, with the heat required by the process being introduced from outside. It differed from puddling basically in that the heat generated on the hearth was so great that the whole charge could be refined without any interference from outside and turn to steel in the liquid state. The bath of molten steel was automatically mixed by the waste gases rising through it, rendering redundant the exhausting work of puddling. In any case, this mixing and, consequently, the refining process took place much more slowly than in the converter and even more slowly than in the puddling furnace, making the whole process last several hours.

The requisite heat was generated by applying the regenerative system first developed by Friedrich Siemens for manufacturing glass and porcelain.[35] His brother Wilhelm adapted this regenerative system for steelmaking, initially in France. He was supported in this by M. Le Chatelier, the *Inspecteur Général des Mines*, who had secured him the cooperation of the firm Boigues, Rambourg & Cie. in Montluçon. It was on their premises in 1863 that Wilhelm Siemens set up a furnace of his own design in which, after protracted experiments directed by his brother Otto, an excellent steel was said to be produced.[36] However, when the furnace melted down and collapsed one night, Boigues, Rambourg & Cie. terminated the partnership, thus obliging Wilhelm Siemens to look around for new premises where he could continue his series of experiments. While he was in Birmingham setting up an experimental works there to produce steel from a mixture of pig iron and iron ore,[37] Pierre Martin in Sireuil managed in 1864 to make perfect steel from a mixture of pig iron and scrap-metal in a furnace which had also been built by Wilhelm Siemens. Both processes were ultimately to prove commercially successful and it was not long before mixtures of all three possible ingredients (pig iron, iron ore and scrap-metal) were being smelted.

The Siemens-Martin or, as it was better known in England, the open-hearth furnace was heated by a mixture of gas and air. The fuel gas was obtained from

35 Ludwig Beck, Zum fünfzigjährigen Jubiläum des Regenerativofens, in: *Stahl und Eisen* 26 (1906), pp. 1421–7.

36 Jeans, *Steel*, p. 90. Maurer, p. 10, on the other hand refers to wholly unsuccessful attempts at smelting, without giving further details.

37 Jeans, *Steel*, p. 90.

coal in what were known as Siemens gas producers.[38] These gas producers were set as deep as possible beneath the foundry base so that the gas, carbon monoxide, could rise unassisted to the furnaces. This gas and the air required for combustion were each passed over a hot regenerator, an open-brickwork chamber (e and f), and conducted to the top of the furnace, where they mingled and combusted producing a flame which swept, as in the puddling furnace, over the charge lying on the hearth (a), smelted it and refined it. The hot waste gases were then conveyed to two more regenerators, which they heated before finally disappearing up the chimney (see Figure 2).[39]

After about half an hour the flow of gas and air was switched around, so that both were heated in the regenerators which the hot waste gases had passed through. These waste gases were now used to re-heat the first two regenerators, which had previously transferred their heat to the gas and air. This 'regenerative heating process' allowed a major part of the heat produced by smelting, which would otherwise have been lost up the chimney, to be used for pre-heating the gas and air, thereby enabling higher temperatures to be reached than in earlier hearth furnaces, for instance in puddling furnaces. These temperatures in the range of *c.* 1,500 to 1,800 degrees centigrade were sufficient to obtain steel in liquid form.

The four regenerators formed the foundations on which the actual open hearth stood. The hearth was similar in shape to the puddling furnace's, except that its grate was set at an angle, allowing the liquid steel to flow off without needing to be lifted out in solid blooms, as was the case with puddling furnaces. Opposite the out-flow opening was a furnace door for supplying the hearth with pig iron, scrap metal and iron ore. The flow of gas and air was switched over by an alternative valve set at the intersection point of the gas ducts.

The open-hearth furnace was constructed mainly from fire bricks; they were used for building the regenerators as well as for supporting the hearth. However, fire bricks were not sufficiently resistant for the actual furnace and hearth. Since the temperatures achieved in the open-hearth furnace equalled those in the Bessemer converter, it required the same sort of fireproof lining; ganister or some other material rich in silicic acid was employed. Using these fireproof materials meant that the open-hearth furnace, like the Bessemer converter, was then firmly set on the acid process, i.e. only phosphorus-free sorts of iron, scrap-metal and iron ore could be processed.

After S. G. Thomas had discovered how to refine phosphoric pig iron in a converter, an attempt was also made to apply his findings to the open-hearth furnace. The hearth was lined with the same tar and dolomite mixture lining as

38 Unless otherwise mentioned, our description of the open-hearth process follows Osann, Chapter 4, Das Herdfrisch- oder Siemens-Martin-Verfahren.

39 Illustration taken from *Gemeinfassliche Darstellung*, p. 111.

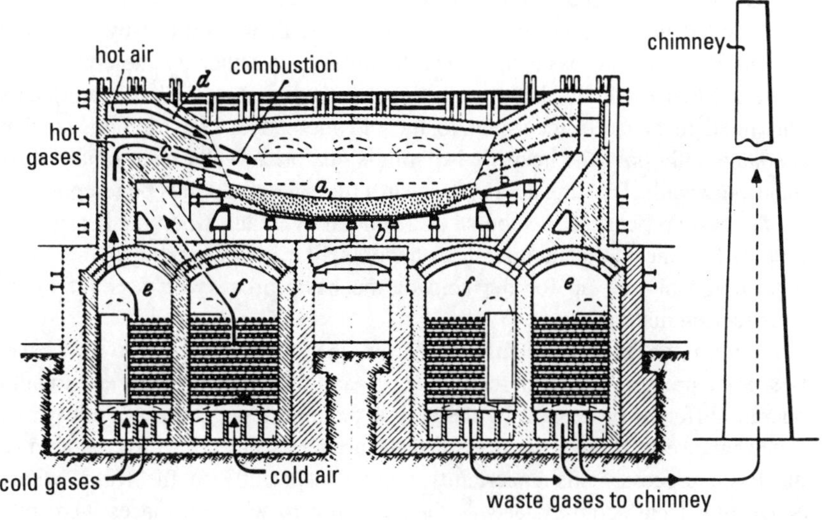

2 The open-hearth furnace

the basic converter, in the hope that adding lime would produce the slag necessary for binding the phosphorus. However, since the tar and dolomite mixture was not suited to the curved furnace walls, acid limestone (Dinassteine) was used as previously.

The phosphorus was indeed removed by proceedings analogous to those in the basic converter, since it did form a slag with the lime which, like basic slag, could also be sold as fertiliser. To that extent, the process ran as expected, but the end result proved disappointing, since the great heat achieved in the furnace turned the dolomite on the hearth into slag along with the other silicious materials (limestone, fire bricks etc.). This led to the rapid destruction of the furnace, entailing expensive repairs. However, it was not possible to dispense with these materials containing silicic acid since only they were capable of withstanding the high temperatures and of ensuring the stability of the hearth and the furnace structure: dolomite was not adequate for this purpose.[40] This had not been an issue with the converter, since its stability was due to its steel casing and, above all, to the fact that the hot gases could escape upwards unimpeded, instead of destroying the structure, as they did in the case of the open-hearth furnace.

It was only towards the end of the 1880s that a satisfactory solution, in both

[40] Osann, p. 350. Beck, *Geschichte*, p. 709. J. H. Darby, The Manufacture of Basic Open-Hearth Steel, in: *JISI* 1889 part I, pp. 89f.

the technological and the economic sense, emerged, when dolomite started to be pounded into magnesite hearths.[41] This was then followed by the rapid dissemination of the basic open-hearth furnace, or rather by the conversion of acid furnaces to the basic process. The regenerators and gas producers continued to be built according to the old method of construction, so that fundamentally only the hearth's isolating silica brick walls were replaced by magnesite walls. Existing acid open-hearth furnaces could thus be re-equipped for the basic process in the course of a general overhaul, when the hearth was rebuilt. This meant that an open-hearth works, like a converter steelworks, was not irrevocably set up for the acid or the basic process on account of its construction material.

As with the gradual changeover from the acid Bessemer to the basic Bessemer process, when the converters in a single plant would occasionally operate different processes, by the end of the eighties it could not always be stated precisely whether the open-hearth furnaces were set up for the acid or the basic process. This uncertainty applied especially to the Ruhr, where converting local scrap-metal was the main use to which furnaces were put. Since scrap steel was in any case low in phosphorus, the question of whether to set up a hearth for the acid or the basic process was determined by the pig iron used in the smelting. In most cases, the wish to use the cheaper Thomas pig iron must have determined the choice of a basic hearth. Apart from which, the basic hearth also permitted the use of puddled scrap metal, which always contained a bit of phosphorus. The director of Phönix works, Herr Thielen, reckoned in May 1889 that 70 to 80 per cent of open-hearth steel in Germany was already coming from basic furnaces.[42]

Apart from the possibility of acid or basic smelting, analogous to the difference between the acid and basic Bessemer processes, two fundamentally different refining processes were available to the open-hearth furnace. With the Martin process pig iron was first melted on the hearth and only then was it mixed with a succession of separate small portions of scrap. The quantity of scrap metal thus absorbed could amount to anything from three to ten times the original quantity of pig iron.[43] In terms of quantity, then, scrap iron was the most important starting ingredient. As in the puddling furnaces, the charge was refined by the residual oxygen left by the combustion gases; in the course of

[41] A Gouvy, Die Flußseisenerzeugung auf basischem Herde in Resicza, in: *Stahl und Eisen* 9 (1889), pp. 397f. Osann, p. 354.

[42] Information provided by Director Thielen, Phönix-Ruhrort, during a discussion at the Iron and Steel Institute, in *JISI* 1889 part I, p. 102. In it he also mentioned that he knew only one firm in Germany which produced Siemens-Martin steel using pig iron, as in Great Britain, instead of scrap iron, which was otherwise usual in Germany.

[43] Jeans, *Steel*, p. 100.

which the slag also served to convey the oxygen into the bath of molten iron.[44] Silicon and manganese were the first to oxidise and turn into slag. The increase in temperature next caused the carbon to oxidise, which, as in the Bessemer converter, produced a fierce boiling period when the charge was thoroughly mixed by the escaping fuel gas (CO). In order to secure enough fuel gas for this automatic mixing process, the charge always had to include pig iron, since scrap iron contained too little carbon for this purpose. This was why coal had to be added when only scrap iron was used in the process. Apart from which, the liquid pig iron curtailed the scrap iron smelt. However, when the proportion of pig iron rose markedly above one third, the charge once again took very much longer because removing carbon in open-hearth furnaces was a slow process.

Unlike in the Bessemer converter, the slag was not an efficient medium for bringing oxygen into contact with the carbon dissolved in the bath of molten iron. So, when large quantities of pig iron were used, iron ore was introduced to the charge to provide more oxygen and to speed up the decarbonising process. Indeed, the greater the proportion of pig iron in the charge, the greater too was the quantity of iron ore which also had to be added, which meant that the whole operation was ultimately carried out without any scrap at all. This then was the pure Siemens process. On the whole, however, operations were kept at the transitional stage, whereby the amounts of pig iron and iron ore were varied according to the amount of scrap metal available. The scope presented by this process was enormous, given that the possible proportion of scrap metal varied from 0 to 90 per cent.

Once the carbon had been removed, the steel, as in the Bessemer process, was de-oxidised by adding Spiegeleisen, and later also ferro-manganese, and it was recarbonised as required. No particular haste was necessary since the heat supply was not interrupted. This meant that after the additional materials had been introduced, a sample could still be taken and assayed, enabling the smelt to be corrected, if necessary, by adding further ingredients. In this way a specific steel could be produced with unerring accuracy, which was not the case with the converter. After this, it was cast in moulds. Open-hearth steel differed from converter steel in that it lay much more smoothly in the forms or ingot-moulds. The slower and gentler refining process and the possibility of allowing the steel to stand in the furnaces meant that when the steel was cast, it contained less waste gases, which produced destructive bubbles in the steel block. The steel did not rise as high in the moulds as converter steel did, a feature otherwise present only in expensive crucible steel. In order to exploit it to the full, the steel would generally be uphill cast to avoid air bubbles.

All this meant that the Siemens-Martin process was especially well adapted

44 Our description of the refining process follows Osann, pp. 424–45.

for producing high-quality steels, and in this it was clearly superior to the two converter processes developed by Bessemer and Thomas. At least as important too was the fact that nearly any kind of pig iron could be processed in the open-hearth furnace. Unlike with the converter processes, a minimum silicon or phosphorus content was not required because the processing heat was introduced from outside. Since most of the available ores contained meagre quantities of phosphorus, the basic open-hearth furnace soon became the commonest refining vessel of all and by the twentieth century it was dominating steelmaking around the world. It thus assumed the position which the puddling furnaces had held right up to the 1870s.

The basic open-hearth process was the only one of the four new steelmaking processes which was capable of replacing the old puddling process, on account both of the steel qualities it could achieve and of the sorts of pig iron it could use. Each of the other three new processes was limited to its own narrow spectrum, although these spectrums overlapped so much that they could have suppressed the puddling process even without the basic open-hearth process. The really universal operational opportunities, which the puddling furnace had offered since the early nineteenth century, were, however, only achieved again with the basic open-hearth furnace, which consequently succeeded it far sooner than the acid or basic Bessemer converter did and took over, to some extent at least, its dominant role in the whole heavy industrial context.

The Bessemer and the acid Siemens-Martin processes cannot, however, be ascribed merely transient significance. They remained in use in places where the raw materials met the limited requirements of these processes. The basic process was especially suited to places where ore with a rich phosphorus content could be mined or obtained cheaply, and it was retained as a cheap refining process for making steel of less stringent qualities until the 1960s, by which time it had been superseded by the basic oxygen process on grounds of price. Nowadays, the basic oxygen process, which involves blowing technically pure oxygen through a water-cooled jet onto liquid pig iron in a converter, is the most common refining process. It has replaced practically all the refining processes presented in our survey, and is about to suppress them entirely.

2 Expanding into the Slump: the railways as major customers of the new steel industry

As the first technology to be ripe for production, the acid Bessemer process dominated the expansion phase of the new steel industry, which was consequently wholly characterised by the process's metallurgical particularities. A new metal had been created in Bessemer steel, one capable of filling the gaps between puddled iron and crucible steel and which robbed both of a good part of their market. It could be cast in any moulds and, unlike the blooms produced in puddling furnaces, in sizes appropriate to the end product, resulting in less wastage. Thus was rendered superfluous the permanently critical practice of welding several pieces together, which the rails manufacturers viewed as the greatest disadvantage of puddled rails and the reason for their lower quality.

In terms of price, Bessemer's steel was far closer to puddled iron than to crucible steel.[1] However, initial ambitious hopes that it would entirely replace crucible steel were not fulfilled, since it never quite achieved the same quality. In fact, the first people to adopt the new process at the start of the 1860s, John Brown in Sheffield and Alfred Krupp in Essen, did still think that they had found a cheaper alternative to their own crucible steel production. Indeed, Krupp, who had been smelting steel made in puddling furnaces down again in crucibles since the 1850s, was already concerned that the new process would render his entire works obsolete.[2] John Brown was about to build a big new works according to Krupp's model when he learnt about Bessemer's process. He came to the same conclusion as Krupp, dropped his plans for a combined puddling and crucible steelworks and went straight over to manufacturing Bessemer steel.[3] However, they both quickly realised that they could not yet

1 The steel from H. Bessemer's first successful charges had already been sold at exorbitant profit in Sheffield as a replacement for crucible steel. However, this market proved to be not nearly as absorbent as the mass demand for railway materials, which in the last analysis promised greater profits in spite of the lower prices. Re Bessemer's price policy during the early years of his process, see Bessemer, p. 176.

2 *Krupp*, pp. 159ff. and 165f.

3 Jeans, *Steel*, p. 63. The view propounded by Carr and Taplin, p. 26, that J. Brown had from the start wanted to manufacture rails, is incorrect.

achieve the high quality of crucible steel and so they immediately procured rail rolling mills for endprocessing the Bessemer steel.

Likewise, puddled steel was not superseded technically, quite apart from the fact that it was still considerably cheaper. The fact was that the Bessemer converter could not produce steel as mild and malleable as the puddling furnace could. Comparable steels would be produced by the open-hearth and Thomas processes only years later, with the assistance of ferrous manganese. However, it was soon established that the new steel was tailor-made for mass demand by the railways. Axles and steel tyres were considerably cheaper when made of Bessemer steel than when using crucible steel, and very nearly as good. It was, however, in the demand for steel rails that Bessemer steel found its biggest market.

It was certainly very fortunate for the further development and dissemination of the new process that steel refined in a converter turned out to correspond less to Bessemer's original quest for a particularly hard steel for artillery than to the requirements of the railways. The very fact that Bessemer steel was remarkably well suited to the demands made of rails explains its rapid expansion and the great significance which it acquired so quickly for the European and North American steel industries. Its expansion phase was almost entirely nourished by the needs of the railways. However, it should be understood at this point that it was not an autonomous demand. Although Bessemer steel was a new material, it could not be used in any really new ways. Steel rails, axles and steel tyres made of Bessemer steel provided exactly the same service as those made of puddled iron or crucible steel, and competed with the latter; in this sense they were homogeneous goods as far as railway managements were concerned. In their case, after all, the market volume had already been determined, unlike the autonomous demand that exists for new products.[4]

It was not the Bessemer process, but the puddling process, already dismissed as moribund by many, which developed a whole new range of applications – especially in shipbuilding and high-rise construction – for its iron during the 1860s and 1870s. This meant that many innovative puddling works could make good the loss of the rails market.[5] Puddled iron production had yet to reach its high point. Indeed, notwithstanding all the recognition due to the new steel production processes invented by Bessemer, Thomas and Siemens-Martin, it should not be forgotten that they simply took over existing markets which had essentially already been created by the puddling process. By opening out a whole range of uses for iron and steel the puddling process had created the

4 Compare Ernst Heuss, *Allgemeine Markttheorie*, Tübingen and Zurich 1965, p. 70.

5 Wilhelm Rabius, *Der Aachener Hütten-Aktien-Verein in Rote Erde, 1856–1913: die Entstehung und Entwicklung eines rheinischen Hüttenwerkes*, Jena 1906, esp. p. 43. Hermann Müller, *Die Übererzeugung im Saarländer Hüttengewerbe von 1856 bis 1913*, Jena 1935, p. 113. For Great Britain see too pp. 124–6.

basic preconditions for the enormous expansion of steelmaking by the new processes. Bound up with this too were the innovatory achievements of the puddling process's later period, which are all too easily forgotten in the light of Bessemer's success story.

1 Sales prospects for the early Bessemer works

It was already becoming apparent by the mid-sixties that future requirements for Bessemer steel would lie almost exclusively with steel rails and to a small extent with axles and steel tyres. The London & North Western Railway's tests had shown that Bessemer rails were substantially more resistant than puddled rails. These were carried out between 1861 and 1864 on especially well-travelled sections and showed that the Bessemer rails of that time were hardly worn at all, whereas comparative rails made of puddled iron had to be replaced several times. Thus the price differential of £7 for puddled rails and £17 10s. for Bessemer rails (1864) was well worth the expense in the case of lines carrying heavy traffic.[6] Further savings were made in labour costs and by reducing interruptions to the service on account of frequent rail replacements.

The heavy wear and tear and, perhaps even more, the frequently unpredictable cracks in puddled rails caused by the material's lack of homogeneity, ended by forcing the railways to engage in the actual pioneering work of introducing the Bessemer process. Robert Mushet, an iron manufacturer in South Wales, wrote the following in a letter to *Engineering* in 1867: 'Indeed, I have often thought that the reason why the public are not allowed to walk on the permanent way is two-fold; first to prevent accidents to those who walk along the line, and secondly to prevent as far as is practicable the public from seeing what a multitude of damaged rails there are in lines laid with iron.'[7]

By 1863 the London & North Western Railway had already taken the excellent results of the tests on their lines as the motive for building their own Bessemer steelworks in Crewe, near Manchester. Two plants, each equipped with five-ton converters, were operating there by 1865.[8] The Furness Railway Co.'s undertakings in the north-west were even more spectacular. In the same year 1863 the railway company founded the Barrow Steel Co. for processing phosphorus-free pig iron from the Schneider and Hannay blast furnaces in Barrow, which was best suited to the Bessemer process. In 1866 Barrow Steel merged with Schneider and Hannay's to become the Barrow Hematite Steel

6 Fairbairn, pp. 196f. Carr and Taplin, p. 29.
7 According to Harrison, *Production*, p. 142.
8 Carr and Taplin, p. 27. Beck, *Geschichte*, p. 134.

Co., the largest Bessemer steelworks of the time, with twelve converters, to which six more were soon added.[9] In Austria, the *K.k. Südbahngesellschaft* (Imperial and Royal Southern Railway Company) followed the example set by the London & North Western Railway and erected a Bessemer works alongside their rolling mill in Graz to produce rails for their own requirements.[10] In the case of France, F. Caron has indicated that the Bessemer process was decisively introduced as a result of pressure from the railway companies, especially the *Compagnie du Nord*. Although these companies did not found their own works as was done in England, they did in some cases provide the necessary capital assistance for erecting a Bessemer works and, by providing long-term and substantial orders, they ensured that the new plant would be viable during its running-in phase.[11] The same applied to the USA, where P. Temin has shown that all the Bessemer works had more or less close connections with the railways, 'usually through the medium of common ownership or directorship'.[12] The three Bessemer works which were wholly new foundations (and thus not based on existing puddling works) were even founded with the direct assistance of the railways and were run by their management.

The discussion about sales prospects during the five years between 1866 and 1870, which featured the end of the diffusion phase of the Bessemer process, at least where decisions to build were involved, consequently revolved almost exclusively around the price differential between iron and steel rails. It was well known, or at least estimated, that there was a growing demand on the part of the railways for new lines and for replacement rails – and that the only point still to be established was how much of the market could be secured for Bessemer steel rails. By 1866 so many companies had already decided to change to Bessemer rails that the demise of puddled rails was already being prophesied. This initial euphoria subsided somewhat when the domestic demand for rails, which was so important to the Bessemer works, subsided,

[9] Ulrich, Aust and Jänisch, Die Darstellung und weitere Verarbeitung von Bessemerstahl in England. Bericht über eine im Jahre 1867 ausgeführte Instructionsreise, in: *Zeitschrift für das Berg-, Hütten- und Salinenwesen in dem preussischen Staate* 16 (1868), pp. 12f. Carr and Taplin, pp. 27f.

[10] Hans-Jörg Köstler, Einführung und Beginn der Stahlerzeugung nach dem Bessemerverfahren in Österreich, in: *Berg- und hüttenmännische Monatshefte, vereinigt mit Montan-Rundschau* 122 (1977), pp. 194–206, here pp. 202f. In any case, the Graz works were not the first Bessemer works in Austria. When they started up in 1864, Bessemer steel was already being produced in Heft and Turrach. Ibid., pp. 198–201.

[11] François Carron, Le Rôle des compagnies de chemin de fer en France dans l'introduction et la diffusion du procédé Bessemer, in: *L'Acquisition des techniques par les pays non initiateurs*, Colloque, Pont-à-Mousson 1970, pp. 561–75, here p. 566.

[12] Peter Temin, *Iron and Steel in Nineteenth-century America*, Cambridge, Mass. 1964, p. 174.

which meant that works, whose order books had still been full at the end of 1866, now waited sometimes in vain for orders.[13]

The sharp competition which had grown up in the meantime drove the price down to less than £12 a ton.[14] In Britain, an annual production capacity of *c*. 300,000 tons was offset by sales of *c*. 150,000 tons only, and some works were already running into considerable problems with payments. For all that, the new industry was still functioning better than the puddling works.[15] Well-run works could even achieve respectable profits in spite of relatively reduced sales. The Dowlais annual report, for instance, could state, when its Bessemer works had been running for three years, that 'The profits already made have recouped the amount of capital invested.'[16] However, it was not possible under these difficult economic conditions to build up a large custom. In spite of all the predictions that the shortage of usable iron ore would force the price of Bessemer steel up, by 1869 it fell to between £9 and £10.[17]

In 1869, after a hiatus of almost three years, the discussion about iron versus steel rails was revived in the wake of rising demand and, above all, rising exports.[18] According to the iron manufacturers who congregated in the newly founded Iron and Steel Institute, prices of £6 10s. for iron rails and £9 to £10 for Bessemer steel rails meant that, for them, the break-even point came in places where iron rails had a life expectancy of 15 years. The puddling works held on to the market for this type of rail; they could survive with this price differential. A reduction in royalties the following year would, however, push the demand back to Bessemer rails. Although only a few voices were prophesying the total disappearance of puddled rails, it was at the same time quite clear that they had no future in the case of lines bearing a heavy traffic load, and that they would be driven back onto side-lines and, for the time being, to new lines which were subjected to moderate usage, especially overseas. As long as this market was expanding, and all the signs were that it was, the large puddling works in South Wales and Cleveland had, with the existing price differential, no grounds for anxiety. They possessed locations along the coast

13 *Engineering* 5 October 1866, p. 255; 16 November 1866, p. 371; 26 July 1867, p. 71. *BITA* 1879, p. 34.

14 At the start of the sixties, when only John Brown was making Bessemer rails in Sheffield, they still cost £18. *Engineering* 11 October 1867, p. 335.

15 Beck, *Geschichte*, p. 242. Beck refers to Tunner's estimates. Others even claimed that the production capacities would have amounted to three times the sales. Carr and Taplin, p. 29. The Lancashire Steel Co. had already gone into liquidation by 1867. *Engineering* 11 October 1867, p. 335; 27 December 1867, p. 586.

16 Dowlais report for 1868, according to Alan Birch, *The Economic History of the British Iron and Steel Industry 1784–1879*, London 1967, p. 359.

17 *Engineering* 20 March 1868, p. 262. *Engineering* 1 October 1869, pp. 22f.

18 Edward Williams, The Manufacture of Rails (Abstract of Paper read before the Iron and Steel Institute at Middlesborough) and minutes of the discussion, in: *Engineering* 1 October 1869, pp. 224f.

which were ideal for exports, and Cleveland in particular had a solid base of the raw material needed for the puddling process. World-wide, the replacement needs of every railway were estimated at 1,000,000 tons per annum, three-quarters of which would fall to puddled rails.[19] In the case of new lines, puddled rails could reckon on an even better share of the market.

The home and neighbouring West European markets had been lost to puddled rails due to the density of the communications network which had been installed in the meantime. On the other hand, the new lines being laid overseas, especially in the conquest of the American West, offered rich sales opportunities, which were only temporarily interrupted by the Civil War. There was not the slightest reason why anyone wanting to equip their works to meet this demand should go over to the Bessemer process. This applied to most of the works in Cleveland and South Wales, which enjoyed sometimes very long-established and close business relations with American railways and could depend on securing the new business too. Richard Crawshay may be considered to have epitomised this connection; he invested the wealth he had created by making railway rails for America back in American railroad stocks. Far fewer Bessemer rails than iron rails were exported to the United States; the market for them was in places where a high level of traffic already existed.

In Germany the same factors applied as in Great Britain, though as yet on a smaller scale. Krupp was the first to start up the Bessemer process in 1862 – at first with two small two-ton converters, to which were added two five-ton converters in the following year. The works were then extended in rapid stages, resulting in the addition of nine more five-ton converters by 1865.[20] Once he had established that the new material was not capable of replacing crucible steel, his chief product, Krupp broke off his agreement with Bessemer, whereby he had been granted the exclusive right to apply the invention in Germany, especially since the Prussian authorities were refusing to issue a patent for the Bessemer process.[21]

There was no longer anything to prevent further works from being built in Germany. At first, however, only a few followed Krupp's example. By 1867 only four further works were operating Bessemer converters, initially only with capacities of two tons. At the start of 1865 the Königshütte, which was then still a state enterprise, had started working plant built according to the English model. Production, however, remained stuck at the expensive experimental stage until its sale was negotiated in 1869, and it was only in 1874, after a lengthy hiatus, that it picked up again to a notable extent.[22] Poensgen started working the Bessemer process in 1863, initially in Oberbilk (Düsseldorf).

19 *Engineering* 21 May 1869, p. 341.
20 *Krupp*, p. 116.
21 Däbritz, pp. 123f. *Krupp*, p. 166.
22 Hermann Illies, Das Bessemerwerk der Konigshütte, in: *Stahl und Eisen* 33 (1913), pp. 231–4.

Since Poensgen's own steel, unlike Bessemer steel made in Sweden, was not suitable for his steel tubes production, he was forced by circumstances to turn it into axles and sometimes rails as well. This proved an unsatisfactory solution and it never acquired much importance.[23] In 1865 the Bochumer Verein started up a regular Bessemer operation. Their works were set up entirely along the lines established by Krupp, their keenest rival, which were already proven by experience, and the steel was milled into rails from the start. In 1868/9 a second plant was built, equipped with larger five-ton converters.[24] At the same time as the Bochumer Verein, the Hörder Verein had obtained Bessemer's plans for extending his rail production and already by April 1864 was capable of blowing its first charge. Parallel to Bochum, here too a second plant was set up in 1868/9 with five-ton converters, by then the usual size.[25]

Thus Krupp, the Bochumer Verein and the Hörder Verein started manufacturing Bessemer rails scarcely any later than in England. Remarkably enough, of these pioneers the Hörder Verein was the only one to be already established as a rails supplier, whereas the other two had hitherto concentrated on crucible steel production, in which they resembled John Brown, the first British user.

Krupp experienced some difficulty in off-loading any samples at all of his new rails. He only managed to do so in February 1863, when the Bavarian State Railway acquired 50 rails.[26] Much more important for his long-term market prospects was, however, the decision taken by the Cologne-Minden Railway in 1864 to carry out comparative tests at Oberhausen on different sorts of rails, much as the Lancashire and Yorkshire Railway had done.[27] As was to be expected, Bessemer steel rails once again proved the best. It was clear too that these rails would make their breakthrough as replacement rails for the railways, especially for the heavily used lines through industrial districts. The number of Bessemer steel rails rose from the 1,540 laid along Prussian lines in 1867, to 1.4 million seven years later.[28]

It seems that Krupp was alone in suffering set-backs during this first period, since his excessively optimistic expectations initially led him to invest far in excess of the market (in 1865 the works were already equipped with two rail rolling lines and thirteen converters). In a rather melodramatic letter to his procuration dated 29 June 1866, Krupp was already wondering whether it would not be better to 'pack in' the business of steelmaking 'and give up

[23] Lutz Hatzfeld, *Die Handelsgesellschaft Albert Poensgen Mauel-Düsseldorf. Studien zum Aufstieg der deutschen Stahlrohrindustrie 1850–1872*, Cologne 1964, pp. 64f. and p. 145.

[24] Werkarchiv Krupp-Bochum. BV 126 00 Nr. 1: General Versammlung, 26 September 1863, p. 6. Däbritz, pp. 113, 121 and 128.

[25] Beck, *Geschichte*, p. 266.

[26] Krupp (1912), p. 169. Däbritz, p. 124.

[27] Jeans, *Steel*, p. 681.

[28] Jeans, *Steel*, p. 702. The Cologne-Minden railway, for instance, was buying Bessemer rails exclusively from 1872. *Iron* 23 September 1876, p. 398.

the whole ruinous show of the works as worthless'.[29] They managed to reassure him, and demand gradually grew to meet his production potential.[30] In 1869 the manufacture of Bessemer steel was already producing a net profit of 400,000 thaler, so that speculations about increasing production by a factor of three started up again. With these profits, the necessary investment could be recouped within six months.[31] However, such plans were never more than speculation. By 1868 more cautious people like Louis Baare of the Bochumer Verein were already warning that the prosperity of the Bessemer works would soon bring the competition onto the scene.[32]

Unlike those in Great Britain, the German works had established scarcely any connections with the overseas markets, which meant that the alternative facing the British (whether to produce Bessemer rails for regions with heavy traffic or to go on producing puddled rails for new lines overseas) was present only to a very slight extent in Germany. Against the background of the rails tests in the sixties and of the accompanying considerations about profitability, previous customers, especially of the Rhineland-Westphalian works, had to be considered as future clients for Bessemer steel rails. This conclusion was only reinforced by studiously following developments in Great Britain. This was the situation when decisions were being made in Germany at the end of the sixties to build more Bessemer steelworks, which were to put an end to the innovation's diffusion phase.

Their buoyant construction of Bessemer works soon earned German steel industrialists the accusation of having been short-sighted and of having built far in excess of the actual demand. It was said that during the speculation boom (*Gründerboom*) which followed the successful outcome of the war with France, they had allowed themselves to indulge in exaggerated hopes of future sales, and were thus to a great extent themselves responsible for their penury during the slump in the seventies.[33]

Unlike the British Bessemer works, which came into being gradually, it is

[29] HA-Krupp, WA IXa172: Abschriften und Auszüge aus der C.R.Akte B3, Allgemeine betriebliche Dispositionen 1864–1883, p. 2.

[30] HA-Krupp, WA III30: correspondence between Fried. Krupp and H. Haass, Paris from 5 January 1869 to 1 February 1870. Fried. Krupp to H. Haass on 13 April 1869: 'the rolling mills are fully occupied'.

[31] HA-Krupp, WA IXa192: Eichhoff to A. Krupp on 18 March 1878.

[32] Werkarchiv Krupp-Bochum, BV 126 00 Nr. 1: General-Versammlung, 19 September 1868, p. 4.

[33] As early as the 17th Congress of German Economists in Bremen, there occurred a vigorous exchange of views on this matter between Baare (for heavy industry) and Philippson (for the free traders). W. Wackernagel, *Berichte über die Verhandlungen des 17. Kongresses Deutscher Volkswirte in Bremen 25., 26. und 28. September 1876*, Berlin 1876, pp. 85–99. Due to the lack of differentiated research into heavy industry during the Grunderkrise, this survey of investment behaviour has gone on far too long. This is the case with Helmut Böhme, *Deutschlands Weg zur Grossmacht, Studien zum Verhältnis von Wirtschaft und Staat während der Reichsgründungszeit 1848–1881*, Cologne 1972, p. 354.

striking that most of the German Bessemer works went into operation in a regular spurt at the beginning of the seventies. This may indeed produce the impression that they were offshoots of *Gründungsfieber* – enthusiasm about the foundation of the Reich. Such an assumption would, however, overlook the fact that these works could not be set up at a moment's notice. There was indeed not a single Bessemer works in Germany on which building had not already started before the *Milliardensegen* and the economic boom of 1872/3. Phönix's management decided on 6 December 1870 to build its own Bessemer steel works, a decision which had already been rumoured the previous year.[34] The GHH works was already under construction in 1870.[35] Leopold Hoesch must have decided to build in 1870 at the latest, since that was the year in which he sent his son Albert to Sheffield to learn about the Bessemer process; the articles of association for his steelworks were signed on 1 September 1871.[36] The Rheinische Stahlwerke was founded in Paris on 27 May 1870; as in the Hoesch case, the decision to do so must have been made at least a year beforehand.[37] The Dortmunder Union's works in the Dortmunder Hütte (iron foundries) were already under construction in 1871, as were those in the Steinhauser Hütte near Witten.[38] In both cases, then, one may assume that the decisions to build were taken in 1869 or 1870, at the latest. The same applies to the Aachener Verein, since its Bessemer works were already operating by the business year 1872/3.[39]

Right from the start, these new works were all fitted out for the production of steel rails. Bessemer works had been present in Bavaria since 1868 and in Saxony since 1866, with guaranteed sales to their respective state railways.[40] Two works had been planned initially only for producing axles and steel tyres; the Daelen, Schreiber & Co. works in Bochum mentioned above in 1868, and the Osnabrücker Stahlwerk near the Georgs-Marien-Hütte in Osnabrück, in order to process its pig iron, which was suitable for the Bessemer process.

34 Mannesmann-Archiv, P 1.25.35: Protokoll der Sitzung des Administrationsrates vom 6 Dezember 1870; P 1.25.11.1: Ordentliche Generalversammlung, Bericht der Direktion, 10 Oktober 1869.

35 GHH-Archiv, 20001/38: Jahresbericht Neu-Oberhausen, 1 Februar 1871. In 1863 Wedding already mentioned plans for a Bessemer works at Jacoby, Haniel und Hyssen, the precursor to GHH. Wedding, *Resultate*, p. 234.

36 Horst Mönnich, *Aufbruch ins Revier – Aufbruch nach Europa, Hoesch 1871–1971*, Munich 1971, pp. 91ff.

37 Anton Hasslacher, *Der Werdegang der Rheinischen Stahlwerke*, Essen 1936, p. 5.

38 *Zeitschrift für das Berg-, Hütten- und Salinenwesen in dem preussischen Staate* 21 (1873), Statistischer Ergänzungsband, p. 239.

39 Rabius, pp. 29f.

40 *100 Jahre Eisenwerk-Gesellschaft Maximilianshütte, 1853–1953* Sulzbach-Rosenberg 1953, p. 42. Ernst Friedrich Dürre, Reisenotizen, betreffend die Anlage und den Betrieb der neu erbauten Bessemerhütte auf dem Arnim'schen Werke Königin Marienhütte zu Cainsdorf bei Zwickau, in *Zeitschrift für das Berg-, Hütten- und Salinenwesen in dem preussischen Staate* 15 (1867), pp. 256–62.

Following the 1867 decision and the associates' meeting in January 1869, the works could be started up in August 1871. At the beginning of 1870, while construction was still in progress, the directors decided to take on rail production too.[41] Finally, there were the Gienanth works near Kaiserslautern, which was, however, an exception in that it was not in any direct competition with the other works, since it used a small converter to produce wrought pieces for machinery, this being carried out from 1871 at the very latest.[42]

To sum up, it may be concluded from the foregoing that the construction of the German steel industry was not materially connected to the foundation of the Reich or to the *Gründerboom*. Furthermore, only Osnabrück and the Rheinische Stahlwerke in Duisburg were real newcomers to the rails business. The other new Bessemer rails works were based on previously established rails works, which, in view of the increasingly obvious supersession of puddled iron by Bessemer steel, were attempting to hang on to their former market by going over to the new process. Their attitude was thus emphatically defensive, which was why they settled for just one plant each, apart from GHH and Dortmunder Union. In their cases, however, the Union acquired its two plants as the result of a merger; they were actually two individual plants. The only works to build more than one plant from the start were GHH and the two newcomers, Osnabrücker Stahlwerk and Rheinische Stahlwerke.[43] During this period, the only Bessemer steelworks to be extended with a further plant was the Bochumer Verein in 1870, thereby closing the technological gap which had since opened between itself and the new works.[44] Consequently the construction of the Bochumer Verein's new Bessemer steelworks preceded the foundation of the Reich and the *Milliardensegen* and this extension cannot be linked to either.

41 H. Müller, *Georgs-Marien-Bergwerks und Hütten-Verein, Osnabrück, die Geschichte des Vereins*, Osnabrück 1896, pp. 67–72.

42 *Berg- und hüttenmännische Zeitung* 31 (1872), p. 167. *Reichsstatistik* 1871, p. 50, in which a production of 250 tons is already mentioned.

43 Osnabrück decided to build the second Bessemer plant, after it had emerged from revised calculations about capacity that the first plant would produce too much steel for their axles and steel tyres manufacture, and too little to operate a rails rolling mill. Müller, *Georgs-Marien*, p. 75. In the case of GHH, the construction of two Bessemer plants with two converters each already meant reducing their original plans by half. It had emerged during the planning stage and from the first running operations, that the rails mills could be fully employed with the production of four converters. Thus the same conclusion was reached there in 1873 as in Osnabrück, although from a different starting position. Nor did it make much sense to make further cuts in the original plans, apart from employing the mill to capacity, on account of the blowing engine which had already been installed and which was conceived for four converters. GHH-Archiv, 2001/36: Promemoria Carl Lueg 1872, betreffend Ausbau Oberhausener Werke; 300108/0: Bericht des Aufsichtsrates und des Vorstandes für die 1. Generalversammlung, p. 25; 320100/0: 1. ordentliche Generalversammlung am 29 November 1873; Bericht des Vorstandes.

44 Däbritz, p. 128.

2 Economic developments during the 1870s

(a) Upswing: 1869–1873

The enormous economic upswing of the years between 1869 and 1873 has been frequently described and is consequently sufficiently well known not to require any further mention. For our purposes, it is enough to establish that it was essentially carried by the strong demand from the railways.[45] This applied to a peculiar degree in the USA and Germany, where both the upswing and the slump were at their most spectacular. In the USA the greatest length of new lines was laid in 1869; in Germany this wave of new construction was temporarily halted, interrupted by the outbreak of the Franco-Prussian War. The first signs of a strengthened demand for rails reached Great Britain at the end of 1869. The American market, which the British were sure of, was capable of absorbing considerably more than the home industry could produce, being then still in its infancy.[46] Besides this, the Indian Railways promised to become the second fat slice of the market when they were taken over by the state. By the very next year it was well and truly possible to refer to railway mania. In the USA 15,000 miles of new lines were planned, and in India a further 10,000 miles. Added to this were a few thousand miles more in Canada and South America.[47] All of which were the British rail manufacturers' traditional markets, which demanded primarily wrought-iron rails, not Bessemer steel rails. Furthermore, the Bessemer works would not have been at all in a position to satisfy such a huge requirement on account of their chronic shortage of suitable iron ore.

It had thus clearly paid to stick to the puddling process and to hold out through the brief dearth of the last years, since the loss of the internal market was more than compensated for by the brilliant prospects overseas. Completely new wrought-iron works in the largest category were even built specially for the American rails market, for instance the Britannia works in Cleveland, with 120 puddling furnaces and a production capacity of 2,000 tons per week.[48] Although the number of Bessemer works increased only slightly (from 18 to 21 between 1868 and 1873), the existing works certainly expanded their capacities.[49] Nor was there any lack of demand here; the trend towards

45 Joseph A. Schumpeter, *Konjunkturzyklen. Eine theoretische, historische und statistische Analyse des capitalistischen Prozesses*, vol. 1, Göttingen 1961, pp. 344–77. Böhme, pp. 320–59. Hans-Ulrich Wehler, *Bismarck und der Imperialismus*, Cologne 1972(3), pp. 53–61.

46 Duncan L. Burn, *The Economic History of Steelmaking 1867–1939*, Cambridge 1961(2), pp. 18f. *ICTR* 24 November 1869, leading article.

47 Burn, p. 19. *ICTR* 27 April 1870, p. 262; 22 February 1871, p. 120; 3 May 1871, p. 284.

48 Harrison, *Production*, pp. 139ff. *ICTR* 1 March 1878, pp. 240, 242.

49 *BITA* 1877, p. 60.

Bessemer rails remained constant on the home market, but it was not possible to obtain as much phosphorus-free pig iron as could have been processed.[50]

The strong upswing clearly identified the characteristic bottle-necks inherent in the two refining procedures: the puddling works suffered from a shortage of coal and from the puddlers' discontent; the Bessemer works were prevented from expanding by the shortage of usable iron ore.[51] Appropriate strategies were developed to surmount these problems: the mechanisation of puddling work on the one hand and the acquisition of their own supplies of low-phosphorus iron ore on the other.[52] In spite of the brilliant economic situation and matching profits, only a few put real energy into pursuing these solutions. The vast majority went on behaving, one might almost say 'as usual', or at any rate as interested observers. Indeed, the actual situation demonstrated quite sufficiently that adhering to tried and proven branches of production was not at all incompatible with entrepreneurial success. The general opinion was that economic developments were no reason for losing one's peace of mind. This was why the large railworks, which had introduced the Bessemer process, had expeditiously retained their puddling works as well. Whitworth, the director of Bolckow, Vaughan & Co., one of the largest railworks, provided an instance of the then prevailing attitude to the question of whether to use the Bessemer or the puddling method when he told the shareholders' meeting of spring 1871 that 'You cannot possibly find such a quantity of ironstone as will enable you to make Bessemer steel to supply all the wants of the world; and I do not therefore anticipate that our regular iron trade will give out while we have anything to do with it.'[53]

50 *ICTR* 18 January 1871, p. 41; 29 November 1871, p. 810.

51 Re the puddlers' strikes, see Chapter 1, note 12. The chronic coal shortage had been worrying British heavy industrialists since the sixties. The notion that British coal supplies were just about to run out had been advanced since 1865 in a book by S. Jevons called *The Coal Question* which appeared that same year. In 1871 a Royal Commission on coal was formed, which was supposed to look into ways of saving energy and consequently coal. In 1873, at the highpoint of the boom, a genuine coal panic then broke out for a brief period, which made prices rocket. Carr and Taplin, pp. 45f. Re the shortage of phosphorus-free iron ores, see *ICTR* 3 May 1871, p. 284; 24 May 1871, p. 334; 29 November 1871, p. 810.

52 The mechanisation of the puddling process was taken furthest in the new iron district in Cleveland, which was suffering particularly from a labour shortage. Commercial success was nevertheless denied to works with mechanical puddlers, since in the meantime a far more rational process had been discovered in the Bessemer process. According to Edward Williams, head of the Bessemer works at Bolckow, Vaughan & Co., these apparatuses appeared 'the day after the fair'. Harrison, *Production*, pp. 141–51, esp. p. 148. The search for low-phosphorus iron ores led in particular to securing the hematite mines in northern Spain near Bilbao, which soon became the backbone of the Bessemer industry in Great Britain and Germany. This is extensively discussed in Michael W. Flinn's MA thesis *British Overseas Investment in Iron Mining, 1870–1914*, Manchester 1952, which has unfortunately not been published. *Idem.*, British Steel and Spanish Ore: 1871–90, in: *Economic History Review*, 2nd ser., 8 (1955–56), pp. 84–90, esp. pp. 87f.

53 BSC-Northern RRC, loc. no. 10490: report of directors, year ending 31 December 1870, p. 16.

In any case, by the beginning of 1871 there already was a general consensus in this branch of industry that this splendid boom would also pass. After reviewing the extraordinary rise in rail production the trade reports of the *Iron and Coal Trade Review* commented that 'Great depression is apparent in this branch of the trade.'[54] The *ICTR* did not need special indicators to reach this opinion, since every boom could be relied on to be followed by a slump almost as a natural law, albeit one based merely on previous experience.[55] However, when the demand persisted unabated and prices went on rising, these warning voices fell silent for a while, until new fears were expressed towards the end of 1872 that it would soon become difficult to find the necessary capital for laying further new lines abroad.[56] At the same time, however, the *ICTR* could see no grounds for concern where the Bessemer works were concerned, since their order books were full for a long time yet, and all they had to reckon with was a future fall in prices.[57]

Although the scarcity of iron ore at the peak of the boom prevented the Bessemer works from meeting the possible demand, on the other hand, the fact that they could rely on a stable base of domestic demand for replacement rails meant they were not as liable to serious set-backs. They had reacted in a very stable manner to the cyclical nature of the rails market, which was primarily induced by exporting conditions – in any direction, as the depression of the later 1860s and the boom in the early 1870s had shown. Given the very regular growth in the size of their production, the economic trend was primarily reflected in their prices, whereas the puddling works also underwent an enormous expansion in the trade.

Germany also experienced an increased demand for rails, in spite of the set-backs caused by the war with France. In addition to the demand for replacement rails, which had backed up, came the extensive projects for new lines, which would enable the German rail network to grow by almost 7,000 km in the years 1872 to 1875 alone[58] – in the wake of the huge wave of railway-building which was taking place all around the world. By April 1871, the *ICTR* was already predicting that 'The German industry will find an important outlet in the *Zollverein* (Customs Union).'[59] Any hopes on the part of British firms that they would be able to participate directly in Germany's railway building, as they had done in the USA and in India, were dashed from the start; the German rail works, unlike the American works, had in the

54 *ICTR* 11 January 1871, p. 24.
55 *ICTR* 20 December 1871, p. 866.
56 *ICTR* 20 November 1872, p. 930.
57 *ICTR* 2 October 1872, p. 788.
58 Max Sering, *Geschichte der preussisch-deutschen Eisenzölle von 1818 bis zur Gegenwart*, Leipzig 1882, p. 156.
59 *ICTR* 19 April 1871, p. 254.

meantime grown sufficiently large to be able to supply most of the internal market. Consequently it was far more a question of taking account of the fact that German heavy industry was expanding beyond the Reich, to the detriment of British works. 'Germany is becoming a formidable rival on the Continent, especially as regards the supply of iron to Russia and the Eastern ports of Europe.'[60] As long as these markets constituted only a fraction of the American demand, and Germany was exporting less than 15 per cent of its anyway meagre rails production, this gave no grounds for concern.[61]

By 1874 German rails consumption peaked at half a million tons, produced almost exclusively within the country. The trend towards Bessemer rails continued uninterrupted. Since many works were still under construction and often suffered further delays due to the war, the oldest Bessemer steelworks such as Bochum, Hörde and Krupp were the ones which could take on the whole brunt of the upswing. By October 1871, Krupp was already writing to C. Meyer, his representative in Berlin, that his firm was completely overwhelmed by orders, requesting him to please bear this in mind in his dealings with the railways.[62] The new works were doing their utmost to complete construction on the plants so as not to miss out on the rails business. A works report to the GHH in February 1871 pointed out in this respect that 'The Bessemer plant is just about a life-and-death issue for Neu-Oberhausen, and as long as this plant is not working at full swing, there is no question of paying interest on the large capital investment.'[63]

According to an Austrian iron manufacturer called Hupfeld, 'The Achilles' heel of the German Bessemer steel industry' was at that time its dependence on English pig-iron.[64] Germany was even worse off than Great Britain where supplies of iron ore suitable for the Bessemer process were concerned. Since basically only the Georgs-Marien-Hütte near Osnabrück could provide the German market with limited quantities of pig iron suitable for the Bessemer process,[65] its phosphorus-free pig-iron requirements had to be supplied by the

60 Forbes, Quarterly report on the progress of the iron and steel industries in foreign countries, in: *JISI* 1871 part 2, p. 113.

61 Re British rails exports, see Table 1, p. 48. German rails production came to 450,000 tons in 1871 and 500,000 tons in 1872. This contrasted with exports of 42,000 tons (1871) and 71,000 tons (1872). *Reichsstatistik* 1871, p. 56; 1872, p. 172. *Statistisches Jahrbuch für das Deutsche Reich*, 2nd issue, 1881, p. 86.

62 HA-Krupp, WA III 12: F. Krupp to C. Meyer from 17 October 1871.

63 GHH-Archiv, 20001/38: Betriebs- und Jahresberichte 1870, Walzwerk Neu-Oberhausen, 1 Februar 1871, Lueg.

64 Wilhelm Hupfeld, Die Ausdehnung der Bessemer- und Martinstahl-Fabrikation in Deutschland, in: *Zeitschrift des berg- und hüttenmännischen Vereins fur Steiermark und Kärnten* 6 (1874), pp. 9–13, here p. 10.

65 While Königshütte in Upper Silesia, Maxhütte in Bavaria and Königin-Marien-Hütte in Saxony also had limited quantities of low-phosphorus iron ores, they obviously processed the entire pig iron obtained from them to Bessemer steel themselves. PEE, Hoesch, p. 240, Müller, *Georgs-Marien*, pp. 26f.

British market, which was already hard put to find enough. The German and British Bessemer works thus came up against the same bottle-neck. Indeed, the situation was even more acute in Germany, where the works under construction reacted to the threat of supply problems by securing huge quantities of suitable iron ore or pig iron, often on credit, which then lay about unproductively in dumps and simply served to exacerbate the shortage.[66] In Germany, fears were already being expressed that the continuing shortage of Bessemer rails, and the ensuing steep price rises would drive the railways back onto puddled rails. This was a rumour which must have cast into great despondency those works which were in the throes of conversion: 'if that is the case, the Bessemer plant won't have been worthwhile, since a splintered production comes too expensive'.[67] However, as the economic situation settled down and prices fell, the problem resolved itself. Though the wrought-iron works experienced their golden age during the founding years of the railway boom, whereas the Bessemer steel works were still having to struggle with a raw material shortage, the whole situation was reversed by the downswing.

(b) Depression: 1873–1879

In order to understand the economic depression of 1873–9 and its effects on the steel industries of Great Britain and Germany, it makes sense to distinguish three levels: (a) the American railroads' financial crises and their effects on Europe; (b) the competition between wrought-iron rails and their substitute, Bessemer rails; and (c) the different structures of the German and British rails manufacturers' markets. On all three levels, the wrought-iron and Bessemer works of both countries were affected in quite different ways. Whereas the financial crisis in the USA had only slight influence on production in German wrought-iron and Bessemer works, in Great Britain it led to the almost total extinction of the largest branch of the iron industry; the iron rails works. On the

66 This presented the exact picture drawn by Spiethoff in his interpretation of the economic trend from the highest point of the upturn, which already carried the kernel of the following downturn within itself. Arthur Spiethoff, *Die wirtschaftlichen Wechsellagen. Aufschwung, Krise, Stockung*, 2 vols. Tübingen and Zurich 1955, here vol. 1, pp. 181–4. Phönix was one of these 'stock-piling sinners'. The construction of the steelworks there did not progress as quickly as planned, so that iron ore from previously contracted deliveries piled up there in the course of the year. These excessive stocks brought Phönix considerable liquidity problems, which could only be surmounted by means of a comprehensive rescue package from the Schaaffhausenscher Bankverein and the Disconto-Gesellschaft. Mannesmann-Archiv, P1.25.35: Sitzung des Administrationsrates am 9 Juni 1874 und 11 Dezember 1876; P1.25.11.1: Ordentliche Generalversammlung am 30 Oktober 1872, Vortrag des Administrationsrates, p. 5; re rescue package P1.25.35: Sitzung des Administrationsrates am 19 November 1875.

67 GHH-Archiv, 20001/43: Betriebs- und Jahresberichte 1871, Walzwerk Neu-Oberhausen, 25 February 1872, handwritten comment by the technical director, F. Kesten.

other hand, its overall effects on Britain's general economic development were very limited, whereas in Germany it hastened the collapse of the wave of national speculation following the foundation of the Reich and so developed into the severest economic crisis of the nineteenth century. While the replacement competition between iron and Bessemer rails was overshadowed in Great Britain by the loss of its greatest export market, in Germany it was overshadowed by the shrinkage of their internal market. Thus, although the phenomenon of a dwindling market for rails may appear at first glance similar in both countries, its origins were very different in each and the counter-strategies adopted by the affected works were also very different.

The turnaround to depression came, exactly as expected and predicted, from the USA with the breakdown of the money market which financed the construction of new railroads. It began with the collapse of the New York banking house Jay, Cook & Co., following the failure of its bid to place new bonds at 6 per cent with Rothschild's in Europe. Jay, Cook & Co. were very heavily involved in the Northern Pacific Railroad, which had been founded with a high level of speculative risk. Of the anyway meagre nominal capital of 2 million dollars for 500 miles of line, only 10 per cent had actually been paid up.[68] The failure of Jay, Cook & Co. to find in Europe the money they needed to complete the lines led to their collapse and, in the end, as part of a regular chain reaction, to eighty-three railroad societies being declared insolvent with outstanding debts totalling 250 million dollars. This wave washed over Berlin with the bankruptcy of the Quistorpsche Vereinsbank, which was followed by 61 other German banks and various enterprises until early in 1874.[69]

This meant that American railroad construction had for the time being lost its European credit, which put an abrupt stop to laying new lines. Nevertheless, there was no longer any question of the British rails manufacturers, who had been the leading suppliers to the new lines, being caught by surprise at the time, although the size and extent of the collapse were unexpected.

In May 1873, with four months to go before the New York Stock Exchange folded, people in Great Britain were already referring to an 'uneasy feeling' about the price of rails, which was bound to affect the railways' orders. No one was yet expecting a real crisis to happen, but rather a drop in prices to stabilise demand.[70] A month later came the first cry that the iron industry all over the country was in a 'critical' state. The dwindling rails market would soon drag the other branches of heavy industry into difficulties as well, leading to the first announcements of works closures.[71] By the end of the year matters had gone so

[68] Schumpeter, *Konjunkturzyklen*, 1, pp. 346ff. Henry Underwood Faulkner, *American Economic History*, New York 1960 (8), pp. 515f. Böhme, p. 342.

[69] Böhme, p. 343.

[70] *ICTR* 21 May 1873, p. 500.

[71] *ICTR* 25 June 1873, p. 620.

far that the *ICTR* published a leading article about the need to revitalise demand, recommending state intervention in the railways, or, better still, the construction of state railways.[72] This was followed by a series of articles throughout 1874 forecasting a quick end to the depression. Such hopes were generally based, as previously when the downswing was being predicted, not on any kind of indicators, but on previous, general experience of economic cycles. Where the wrought-iron works were concerned, it was only a matter of lasting out yet another lean period, and new orders would surely come sooner or later.[73] This time, however, they did not.

Where British heavy industry was concerned, the end of the American railroad boom entailed the loss of their largest single market. Given that the accompanying financial crash had fundamentally shattered all confidence in overseas railroad stock, it was extremely difficult, not to say impossible, to find the capital for any planned projects, and not only in the USA.[74] Consequently British heavy industry felt that the main reason for its penury was the reluctance of its fellow nationals to risk their money once more in railways overseas. For Great Britain was not only the workshop of the world, but also its leading money-lender, which meant that it had to extricate itself from its predicament. 'When we stopped the loans the countries ceased to make their railways, and to take our iron',[75] ran the lapidary explanation for the sales crisis in heavy industry. There would be no new upswing before the capital had first of all been raised in the countries that previously had borrowed money. As it was, however, the rails-exporting business collapsed and with it a whole row of wrought-iron works, which had been producing almost exclusively for it, whereas the internal demand, for Bessemer steel rails at any rate, remained stable (see Table 1).[76]

The Bessemer works had not only survived this depression, as they had the previous one in the late sixties, but had also improved their results. At the same time, they could rely on the secure base of the domestic market for replacement rails, which practically stopped using iron rails after 1877.[77] During 1874 and 1875, the first two slump years, most of the Bessemer works flourished as in the most favourable economic climate. They were overwhelmed with orders

72 *ICTR* leading article of 31 December 1873, p. 1317.

73 *ICTR* 7 January 1874, p. 9; 14 January 1874, p. 36; 28 January 1874, pp. 94f.; 20 May 1874, p. 572; 17 June 1874, p. 698; 15 July 1874, p. 72; 5 August 1874, p. 166; 26 August 1874, p. 263; 16 September 1874, pp. 358f.; 23 September 1874, p. 390.

74 This applied just as much to South America. *ICTR* 15 February 1878, p. 183.

75 Mundella MP at the annual dinner of the Sheffield Trades Council on 20 November 1877, in: *ICTR* 23 November 1877, p. 579.

76 Table 1, in: *BITA* 1879, p. 34 (Production and Export) and *B.T.* 1871, p. 94; 1872, p. 94; 1873, p. 94; 1874, p. 92; 1875, p. 92.

77 *ICTR* 12 January 1877, p. 43.

'to utmost capacity'.[78] Since the price drop was accompanied by a comparable fall in costs, profits remained at a satisfactory level. Even works like Samuel Fox, which was unfavourably sited with no railway link until the end of 1876, were able to pay out their required 10 per cent dividends.[79]

It was only when the unremitting expansion of production led to sales being diversified to the export markets and when the continental steelworks, which had since grown stronger, started to compete with the British, that the works around Sheffield with their less adequate freight facilities, started to lag behind.[80] As the prices of rails and the cost of materials fell sharply, the proportion of freight costs in the overall expenses naturally rose, which meant that these works were at a disadvantage when competing with coastal works. The two largest steel manufacturers in the Sheffield district, John Brown and Charles Cammell, attempted to alleviate this effect by manufacturing steel for high-quality products, namely armour plating.[81] Although these were shipbuilding materials, the value of the product was so high that the proportion of freight costs no longer played a major role.

In 1876 the price war between the Bessemer works developed for the first time into a serious danger that prices would be agreed which, in the opinion of outside observers, would necessarily make a loss. Such prices could only be justified if they ensured that the works could run at full capacity. Nevertheless, the general tenor of trade reports about the Bessemer industry remained positive, compared with all other spheres of heavy industry.[82] It was the only branch which could show uninterrupted growth.

Wrought-iron rails were driven out of the German market in much the same way during the economic downswing. Indeed, this had not been unexpected; the large railworks had begun adapting their plants years ago, but most of them were too late for the rails boom which came after the war, and construction work had anyway been partly delayed by the war itself. The ensuing anxious period of shortage of Bessemer rails and the prices charged for them, which were clearly higher than British or French levels, had secured the old wrought-iron works yet another unhoped-for upswing.[83] Subsequently, however, wrought-iron rails began to disappear from the German market at the same rate

[78] *ICTR* 26 February 1875, p. 234.

[79] BSC-East-Midlands RRC, loc. no. 42129: minutes of general meetings, 30 August 1875 and 30 August 1876.

[80] This tendency was first remarked on in *Iron* 10 October 1874, p. 462. And again in *Iron* 5 December 1874, p. 518, and *ICTR* 4 June 1875, p. 681.

[81] These traditional steel producers enjoyed some protection from imitators in that the production plant for armour plates cost £250,000 in the first instance. *ICTR* 8 June 1883, p. 665.

[82] *ICTR* 1 September 1876, p. 239; 29 September 1876, p. 355; 22 June 1877, p. 692; 24 August 1877, p. 213; 20 July 1877, p. 71; 9 November 1877, p. 523; 16 November 1877, p. 551; 6 September 1878, p. 267; 27 September 1878, pp. 360 and 361.

[83] *BITA* 1879, p. 41.

Table 1. *Production and export of rails: Great Britain 1871–1878*

Year	Production (in tons)	Exports (in tons)	USA exports (in tons)
1871	1,365,367	979,057	505,920
1872	1,276,637	945,420	497,783
1873	1,104,329	785,014	178,051
1874	1,105,243	782,665	91,993
1875	865,446	545,981	17,013
1876	852,210	514,654	discontinued
1877	871,406	498,256	discontinued
1878	779,534	439,392	discontinued

as the Bessemer works were stepping up production. Unlike in Great Britain, however, the German domestic demand for rails was not at all steady, being more in line with developments in America, although Germany was largely spared the dramatic repercussions of the recession there.[84] It was only after 1874 that the construction of new lines and renovation projects were past their peak. Once past it, however, the domestic consumption of rails fell to less than half in just two years, from 1874 to 1876.[85] In heavy industry, the wrought-iron railworks bore the brunt of this development, whereas the Bessemer works were in the same period able to raise production by a third. By this time, however, the precondition for this sort of expansion lay in stronger rails exports, which had not previously played an important part.

As in Great Britain, in Germany the production capacity of Bessemer works had grown faster than demand. Complaints were made about the plants' lack of work and the number of converters actually in use fell for the first time in 1876.[86] Unlike Great Britain, the German works were not sheltered at all from the severest effects of the slump. No dividends at all were paid in the Ruhr between 1875 and 1877, let alone the 10 to 18 per cent announced by British

[84] All the same, four railway companies had gone bankrupt by the beginning of 1874. Böhme, p. 343.

[85] Construction of new lines: 1873 – 1475 km; 1874 – 1615 km; 1875 – 2496 km; 1876 – 1409 km; 1877 – 1138 km; 1878 – 775 km according to Sering, p. 156. For consumption of rails, see Table 11 on p. 112. Consumption amounted to 524,000 tons in 1874, and only 247,000 tons by 1876. See too Rainer Fremdling, *Eisenbahnen und deutsches Wirtschaftswachstum 1840–79. Ein Beitrag zur Entwicklungstheorie und zur Theorie der Infrastruktur*, Dortmund 1975, pp. 13–34.

[86] From 63 (1875) to 44 (1876) and finally to 32 (1877). According to *Reichsstatistik* 1875, pp. 91 and 93; 1876, pp. 91 and 93; 1877, p. 122. These absolute numbers should not be attributed too much significance, since the Statistics Office worked partly on the basis of estimates. The trend, however, is so clear that this does not affect it.

Bessemer works.[87] The situation had not substantially improved by 1876, although production was buoyant and was climbing again. Although we have already seen that the construction of works had not been determined by the *Gründerboom*, the accusation that management had overestimated future sales and built too big, still holds good. In any case, the outward signs, under-capacity production in spite of rising sales, indicate that the malaise was self-inflicted and that its origins lay in over-investment at the beginning of the decade. This conclusion can be tested, in that projected sales as well as actual sales, were compared with what were known at the time to be the new plants' production capacities. The starting point for this can only be the projected sales and the production capacity of the plants at the moment when the decision to invest was made. It is only when it can be shown that the production capacities of the Bessemer installations, when they were under construction in 1870–1, were already greater than the subsequent demand, that the accusation of having invested in excess of actual requirements holds good. If this was not the case, the factors which determined the slump should be sought after the construction phase and not during it.

3 Supply potential and market absorption

Determining production capacities and the adjustments to them requires taking a further more precise look at the Bessemer plants of the late 1860s and early 1870s. Like most resources, they did not exist in quantities or sizes which could be pooled to demand. Charles Babbage's well known principle of multiples applied here too,[88] so that the expansion of steelmaking could only happen in rough stages and not continuously.

The standard layout for Bessemer plants, as outlined in 1860–2 by Bessemer and his partner Longsdon and the machine construction firm Galloways in Manchester, had proved extremely long-lived and was retained in its basic principles for several decades. The similarity of the plants was also based to a major extent on the fact that Galloways was initially the only firm licensed to produce the machines according to Bessemer's patents. Although its apparatus was not always purchased, it was always faithfully copied.[89]

A Bessemer plant generally consisted of two or three converters, grouped

[87] For instance, 10 per cent with S. Fox, see note 79; 18 per cent with Brown, Bayley and Dixon for 1878; see *ICTR* 8 March 1878, p. 267.

[88] The term 'principle of multiples' is derived from Philip Sargant Florence, *The Logic of British and American Industry*, revised edition, London 1961, p. 51. In it, Sargant Florence expressly mentions that the content of the term had already been described by Charles Babbage in his book *On the Economy of Machinery and Manufactures*, London 1832, p. 175.

[89] Carr and Taplin, p. 26. Königshütte in Upper Silesia, which was then still government property, copied a Galloways plant in the 1860s after an earlier construction of their own had completely failed to work. Hermann Wedding, Versuche zur Entphosphorung des Roheisens

around a circular casting pit (see Figure 3). The converters were canted inwards to allow the steel to be poured into a ladle suspended from the central casting-pit crane; it was then swung round and poured into individual ingot-moulds, where it solidified into ingots. Once filled, the ingot-moulds were lifted out of the casting pit by small cranes, and taken to the smiths' shop or the rolling mills.

The crucial element in the plant was its tippable converter, in which pig iron was refined to steel. Its average loading capacity gradually settled down during the sixties at around five tons. The smaller converters of the early days, with a loading capacity of 1½ to 3 tons were soon replaced by the bigger versions, once it had been established that the refining process occurred more reliably over the same length of time when larger charges were involved.[90]

The converter bottom was the most susceptible part of the whole plant; it was subjected to heavy wear and tear because the reaction between iron and air was most violent at this point. During the 1860s it would have been most unusual to make more than six charges using only one bottom.[91] By then it would be so badly burnt that there was a danger of the steel breaking into the wind chest. If the bottom had been burnt to within minimum safety levels or if the tuyeres had been destroyed, then the converter was allowed to cool, new tuyeres were fitted and a new bottom was pounded into the converter from inside. Bessemer had already provided replaceable bottoms to make this frequent task easier; however, once they had been used these bottoms, like the usual ones, could only be replaced when the converter had cooled down, and then they too had to be pounded down from inside. While this had little effect on the actual time required for repair work, it did allow the bottoms to be repaired more carefully and so provide the tuyeres with a better fit.

On account of the hours spent every day on repairs to the converters during the course of normal operations, Bessemer had prescribed a minimum of two converters per plant (see Figure 3).[92] The idea was to ensure that one converter

in Königshütte, in: *Zeitschrift für das Berg-, Hütten- und Salinenwesen in dem preussischen Staate* 14 (1866), pp. 155–9; Illies, pp. 225–9. Krupp also copied the Galloway plant when he extended his Bessemer works, after buying four converters there. HA-Krupp, WA IX n 294: V. Nida, Neuanlagen und Reparaturen 1853–62, p. 5. Further references to the purchase of Galloways plant can be found in *Engineering* 4 January 1867, p. 19; 8 February 1867, p. 125; Däbritz, p. 113; Beck, *Geschichte*, p. 615.

90 The 5-ton converter was long considered the optimum size, since in it Spiegeleisen (which was added during the process) was distributed best in the steel, whereas with larger converters this was not always the case. Samson Jordan, *Métallurgie du fer et de l'acier*, 2. *Études pratiques et complètes sur les divers perfectionnements apportés jusqu'à ce jour dans la fabrication des deux métaux*, Paris 1872, pp. 352f.

91 Beck, *Geschichte*, p. 130. For greater detail see Chapter 3, p. 000.

92 Illustration from Louis Grüner, *The Manufacture of Steel*, translated from the French by Lenox Smith with an appendix on The Bessemer Process in the United States, by the translator, New York 1872, pl. VIII.

would always be in running order while the other was being repaired. A third converter would simply provide an additional fall-back in case the converter in use should break down before repair work on the other was finished. The basic idea behind the whole arrangement was to have one converter in use at any time, allowing the steelworks to engage in constant production. The important thing about this was that it was not the number of converters but the number of casting pits which determined the productivity of a Bessemer works. The decision whether to fit out a plant with two or with three converters depended simply on the degree of reliability required – in other terms, on the works manager's organisational skills. Irrespective of the number of converters in a works, only one would be operating at any given time per casting pit.[93] A Bessemer casting pit of this kind formed a closed unit, which could not be subdivided, and the works were made up of such units.

Reminders of the puddling process were to be found only in the stages immediately prior to and after the actual conversion process: smelting the pig iron and processing the castings into ingots. Since the converters, unlike the puddling furnace, were loaded with liquid pig-iron, this had first of all to be smelted; the same applied to the Spiegeleisen which was added after the refining process. The smelting was done in what were known as 'reverberatory furnaces', the construction of which was very similar to that of puddling furnaces. The most important feature common to both furnaces was the way the fuel was kept separate from the iron, so that the latter was exposed only to the heat of the flames and not to the impurities in the coal. Basically, the first stage of the puddling process was replicated by the smelting process in the reverberatory furnace, but the actual refining process was left to the converter and not to a workman.

The reason behind this laborious and 'cleaner' method of smelting pig iron was to be found only partly in the desire to retain as many tried and trusted elements as possible in the new process, although this certainly did play a part. The great sensitivity of Bessemer steel to the slightest quantities of phosphorus had in the initial stages given the process the reputation of being very sensitive to the raw material employed, so that its early users exercised excessive caution in this respect because they lacked precise means of analysis. The reverberatory furnace was attributed quite incorrectly with having a cleansing effect on the pig iron.[94] For this reason, too, no attempt was made to run the liquid pig iron directly from the blast furnaces to the converters, and thus avoid re-smelting it. Given that Bessemer steel had penetrated the market in the first place as a new high-quality material, there was little inclination to take any

[93] Fairbairn, p. 175, mentions the frequent need for repairs as a reason. He considers running two converters simultaneously as an exception for casting particularly large pieces. Ibid., p. 172.

[94] Beck, *Geschichte*, p. 160.

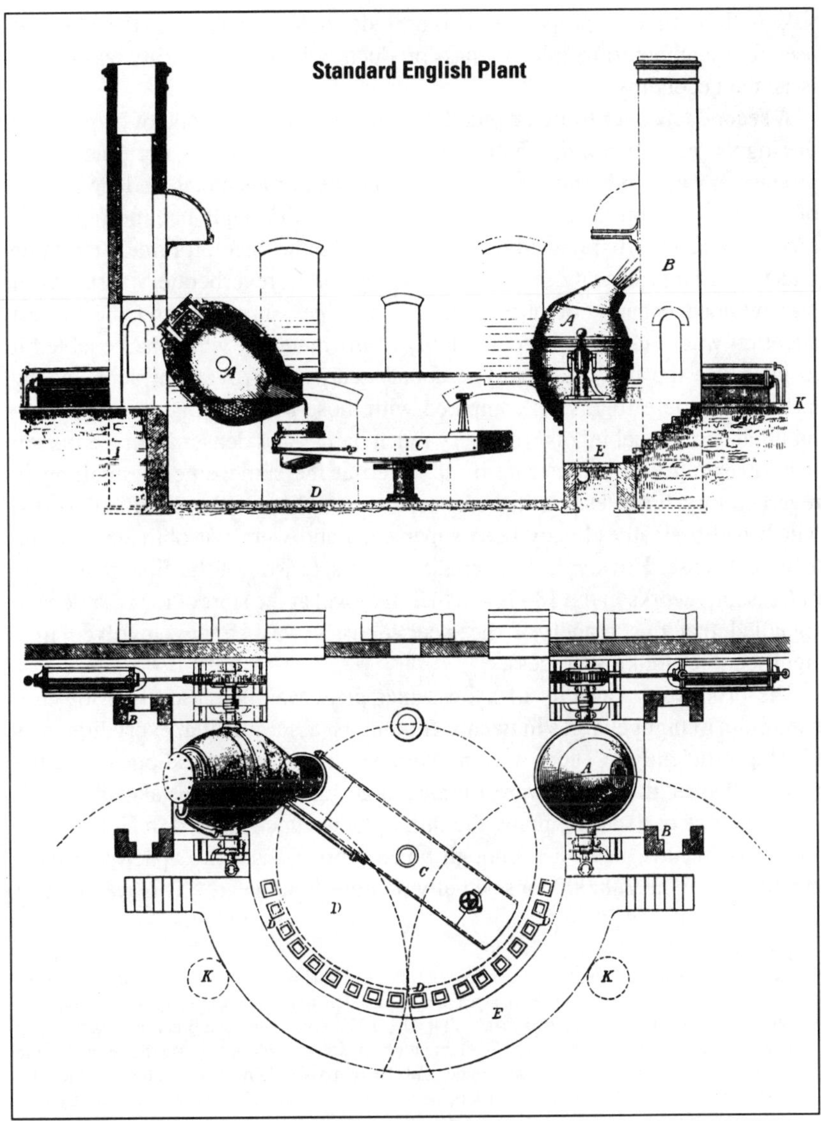

3 A Bessemer plant in the 1860s

risks with its material properties. This attitude changed only when the steel was used for manufacturing uniform mass products, when select quality was less an issue than economy.

A second left-over from the puddling process was the method of hammering the ingots prior to rolling. With the puddling process it was not practical to produce any sort of homogeneous ingot from the porous metal, and the quality of the finished product depended basically on thorough hammering. This practice was initially retained in the case of Bessemer steel, for much the same underlying reasons as those for smelting pig iron in reverberatory furnaces. It was retained because of uncertainty about a new process, and the utmost attention was paid to the special quality of the material, which had enabled it to penetrate the new market. The actual quality improvement achieved by hammering these ingots, as compared with those made of puddled iron, was minimal, since steel ingots were very much harder and denser from the outset. It is revealing that hammering disappeared at the same time as smelting in reverberatory furnaces. Both these steps were relics of the puddling works, which had basically already been superseded and were not obligatory in the technical sense. However, they remained characteristic of the first generation of Bessemer works in the 1860s and their removal at the start of the next decade signalled that this branch of the steel industry was already involved in a significant re-thinking process.

The production capacity of a Bessemer plant was increased in 1869 to a maximum of eight charges in twenty-four hours, assuming that everything was running efficiently.[95] There were no national or regional variations, since the plants all over the world were by and large uniform. If we take this daily production as our starting point, the thirty plants which existed in Germany in 1876 would have been just enough when working at full capacity to have manufactured the 288,000 tons actually produced that year.[96] This would have

[95] *Engineering* 15 October 1869, p. 259.

[96] According to a statement by the director of the Osnabrücker Stahlwerk there were 26 Bessemer plants in Prussia in 1876. Hermann Brauns, Die nordamerikanische und die deutsche Flußstahlerzeugung, in: *Zeitschrift* des VDI 21 (1877), col. 91–4. The other six plants were: Königin-Marien Hütte in Saxony (2), Gienanth near Kaiserslautern (1), Maxhütte in Bavaria (1) and the two Lorraine works, de Wendel and Dietrich & Co. However, this list also includes old plant which had by this time been superseded technically and closed down, as, for instance, in the case of the first small plants in the Hörder Verein, the Bochumer Verein, the Königshütte and Poensgen. It was a long time since these firms had considered operating these plants. Ibid., col. 91f. It is very doubtful whether these plants were at all capable of being worked. Indeed, the assumption that altogether 30 plants were in working order is not set too low at all. If one assumes that the average size of the converters was 5 tons, this gives a possible annual production of 5 tons × 8 (charges per day) × 5.5 (24-hour working days per week) × 50 (weeks) × 30 (plant) = 330,000 tons. Since the *Reichsstatistik* recorded the production of finished steel products and not of ingots, we have to subtract a further 10–15 per cent processing losses, which works out at almost exactly the actual annual production of 288,000 tons.

required scarcely any degree of over-capacity and certainly no whole plant stoppages, such as actually occurred, and the quantities produced from 1877 onwards would have been completely impossible. Since the plants were all derived from the period prior to the collapse of the boom (*Gründerkrach*), the technical and organisational improvements to the existing works during the years after 1873 must have played a considerable role, and the figures relating to the capacity of these works were consequently being constantly corrected and upgraded during the course of these years.

A widely read article appeared at the beginning of 1874 in the *Zeitschrift des berg- und hüttenmännischen Vereins für Steiermark und Kärnten* (Journal of the Society of Miners and Ironmasters in Steiermark and Kärnten), which calculated the capacity of the German Bessemer works and gave vent to the first considerations about the feasibility of finding a sufficient outlet on the market. Assuming that 60 of the 71 converters available could be worked continuously, a possible yearly production of 450,000 tons was mooted.[97] Taking two converters per plant as the starting point, this would give ten charges a day per plant. Only a year later, the *Moniteur des intérêts matériels* provided a new estimate of German Bessemer steel production, amounting to 600,000 tons a year from 69 converters.[98] This was already assuming twelve charges per day per plant. A year later again, in June 1876, a Belgian calculation of the Bessemer steel capacities in the whole world was published. It came to a possible production of over 2.5 million tons from 270 converters, of which 700,000 tons were attributed to Germany.[99] The whole calculation was based on the assumption that 10,000 tons could be produced per converter, involving an average rate of more than thirteen charges per day per plant.

Definite progress was reported in the survey which Patrick Peterson conducted in Great Britain during the winter of 1876–7. It was first published in *Jernkontorets Annaler* and soon afterwards was given a wider readership by other trade journals.[100] He reckoned on a capacity of five to eight tons per converter, and on a weekly production of 500 to 1,000 tons per plant with two converters. If one bases his lower figure on the then usual five-ton converters and his higher figure on the then newest eight-ton converters, this entails a daily rate of 18 to 23 charges. Somewhat higher still were the figures provided by H. Brauns, the director of the Osnabrück Sahlwerk. He considered in 1876 that most German works could 'easily' achieve 22 to 26 charges per day per plant. His article was prompted by reports from the USA that over there 30 charges had already been common in 1875.[101]

97 Hupfeld, *Ausdehnung*, p. 11.
98 According to *ICTR* 22 January 1875, p. 77.
99 According to *ICTR* 30 June 1876, p. 735.
100 According to *Berg- und hüttenmännische Zeitung* 37 (1878), p. 397.
101 Brauns, col. 93.

Given that these capacity increases in Great Britain, Germany and the USA applied not only to new works but to the industry as a whole, the alterations which were effected to increase capacity must have been undertaken from the beginning of the seventies onwards, and they were particularly effective by the middle of the decade. Hence the well-nigh perfect equilibrium between productive capacity and market volume in Germany on the basis of the starting data could not materialise. It cannot, however, be claimed at this stage that the industrialists would have anticipated the cumulative effect of their investment decisions with this degree of precision. There are, however, a few unmistakable indications that considerations of this nature did at least play a role.

When one looks at the steelworks in Rhineland and Westphalia, it is once again noticeable that the works which came late onto the market generally restricted themselves, unlike already existing works, to a single Bessemer plant, adopting in this way a minimal strategy. Exceptions to this were the total newcomers to the scene, Rheinische Stahlwerke and Osnabrücker Stahlwerk, which could reckon on having a competitive edge since they possessed their own iron-ore supplies, and the previously established GHH works. GHH's original plans had provided for four plants and the money for them had also been agreed.[102] They were each to have been built in two stages. The first stage advanced only very slowly, which meant that they lost their initial lead. Consequently, the second construction phase was abandoned in 1873. It did not appear economically sound to reduce the two plants to one since a large twin steam engine for a total of four converters had already been provided. Apart from which it has to be borne in mind that GHH, like most of the other new Bessemer works, was already established as a major rails supplier and disposed of considerable 'good will' in this business, which made it a primary consideration to maintain their position by keeping supplies steady. Krupp too, otherwise so expansion-happy, behaved in much the same way as GHH in this respect. Krupp's Bessemerwerk III was planned to provide a further five converters by the end of the sixties, after which it turns up in various sources although it was never actually built.[103]

[102] GHH-Archiv, 2001/36: Promemoria Carl Lueg betr. Ausbau Oberhausener Werke vom 1 Juni 1872.

[103] Brauns, col. 91f., was already complaining that the number of Krupp converters had been set far too high. The construction plans for Bessemerwerk III had been abandoned by October 1873, after the building for it had already been erected. A Siemens-Martin works was later set up in this structure. HA-Krupp, FAH II B 313: A. Krupp to Procura on 8 October 1873. See too pp. 171f. The confusion about the number of converters at Krupp's and the works extension there has its origins in the firm's obsession with secrecy: the Statistics Office could never do more than estimate its production figures and means of production, because Krupp never produced any information about these. *Zeitschrift für Berg-, Hütten- und Salinenwesen in dem preussischen Staate* 21 (Berlin 1873), *Statistischer Ergänzungsband*, p. 239. An article by Hermann Wedding in the *Reichsanzeiger* 4 October 1876 was the first to publicise the

On the basis of the conclusions reached above, it can already be stated that the accusation which was levelled at the Rhineland and Westphalian steelworks in particular, that they had blindly overestimated the sales possibilities and had provided far too large capacities, cannot be sustained. Rather, the origins of their over-capacities must be sought in the technical and organisational development of the Bessemer process *after* the fundamental decisions to invest had been taken.

'These plants were fitted out in accordance with experiences at the time of actual consumption; however, the technology of this new process very soon developed to such an extent that four or six times the previous quantity could be made using the same apparatus, meaning that production capacity was increased to superfluity.' This was how the firm of Krupp viewed this progression during the seventies.[104] The Bochumer Verein's management spoke of a 'technical crisis' in this connection, which was supposed to have introduced the oppressive over-production.[105] The secretary of the Iron and Steel Institute J. S. Jeans characterised the development as follows in his handbook on steelmaking, which appeared in 1880: 'concerning the large yields obtained in modern steel works. It is in this respect, more, perhaps, than in any other, that the Bessemer system has advanced within recent years. In its main mechanical features that process is much the same now as it was when originally established on a working scale by its distinguished inventor. But by abolishing the pit, by using improved bottoms, and by sundry other improvements of detail that are referred to in the text, the output of steel from a given plant has gradually been raised to a figure that would have been deemed almost incredible a few years ago.'[106]

The data which were available at the time when the decision to build was made does not allow one to conclude that the potential sales markets had been overestimated to a significant extent. The behaviour of those works which came late onto the scene even allows one to assume that they had indeed perceived the limits of the market and by making a minimal entry into the new

excessively generous number of eighteen converters in working order,which would in fact only have been achieved had Bessemerwerk III been ready for operation. It proved particularly effective when Wedding's figures were adopted by Franz Reuleaux, *Briefe aus Philadelphia*, Brunswick 1877 (2), p. 83, a work which won considerable renown by attempting to demonstrate how backward German industry was compared with American. From there they found their way to Ulrich Troitzsch, Die Einführung des Bessemer-Verfahrens in Preussen – ein Innovationsprozess aus den 60er Jahren des 19. Jahrhunderts, in: Frank R. Pfetsch (ed.), *Innovationsforschung als multidisziplinäre Aufgabe*, Göttingen 1975, p. 239. The converters intended for Krupp's Bessemer works had apparently already been finished by the seventies, but had not yet been set up. HA-Krupp, W. A. VIII 130: Erinnerungen von Georg Zanders vom 10 Februar 1909.

104 PEE, Meyer (Krupp), p. 77.

105 PEE, Baare (Bochumer Verein), p. 807.

106 Jeans, *Steel*, p. IX.

industry with only one plant, they simply wanted to avoid being totally squeezed out of their familiar market. This was the explanation which Louis Baare gave to the Bochumer Verein's shareholders at their AGM in 1875: 'One produced Bessemer steel rails with twice and three times the resistance of iron ones although at only a quarter more cost. What was the inevitable result? Even the iron works had to come to the conclusion, after lengthy resistance, that Bessemer steel would drive iron out of the railway market, as has already happened to a great extent today, and consequently, there is only one answer to the question about the survival of the major German iron industry! It is: go into it with new and heavy sacrifices, so as not to go under!'[107]

Although extensive plans for expansion did exist around 1870 for a few Bessemer works in the Ruhr in order to equip them for an export trade which was thought would develop shortly, for the time being not one of them left the drawing board. Take for instance Krupp's plans for his Bessemerwerk III, which are mentioned above. Moreover, he owned a further site outside his Essen works on which he meant one day to build an integrated blast-furnace and steelworks system with several Bessemer plants.[108] The Hörder Verein had laid out its new blast-furnace works precisely so as to allow a further steelworks to be inserted alongside, which could be fed pig iron directly from the blast furnaces.[109] The Bochumer Verein was planning to build a blast-furnace plant within its precincts, equipped with six blast furnaces of the largest dimensions available at the time, to make pig iron exclusively for processing to steel. Of these, two were built during the seventies, the second of which was started up with scant enthusiasm only in 1878. In contrast with these ambitious plans, the actual level of investment was characterised by extreme reluctance and, as far as individual works were concerned, permitted only a minimal programme. Why, in spite of all this, it was soon necessary to complain about 'massive over-production' has already been intimated in the foregoing quotations from J. S. Jeans and the Krupp representative and will be further investigated in the following pages.

[107] Werkarchiv Krupp-Bochum, BV 12600 Nr. 1: General-Versammlung am 28 September 1875.

[108] HA-Krupp, WA IV 1429: Meyer to Goose on 8 March 1876. Their plans in Duisburg and northern Spain are discussed in this letter.

[109] *Hoerder Bergwerks- und Hüttenverein, 50 Jahres seines Bestehens als Aktiengesellschaft, 1852–1902*, Aachen 1902, pp. 22f., 110. Däbritz, p. 172.

3 Surmounting the Slump: the individual strategies of firms

1 Rationalisation strategies

By rationalisation strategies we mean here 'all measures which are aimed at optimising the labour input in the means of production, the material input in the means of production and the length of time required for the production process'.[1]

It became apparent while researching Bessemer steelworks that these particular three measures were not always pursued with the same degree of intensity. It is much more a case of distinguishing between trends in different periods and places. Until the outbreak of the economic crisis of 1873 efforts at optimising the material input predominated, and were initiated chiefly in Great Britain. After which, during the slump years between 1871 and 1879, people's interest turned more towards cutting the time required for the production process. This strategy was undertaken primarily by American and German Bessemer works. The optimisation of the labour input, on the other hand, did not produce any independently formulated strategy during the period in question. The essential advance in the manufacture of steel had already been achieved by the changeover from puddling to the Bessemer process. The rationalisation of labour input never amounted to more than the natural effects of speeding-up the refining process in the steelworks, which took place anyway even without being specifically intended. This was certainly due to the low proportion of wages to overall costs.

(a) The optimisation of the material input

This was mainly a matter of eradicating the remnants of the puddling works, referred to above. Once the first-generation Bessemer works in Great Britain had gained some experience with the new process, it was recognised that both

[1] Gerhard Brandt, Bernard Kündig, Zissis Papadimitriou, Jutta Thomas, *Sozioökonomische Aspekte des Einsatzes von Computersystemen und ihre Auswirkungen auf die Organisation der Arbeit und der Arbeitsplatzstruktur, Bundesministerium für Forschung und Technologie: Forschungsbericht DV77-04*, Frankfurt am Main 1977, p. 108.

smelting pig iron expensively in reverberatory furnaces and hammering the steel castings prior to rolling were superfluous and extravagant procedures which no longer seemed necessary in technical terms. Efforts at removing them were initiated at almost the same time in 1866/7.[2]

Analyses had shown that fears about the pig iron being exposed to impurities in the cupola furnace through contact with coke were unfounded.[3] At the same time, cupola furnaces were being used in foundries for smelting pig iron. This sort of furnace was basically similar to a small blast furnace and was distinguished above all by its ability to smelt pig iron in an uncomplicated and cheap way. English steelmakers had further established that during the unavoidable oxidation of the pig iron in the hitherto usual reverberatory furnaces, a considerable part of the silicon was already turned into slag.[4] This was, however, thoroughly undesirable since silicon was desperately needed to generate heat in the Bessemer converter, and its presence in pig iron was promoted in blast furnaces with a high fuel consumption, and correspondingly high costs. On the other hand, no observable loss of silicon occurred in the cupola furnace.[5] Thus, by the end of the sixties, these two factors were the principal reasons why the cupola furnace replaced the reverberatory furnace. According to Dürre, who propagated it in Germany on the basis of British experience, the fuel consumption of the cupola furnace was only 45 per cent that of the reverberatory furnace.[6] At the same time, cupola furnaces were far less liable to need repairs.

Cupola furnaces were rapidly established with the construction of new Bessemer plants and even the older works had replaced the reverberatory furnaces in their operating Bessemer plants with cupola furnaces by 1875 at the latest. A French ironmaker, who had gone on an extended journey through the iron districts of Britain in 1875, emphasised in his report that he had not seen a

2 Beck, *Geschichte*, p. 160. Ulrich, Wiebmer und Dressler, Reisenotizen über den englischen Eisenhüttenbetrieb, in: *Zeitschrift für das Berg-, Hütten- und Salinenwesen in dem preussischen Staate* 14 (1866), p. 327. Ulrich, Aust und Jänisch, pp. 6 and 13.

3 Beck, *Geschichte*, p. 160. Ernst Friedrich Dürre, Notizen über das Bessemerwerk zu Seraing (Aktiengesellschaft John Cockerill) mit besonderer Berücksichtigung einer späteren Verwendung des fabricierten Stahls, in: *Zeitschrift für das Berg-, Hütten- und Salinenwesen in dem preussischen Staate* 18 (1870), p. 268.

4 Knut Styffe, Über Bessemern und Gusstahlfabrikation, in: *Berg- und hüttenmännische Zeitung* 27 (1868), p. 169.

5 William Hackney, The Manufacture of Steel, in: *Minutes of proceedings of the Institution of Civil Engineers* 42, session 1874–5, part IV, p. 21. Re the increased cost of making silicon-rich pig iron see *Zeitschrift des VDI* 32 (1888), p. 457.

6 Ernst Friedrich Dürre, Über die relativen Vorzüge des Cupoloofen und des Flammofenbetriebes für das Einschmelzen der Bessemerchargen; mit besonderer Beziehung auf die Bessemerwerke von Rheinland-Westphalen, sowie die Cupoloofenprofile der Neuzeit, in: *Berg- und hüttenmännische Zeitung* 28 (1869), p. 422.

reverberatory furnace still operating in a single Bessemer works.[7] In Germany, Krupp was the last works to complete the changeover to cupola furnaces.[8]

The cost of altering existing Bessemer plant was not too exorbitant. The 33,000 marks set aside by the Bochumer Verein in 1875 for this purpose may serve as a pointer. The firm reckoned with 2,500 marks for a single cupola furnace.[9] The actual Bessemer plant, with its converters and casting pits, was generally unaffected by the alteration work and did not need to be adapted.

As well as the proportional savings in the cost of smelting pig iron, the introduction of cupola furnaces also secured additional capacity which was much more important for the development of the Bessemer process. As we have already seen, the production capacity of a first-generation Bessemer plant was limited by the protracted smelting procedure and the high cost of repairs and delays associated with reverberatory furnaces. This did not necessarily interrupt works operations because the number of reverberatory furnaces per plant could in principle be increased – and indeed, this is what happened in Dowlais. Here every converter was served by, not one but two reverberatory furnaces, which effectively doubled the plant's production potential. Exactly the same thing was done in the Barrow works, where three reverberatory furnaces were run for every two converters.[10] The production capacity of the Dowlais Bessemer works was accordingly set at a daily rate of 42 charges, i.e. 14 per plant. The maximum output for an individual plant was set at sixteen charges a day. However, such quantities were only calculations and they were not achieved in the course of actual operations. They required completely uninterrupted and continuous operation on the part of the reverberatory furnaces, which were not in general particularly distinguished for this. In any case, the available blowing engines were not capable of an accelerated operation of this kind. Thus the high capacity figures published by the Dowlais works, which were very exceptional, demonstrate the development possibilities envisaged in 1867 rather than the actual practice. The interesting thing about them is that the productivity limits of a Bessemer plant were already understood to lie, not with the actual conversion plant, consisting of converters and casting pit, but with the supply of liquid iron ore.

In 1867 two new Bessemer plants were set up for the purposes of comparison in Barrow, one containing the usual reverberatory furnace and the other a cupola furnace, enabling a decision to be made in the course of trial

7 L. Pelatan, Eisenhüttenmännisches aus England, in: *Berg- und hüttenmännische Zeitung* 37 (1878), p. 63.

8 The Krupp case study examines this in detail, pp. 51ff.

9 Werkarchiv Krupp-Bochum, BV 12900 No. 11: Protokolle der Verwaltungsratssitzungen vom 21 November 1875 und 29 Januar 1878.

10 The following account of the rationalisation measures taken at Dowlais and Barrow follows Ulrich, Aust und Jänisch, pp. 1f. and 13.

operations in favour of one or the other working method. They established that the cupola furnace could smelt its charges in half the time required by the reverberatory furnace, which meant that it had the same capacity potential as twice the number of reverberatory furnaces. Indeed, one unforeseen side-effect of the cheaper smelting method was that the exploitation of these newly acquired opportunities presupposed further developments in the organisation of production in the conversion process. In itself, removing a technical hindrance would not increase the production capacity of a plant. It had already been pointed out during the discussion about the high theoretical production capacity of the Dowlais works that 'The estimated production capacity of this plant depends on the possibility of carrying out the work necessary for casting, of removing the ingots after casting and of putting the ingot moulds back for the next charge within the shortest possible time.'

For the time being, however, there was apparently no compelling need to change the proven organisation of production in this sense. Bessemer steel production was exceptionally profitable, and its immediate bottle-neck lay not in its productive potential but in the supply of phosphorus-free pig iron. Figures relating to the capacity of new works, which were fitted from the outset with cupola furnaces, did not once attain levels which had already been considered possible at Dowlais in 1867.

(b) The optimisation of the time taken by the production process

The critically deflationary conditions in the years between 1873 and 1879, when the raw material shortage affecting the Bessemer works ended and the ensuing price competition was exacerbated by plentiful supplies, induced the works to turn their attention to more rational, that is to say more intensive, use of their available resources. The management of Phönix, for instance, formulated this strategy as follows: 'Given the extraordinarily low sales prices it was, if one did not want to pack in the whole concern, only possible to drive production costs down and to limit losses as far as possible, so that the necessary arrangements for a rational operation could be exploited as intensively as possible.'[11]

Given that the organisation of production enhanced the technical potential of existing plant, it was possible to achieve higher production in the Bessemer works without requiring extra capital investment. Speeding up operations is supposed even at that time to have achieved the effect of the fixed costs degression resulting from rising plant capacity use, which was demonstrated half a century later by Schmalenbach, who used the heavy industry of the

[11] Mannesmann-Archiv, P1.25.11.1: Ordentliche Generalversammlung, 27 November 1877, Bericht der Direktion.

Weimar Republic as his example. At the time, this was expressed in the following terms: 'Under certain circumstances this large number of charges is in any case associated with conspicuous advantages and one or other of our Bessemer works could possibly derive benefit from it, if it were to be capable of reaching a greatest possible production with its actual apparatus. In any case this would ensure that the interest payable on the investment capital, relating to production per hundredweight, would be reduced to a minimum.'[12] Although the changeover to cupola furnaces was effected from quite different motives, it served to remove the bottle-neck at the smelting stage and thus provided new openings for this rationalisation strategy.

The removal of one bottle-neck only made the next one all the more obvious. While there was now a continuous and constantly increasing amount of liquid pig iron available for the converters, a truly continuous refining process, during which all the equipment could have been in constant use, broke down on account of the converters' constant need for repairs. Further alterations to the existing plant were required before the intensification of the refining process as envisaged could be fully implemented. Unlike in the preceding rationalisation phase, particular technical problems (namely the high level of repairs required by the converter and lack of space) underlay these efforts towards achieving productive and organisational goals, which had to be solved before the latter could be realised.

Conditions in the Bessemer works had been turned around because the advances made in smelting the pig iron had overtaken the capabilities of the converter. The converter had itself become a hindrance and efforts were now being made to improve it, which followed two simultaneous courses. One strove to improve the fireproof lining so that the intervals between repairs could be extended. The other attempted to speed up the process of changing the converter bottom in order to curtail repairs. The common denominator to both attempts at solving the problem was the wish to extend the converter's effective operational life, and so to adjust it to the capabilities of the other apparatuses. Improving its fireproof lining involved a further motive, that of cutting costs in proportion to its greater durability. After all, preparing a bottom cost on average 40 marks in Germany, which was just as expensive as refining one ton of steel in the converter. Therefore it already made a perceptible difference to running costs if such a bottom could hold twenty instead of the previous five charges.[13]

A feature common to all the previous methods of repair was that the converter lining was patched from inside. The duration of this job depended in

[12] Peter Tunner, Amerikanische Verbesserungen am Bessemer-Converter, in: *Zeitschrift des berg- und hüttenmännischen Vereins für Steiermark und Kärnten* 7 (1875), here p. 236.

[13] Brauns, col. 93.

the first instance on the quality of the repairs required, but the bottom line lay in the time it took for the converter to cool, since the work had to be done inside it. Especially careful repairs, such as were, for instance, undertaken in the state's Königshütte, took up to thirty hours for one bottom, which would then last for ten to twelve charges. If the process was curtailed to nine or ten hours, the resistance of the bottoms was reduced to two or three charges.[14] Given a daily rate of six to seven charges, as was usual in the 1860s, these time-scales did not as yet give rise to any problems.

As might be expected, the shortest repair times in this period were achieved by Dowlais, the pioneer of rapid operations in Europe. There, a converter bottom lasted for a maximum of eight charges. This description of the job of changing the bottoms has been taken from an account by touring Prussian ironmakers:

> As soon as the tuyeres have been worn down to six or seven inches, the bottom is replaced; they turn the converter on its head, open the wind chests and push the old tuyeres and bottom out, after which it is set horizontally and sprayed intensively through the tuyeres until it is cold enough for someone to survive in there as necessary, and then they move on at once to installing the new tuyeres and to ramming the bottom. The tuyeres are inserted from below, at which the converter is set on its head again. The tuyeres are firmly stuck into the bottom plate and wooden props are set between the tuyeres and the closing plate of the wind box, after which they set the converter upright again and begin ramming. Although they are completely naked, the workers can endure only five minutes inside the converter, on account of the great heat, and they are then relieved. As soon as they have finished pounding the bottom, fire is introduced into the converter, and soon after some wind is blown into it and the bottom is quickly dried. The whole job takes only two to three hours and the converter is often worked again on the same day as the bottom was made, although repairs generally take place towards evening at the end of that particular work process.[15]

Whether greater emphasis was put on accelerating repairs or on improving the lining, depended in the first instance on the availability of fireproof materials. Dowlais in South Wales, for instance, procured its ganister from the Sheffield region, several hundred miles away. The biggest advances in improvements to the lining were made in Austria. In Ternitz the resistance of the masonry bottoms is supposed to have been gradually increased from *c.* 5 charges at the start to more than 25 charges and finally to 70–80 charges per bottom during the course of the seventies.[16] However, this was an

[14] Hasenrnohrl, Das Bessemern zu Königshütte in Oberschlesien, in: *Zeitschrift für das Berg-, Hütten und Salinenwesen in dem preussischen Staate* 16 (1868), pp. 213f.

[15] Ulrich, Aust und Jänisch, pp. 2f.

[16] Bleichsteiner's statement in: Protokoll über die am 6. Juni 1880 im städtischen Rathaussaale zu Leoben abgenhaltene Jahresversammlung der Section Leoben des Berg- und hüttenmännischen Vereins für Steiermark und Kärnten, in: *Zeitschrift des berg- und hüttenmännischen Vereins für Steiermark und Kärnten* 12 (1880), p. 343.

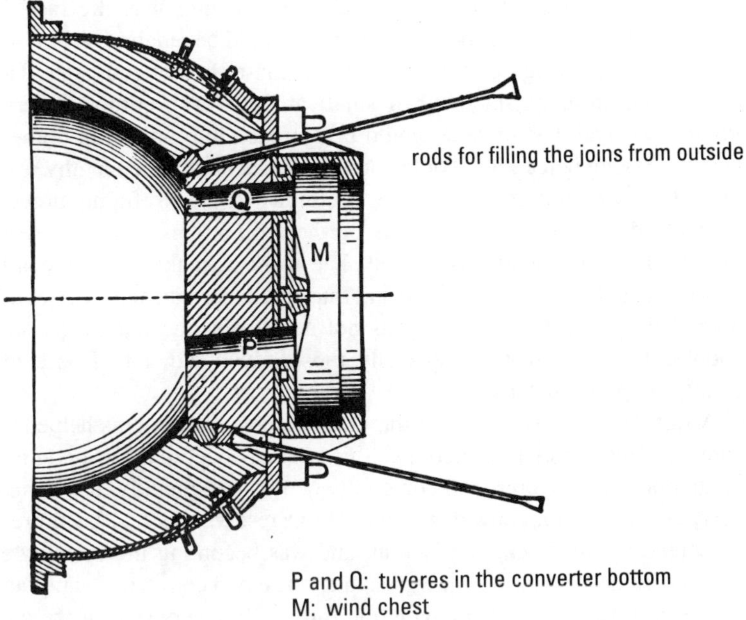

4 Holley's removable bottoms

exception. In 1880 changing the bottoms every 12 to 18 hours after 20 to 30 charges respectively was considered normal practice in Germany.[17] It should be mentioned that this calculation is already based on a rate of 40 charges in 24 hours. Thus the further development of fireproof materials almost kept pace with the increase in productive capacity.

In places where fireproof materials of a high quality were not available at affordable prices, however, the rate stayed at five to ten charges per bottom. This applies especially to the USA, where, as before, little more than six charges could be achieved. Effective relief was obtained only with the development of a new method, which simplified the task of changing the converter bottoms significantly and thus had the same effect on production capacity as the extremely resistant bottoms. The solution was patented by A. L. Holley on 17 December 1872, having been fully tested by him since 1869; unlike previous methods for changing bottoms it enabled the joins between bottom and converter lining to be filled in from outside (see Figure 4). Changing a bottom this way could be effected in less than an hour, which, given two converters operating alternately in one casting pit, meant that delays were

[17] *Berg- und hüttenmännische Zeitung* 40 (1881), p. 285.

almost eliminated.[18] Its greatest time-saving feature was the fact that the converter did not have to cool down before it could be repaired. A bottom could be heated to glowing point before it was set into the hot converter. The joins were filled from outside with a small amount of fireproof material (see Figure 4),[19] and the converter could be put back into operation immediately, without having perceptibly cooled down. Consequently the lengthy re-heating procedure was not required. The continuity of the refining process was guaranteed by a supply of finished, pre-heated bottoms, which were brought into the Bessemer works as needed. The work of making and preparing the bottoms could be done elsewhere and thus no longer interfere with continuous operations. Since this solution was not dependent on materials, it could be adopted by every works; especially those which did not dispose of high-quality fireproof materials.

A further aspect was the fact that these removable bottoms helped to overcome lengthy works interruptions. Since defects occurred most commonly in the bottoms and tuyeres, and could now be repaired quickly, it was no longer necessary to keep a converter constantly in reserve. As had been previously attempted by the Bochumer Verein, and was becoming increasingly general practice in the USA, it was now possible to use two converters simultaneously in one casting pit, thus employing the further 'free services' of the converter which had previously been held in reserve. The end result was that the Bessemer pit with two or three converters was now no longer the smallest viable unit. From now on, the efficiency of the single converter came to the fore.

Whereas previous advances in the practice of Bessemer works had mostly originated in Great Britain and had been associated with solving the problem presented by the materials, the American works now took the lead in technological matters with a series of technical and organisational changes, all of which pursued the same object: speeding up the conversion operation and making it more continuous by reducing the intervals between charges as far as possible. The time taken between one charge and the next was to be reduced to the minimum specified by the process, the actual refining period in the converter. The first time this was approximately achieved was in 1875 in the USA, where 50 charges had already been processed in 24 hours by one converter.[20] By 1877, a total of 73 charges in 24 hours had been achieved by one plant.[21]

18 Lenox Smith, Der Bessemerprozess in den Vereinigten Staaten, in: *Berg- und hüttenmännische Zeitung* 31 (1872), p. 424.

19 Illustration taken from Henry Marian Howe, *The Metallurgy of Steel*, 1, 2nd edn, revised and enlarged, New York 1891, p. 350.

20 *JISI* 1874 part 1, p. 231. *Berg- und hüttenmännische Zeitung* 33 (1874), p. 283.

21 *ICTR* 23 March 1877, p. 333.

Continuous operations had thus been achieved and the hitherto prevailing principle, that only one converter per plant could operate at a time, was dropped. When 73 charges were being made within 24 hours, only 20 minutes were left for refining each charge, with the blowing engines serving both converters in the plant alternately. This, however, was also the sheer time required for refining a Bessemer charge, meaning that the blowing engine, by far the most expensive part of the whole plant, was operating without a break. By then there can hardly have been any idle periods left. The same applied to the other appliances. While refining was going on in one converter, the finished steel in the other was simultaneously being de-oxidised with Spiegeleisen and subsequently cast. Immediately afterwards liquid pig iron had to be removed from the cupola furnace and charged into the converter, so that it was completely ready to go into operation again once the other converter had completed its twenty minutes' refining time, and the blowing engine had simply to be switched over. The same pressure was applied to teeming the ingots; since there was only one casting pit crane per plant, this too had no more than 20 minutes in which to teem a whole charge of five tons. According to R. M. Daelen the maximum speed at which Bessemer steel could be teemed was 300 kilograms per minute,[22] which meant that casting a five-ton charge took about 17 minutes. The last three minutes were surely needed for shifting and filling the ladle.

Thus it was clear that this sort of accelerated operation meant that the vital parts of a Bessemer plant – the converter, casting pit crane and blowing engine – were all being used at the same rate and without a pause. Compared to the changes in the organisation of production which underlay this ten-fold increase in the number of charges, the technical changes to the plants which made it possible – cupola furnaces and Holley's replacement bottoms – seem truly modest. As for the concrete organisational steps taken with regard to the labour input which had, by ensuring constant attendance on a Bessemer plant, made this enormous advance in productivity possible, these have not been recorded. Information about the organisation of labour was conveyed by means of personal experience and observation and these, unlike the costs involved and the results, were not noted down in the firms' records. There are, however, references to works managers looking around other works both at home and abroad, in order to deepen their organisational know how and to seek elsewhere hints about how to solve acute organisational problems; the same thing applied to journeys to the USA to study rapid driving there.[23] As far as the

22 R. M. Daelen, Über die Fortschritte in der maschinellen Einrichtung der Bessemer-Stahlwerke, in: *Zeitschrift des VDI* 25 (1881), col. 388.

23 Snelus, director of the West Cumberland Iron Co., went to the USA as early as 1872, according to his own account, in order to view the 'rapid driving' there. *JISI* 1880 part 1, p. 90.

managements of those firms whose written records are retained in their archives were concerned, the problem was reduced to the search for a suitable works manager, one capable of keeping abreast of technical and organisational developments within the industry.

The works manager had to present monthly cost accounts to the directors, thus allowing them to check up on his performance. They used these accounts to determine whether his efforts at rationalisation had achieved the desired results. For it was not enough simply to raise the number of charges and thus employ the plant more intensely, if at the same time the quality of the steel dropped below that required by the customers and more rejects were produced. During a discussion with British steel industrialists in 1874 Holley revealed that, although 50 charges a day were already being achieved in the USA, for the moment the economic optimum stood at only 30 to 35 charges a day. Charges over and above this number could be ascribed to rivalry between a few works managers, and in any case did nothing to improve profits.[24] This kind of brief spurt to test the technical and organisational potential of a Bessemer plant had also been undertaken in Great Britain at the same time. It is known that fourteen charges were once made with one plant in twelve hours at John Brown's in Sheffield towards the end of 1873. In Sheffield too, and at the same time, Wilson, Cammell & Co. made as many as 48 charges in 24 hours.[25]

This high tempo of production only became general practice in Great Britain in the 1880s and even then not everywhere. It was not only the organisation which had to be mastered but the qualitative requirements of the steel user, primarily the railway companies both at home and abroad, which also had to be met under these conditions. This was the main problem, which the American Bessemer works circumvented the more easily in that American railroads set less rigorous conditions than European companies, a fact which the German and British works had to take into account.[26] Blasting thirty charges of steel a day for milling into rails for a Prussian or English railway line made higher demands on the organisation of production in the steelworks than were made by the same quantity for use in America. Still, there was no doubt in Europe that the American method of 'hard driving' represented the most promising rationalisation strategy for the immediate future. Holley's credo was publicised in many trade journals and ran as follows:

24 Alexander Lyman Holley during the discussion about his lecture to the Iron and Steel Institute on 'Setting Bessemer Converter Bottoms', in: *JISI* 1874 part 1, p. 372.

25 B. Walker during the discussion about his lecture to the Iron and Steel Institute on the 'Description of the latest improvements in appliances for the manufacture of Bessemer Steel', in *JISI* 1874 part 1, pp. 374–83, here p. 382.

26 C. P. Sandberg, De la spécification et de la réception des rails en Europe, in *Revue universelle*, 2nd ser. 10 (1881), here p. 531.

A very large and regular product is essential to commercial success. The same engine and boiler capacity, the same vessels and accessories, the same quality and nearly the same extent of hydraulic machinery, melting apparatus, and buildings are required – in all, say 200,000 dols.' worth of plant – to make six 5-ton heats per day, as to make sixteen 5-ton heats per day. Six heats was the maximum work in England three or four years ago, and still is in some foreign works, ten or twelve being the average abroad, while eighteen to twenty-four heats are the standard practice in America. This additional work, got out of nearly the same capital, is the result of these mechanical refinements. A 40000 dols. blowing engine must be kept at work to pay dividends, in this country at least.[27]

His views met with an enthusiastic response in England. For Peter Tunner, teacher to many successful ironmasters, the value of Holley's improvement lay above all in the fact that 'this American achievement has recently identified a very common problem in our ironworks, namely that the size of production in relation to the available apparatus and to the investment capital employed, is much too small'.[28] In France, his colleague Jordan also welcomed the news that the tedious task of repairing bottoms, 'this serious obstacle to the commercial success of the process', had now been overcome and he stressed that the larger quantities, which could now be produced, were 'obtained with an almost equal capital [outlay]'.[29]

Holley's bottoms spread gradually over Europe in the years following 1873, when the first usable descriptions and drawings were published. Since the resistance of converter bottoms was less of a problem in Europe than in the USA, and the increase in production made possible by the cupola furnace was on the whole not fully exploited, there was no particular compulsion to adopt them until about 1876. Around this time the production capacity of the Bessemer works had already been raised to such an extent by the previous improvements that they were hardly able to off-load the produce of one single plant. Cockerill's new Bessemer works in Seraing/Belgium of 1873/4 is supposed to be the first to be built according to American plans and fitted with Holley's bottoms. At Cockerill's the results of both ways of improving the

27 Alexander Lyman Holley, Bessemer Machinery (lecture delivered before the students of the Stevens Institute of Technology), reported in: *Engineering* 20 December 1872, p. 416. This lecture, or statement, was also widely disseminated in the German-speaking sphere, as for instance in: *Dinglers Journal* 207 (1873), pp. 394–400, *Polytechnisches Zentralblatt* 1873, p. 410.

28 Peter Tunner, Amerikanische Verbesserungen am Bessemer-Converter, in: *Zeitschrift des berg- und hüttenmännischen Vereins für Steiermark und Kärnten* 7 (1875), p. 236.

29 Samson Jordan, *Notes sur la fabrication de l'acier Bessemer aux Etats-Unis d'après MM. Holley, Smith, etc.*, Paris 1873, pp. 20 and 33f. Louis Baare of the Bochumer Verein also argued along the same lines as Holley at the Congress on Political Economy (*Volkswirtschaftlicher Kongress*) in 1876, in that he pointed out that the use of steam and blowing engines with the Bessemer system permitted only 'bulk production'. Wackernagel, p. 87.

converter bottoms were already being combined. On the one hand, they could be replaced quickly and on the other they could last for up to twenty-two charges due to the improved fire-proof lining.[30] In Great Britain, following a lecture given by Holley at a meeting of the Iron and Steel Institute in 1874 in Barrow, the director of the West Cumberland Steel Works, Snelus, offered to show them around his works, where the new bottoms had just been introduced. The following year, when the Institute met again, he was able to convey his satisfaction to them, since his expectations had even been surpassed.[31] In Germany, the Hörder Verein was probably the first to introduce the new bottoms in 1875 and they were used increasingly from 1876 onwards.[32] That Holley's replacement bottoms were in common use by the end of the 1870s may also be inferred from the fact that in the meantime the absence of these bottoms was described as exceptional in works guidelines.

The cost of introducing these replacement bottoms has not been established. However, given that they required only a slight alteration to the lower half of the converter, costs cannot have been excessively high. In Dowlais, for instance, costs for improved converters were not singled out, but, as was usual with repairs, were simply added to the overall production costs for steel.[33] Only greater expenses, which had to be approved, appear in the directors' minutes and are thus perceptible to us. The fact that Holley's bottoms were generally introduced but feature hardly at all in the records, indicates that this innovation cannot have been very expensive and was classed as a running improvement, for which the works managers were responsible.

Once both the bottle-necks of the first generation Bessemer works, the tedious smelting of pig iron and the frequent repairs to the converter bottoms, had been overcome, the potential of existing plant had been exhausted. As it was, most works were already finding it impossible to exploit both improvements to the full. They came up against a barrier, consisting of the impossibility of removing the steel from the shop as quickly as it could be produced in it. They were in real danger of choking on their own product. Their production capacity was increasingly dependent on structural factors, which had not played a part when the works were planned. Against this background, enlarging the converters, as was done in the case of new constructions during the 1870s, was not always the successful course in the case of existing works. On the one hand, the usual 5-ton converters could take up to seven tons when required, which meant that there was not much difference between them and

30 *Iron* 26 September 1874, p. 386.

31 *JISI* 1874 part 1, p. 371 and 1875 part 1, pp. 204f.

32 Hermann Wedding, *Die Darstellung des schmiedbaren Eisens in praktischer und theoretischer Beziehung*. Erster Ergänzungsband. *Der Basische Bessemer- oder Thomas-Process*; Braunschweig 1884, p. 165. Beck, *Geschichte*, p. 618.

33 Glamorgan CRO, D/DG E 3; report for the year ending 2 April 1881, p. 5.

the new eight-ton converters.[34] On the other hand, however, these larger quantities would have required longer intervals between charges to allow for teeming into the ingot moulds and removing them. This, however, was at variance with the fuel and operational savings effected, since the advantages of continuous operations compared with earlier sporadic workings would thus have been counteracted. 'Efforts today are directed much more towards speeding up operations, not towards increasing the charge' was how the forementioned French iron manufacturer summed up operational strategy in the British Bessemer works.[35]

There were very few opportunities subsequently to make casting and removing the steel easier in existing plants. The Bochumer Verein managed to achieve this at extremely modest costs of 13,000 marks by improving its lay-out.[36] The bottle-neck here was the circular casting pit, a feature of every Bessemer works until the 1870s. It was not capable of taking the charges at the same tempo at which they were turned out by the converters. Consequently it was enlarged by adding two extended casting trenches of the same depth, over which a waggon ran bearing a casting ladle taken from the central pit crane (see Figure 5). By means of this adjustment, the capacity of the central pit crane and the casting pit could be subsequently increased.

However, the point of these measures was not to effect an overall rise in production (indeed Bochumer Verein had only just stopped working a night shift) but to use the plant more intensively while it was in operation.[37] Overall production was stagnating at around 55,000 tons a year.[38] Working day and night shifts would have resulted in an annual production of 100,000 tons from one plant; I am not aware that a comparable production capacity was achieved anywhere in Europe at this time. By the second and third month following this alteration the cost of operating a converter had already fallen by a full three marks per ton.[39] When this is set against annual production and the level of costs, the potential significance of these minor improvements is revealed.

At Hoesch in Dortmund in the neighbourhood of the Bochumer Verein, an elongated casting trench was also laid on to extend the pit crane (see

34 Pelatan, p. 63. Obviously this applied by analogy to larger converters as well. The Dortmunder Union, for instance, was filling its 7.5 ton converter with 9 tons in 1879. However, this involved a greater risk of metal wastage because more metal was thrown out during the boiling period. Jeans, *Steel*, p. x.

35 Pelatan, p. 63.

36 Werkarchiv Krupp-Bochum, BV 12900 Nr. 11: Protokoll der Verwaltungsratssitzung am 16 Mars 1878.

37 Werkarchiv Krupp-Bochum, BV 12900 Nr. 11: Protokoll der Verwaltungsratssitzung am 20 November 1877.

38 Däbritz, p. 175. PEE, Baare (Bochumer Verein), p. 790.

39 Werkarchiv Krupp-Bochum, BV 01000 Nr. 1: a comparative survey of production costs for various products. These three marks corresponded to *c*. 15 per cent of the conversion costs in the steelworks, and thus represented a considerable saving.

Figure 5).[40] In consequence this plant, which had been built in 1872/3 and had subsequently been fitted with removable bottoms, achieved a daily rate of 35 charges. Even the old plant of the Bochumer Gesellschaft für Stahlindustrie, which had been built very compactly, was provided, insofar as its restricted area permitted, with an extra casting pit and casting waggon according to the Bochumer Verein's example.[41] Furthermore, the plant was equipped with cupola furnaces instead of the original reverberatory furnaces, and with Holley's replacement bottoms. Although a daily rate of two charges had been envisaged in 1869, by the beginning of the 1880s a daily rate of 30 charges was being achieved under exceptionally tight conditions. Comparing the modest level of equipment in these works with that of the new Bessemer works in the late seventies and early eighties, most of which did not produce any more rapidly, makes it clear how highly abilities relating to the organisation of production are to be esteemed, compared with the works' technological equipment.

Where it was not possible to extend casting capacity, which was generally due to compact construction, the only solution was demolition and reconstruction according to the 'American plan'. The chief notion behind the American plan was to improve all the transportation paths in the steelworks in order to master the increased production quantities. Since the American works had, in the course of the seventies, overtaken the production capacity of the Bessemer plants they had also been the first to encounter this problem. Their most vital innovation involved abandoning the deep casting pit and stopping all work beneath the foundry floor. More space for casting ingots was created on the one hand by increasing the radius of the casting pit crane, and on the other by placing the two converters close together. Since casting the ingots now took place above ground, they could be removed easily and above all quickly by the railway which ran through the shop. On top of this there was enough space around the casting crane to enable the ingots to be shifted temporarily by using other cranes. A necessary result of casting at ground level was that the converters too had to be set higher, which meant that it was now easier to get at them to work on repairs. In particular, the work of replacing bottoms was simplified since there was now enough space beneath the converter to allow a new bottom to be moved into place on a waggon. Behind the converters on a higher level still stood the cupola furnaces, whence the smolten pig iron was conducted along channels to the converters.

Cockerill in Belgium was the first European works to be built according to

40 Illustration taken from Macar, planche 18. On the day the works were inspected by the gathering of the Iron and Steel Institute in 1880, 52 charges were made in 24 hours in this plant. *Stahl und Eisen* 2 (1882), p. 154. Compare R. M. Daelen, *Fortschritte*, cols. 388–99.

41 Macar, pp. 164–5 and planche 15.

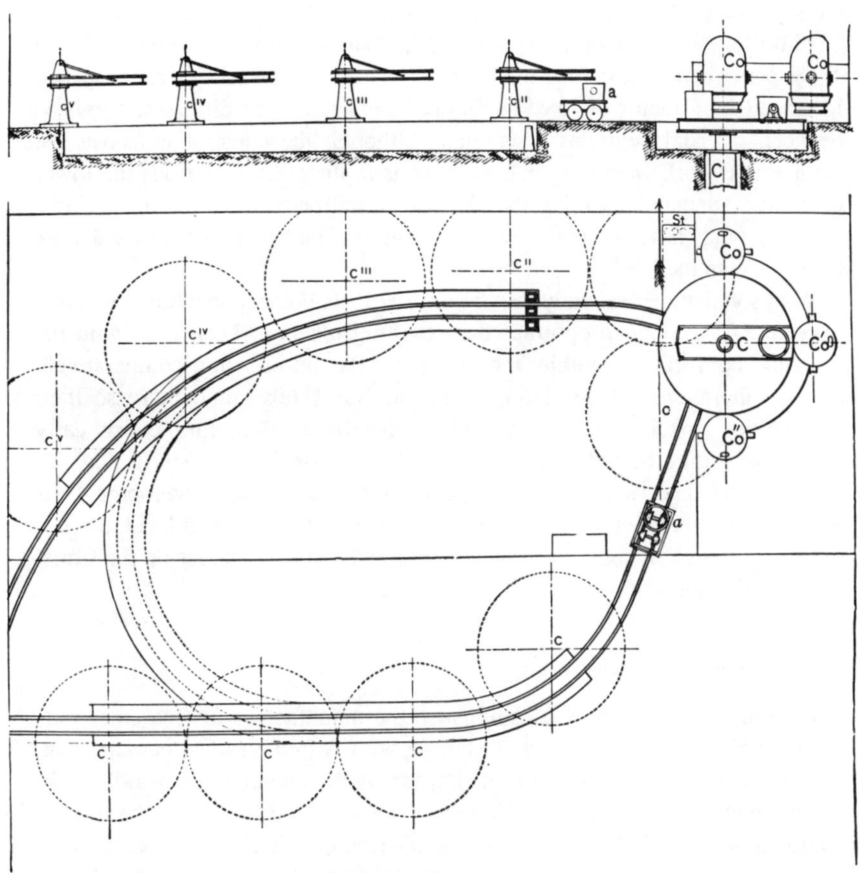

Co = Converter, a = casting waggon, c = ingot crane, C = casting pit, St = control platform

5 The Bochumer Verein's Bessemer plant

this principle in 1874.[42] Figure 6 clearly shows the difference between the old (left) and the new (right) arrangement of the converters at Cockerill's. Some of the plants which came into operation in Great Britain after 1874 followed the same course: these were Bolckow, Vaughan & Co. in Eston; Rhymney Steel Works in South Wales and the rebuilt Moss Bay Steelworks.[43]

Naturally, spacious plant and greater recourse to mechanical aids had to be paid for as well. In the USA, Edgar Thomson Steel Company's Bessemer

42 *Iron* 26 September 1874, p. 386. Illustration taken from Macar, planche 12.
43 Jeans, *Steel*, p. 421 and *Iron* 19 May 1877, p. 622.

works, which were completed in 1877 with two 5-ton converters, had cost $222,000 and were intended to produce 50,000 tons annually.[44] Compared with 1868, construction costs had almost doubled, but its productive capacity had increased by a factor of five. In Europe, the same price differential existed between old and new Bessemer plant, although the construction costs for comparable work were only half as great as in the USA.[45] Besides the lower qualitative demands set by the American railroads, one reason for this difference may have been that investment capital first began to be utilised more intensively in the USA.

Works which were already standing in 1873 – and they constituted the great majority – had only to optimise their existing plants to keep pace with the growing competition within the industry. By introducing comparatively modest alterations, plants dating from the late 1860s could increase their production capacity many times over, thereby confounding every early prognosis within the industry. Bochumer Verein, Bochumer Gesellschaft für Stahlindustrie and Hoesch are all eloquent instances of this development. Its main features will be presented in a case study set at operational level, prior to our final discussion about the effects of rationalisation strategies on the firms' market behaviour.

2 The Krupp case study

The discussion about how to rationalise the manufacture of Bessemer rails began at Krupp's in March 1870, starting with Alfred Krupp's demand that: 'We must summon up all efforts at simplifying the manufacture of rails without the hammering procedure. Hopefully we will find the means before the competition does.'[46] The idea of not hammering cast steel, as was done in the puddling works, but to roll it straightaway, was not a new one, and had already been tried many times in Great Britain. However, the usual grooved rolling mills had turned out to be unsuitable for this purpose. Rails produced in this way were so inadequate that the works had soon reverted to hammering.[47] This was also what Eichhoff, Berta Krupp's cousin and the Bessemer works manager, emphasised in his reply rejecting Alfred Krupp's directive. Eichhoff considered thorough preparatory hammering to be indispensable, in order to meet the high quality demanded by the railways. Economising on the

[44] Jordan, *Note*, p. 56. Jeans, *Steel*, pp. 834f.

[45] Re the USA see Temin, p. 123. Re Europe see Sidney Gilchrist Thomas and Percy C. Gilchrist, The Manufacture of Steel and Ingot Iron from Phosphoric Pig-Iron, in: *Iron* 12 May 1882, p. 375.

[46] HA-Krupp, WA IX a 192: A. Krupp on 16 March 1870.

[47] In 1867 Dowlais already possessed a cogging mill, but could not use it in every instance because many customers were still insisting on the preliminary hammering procedure. Ulrich, Aust and Jänisch, p. 7.

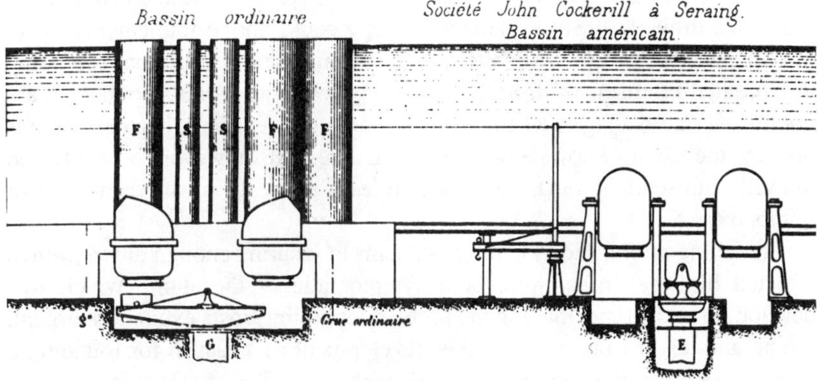

6 The American lay-out for converters

hammering process did not bring any advantages, as was demonstrated at Bochum and Hörde by 'the heaps of rejects which these works make'.[48] For the moment, then, the notion was dropped. It turned up again on several occasions, including in November 1873, in connection with Krupp's gigantic expansion plans. The maxim was now cheap mass production of Bessemer rails. Now that independence in the critical question of raw materials had just been gained, the aim was:

> to be able to create the greatest possible quantity of usable material more cheaply than any other manufacturer – otherwise we would have to abandon our designs on the future world market. Only thus will we achieve this position and then it will be an advantage to have acquired the Spanish mines and the steam ships, and the German mines and iron-works. We must make as much as the entire German competition. Making rails steel of such excellence that it surpasses the conditions would be a waste – with steel for tools or artillery every improvement is valued, but not in this case.[49]

Krupp's strategic decision to seek future salvation no longer in superior quality but in the cheapest possible mass production was at the same time the beginning of the end of the Eichhoff era in his Bessemer works. Krupp replied to one of Eichhoff's sceptical memos about these plans by pointing to the example of 'English operations', which achieved the same rate of production with fewer converters.[50] However, Eichhoff remained a stubborn adherent of the old methods using reverberatory furnaces and hammering. He was deeply suspicious of simplified and speeded-up mass production, and was still too tied

[48] HA-Krupp, WA IX a 192: Eichhoff on 18 February 1870.
[49] HA-Krupp, FAH II P 112: A. Krupp to Procura on 10 November 1873.
[50] HA-Krupp FAH II P 112: A. Krupp on 11 November 1873.

up with the material problems of early Bessemer production to consider even a trial worth-while.[51] Since the new strategy could not be implemented with him in place, he had in the end to go. For a while Krupp considered enlisting an 'initiated Englishman' from a flourishing Bessemer works, 'so that we are not misled into having to let everything pass as the very best, because there is nobody there who knows even better'. In the end, however, he contented himself with sending a man to Britain to learn about the most recent state of affairs there.[52]

The falling profits and mounting problems in securing enough orders, which affected his steelworks, were both symptomatic of the slump which was developing at the time, making the problem of cutting costs even more urgent. Krupp and his procuration used this development as a reason for founding a *Commission für Bessemerwerksangelegenheiten* (Commission for matters relating to Bessemer works), which was able to evaluate information from other works, especially from Britain, and make it the basis for the reorganisation of the whole operation.[53] Shortly afterwards, the situation was further dramatised by the discovery of faulty accounts, which had presented far too rosy a picture of the profits and had misled Krupp into engaging in inappropriate plant and expenditure.[54] It was only the state's at the time extraordinary intervention through the Prussian Seehandlung (state bank), which saved Krupp from ruin.[55]

The Bessemer Commission's duties were of vital importance. Nevertheless, the Commission did not possess real powers of decision since Krupp had made it clear right from the start that he would insist on removing the reverberatory furnaces and the hammering procedure. He advised against any form of autonomous innovation, since functioning processes were to be copied from others, in order to proceed safely. Dowlais was to set the standard in every-

[51] HA-Krupp FAH II P 112: Eichhoff to A. Krupp on 20 October 1873.

[52] HA-Krupp FAH II P 103: A. Krupp to Procura on 15 May 1874.

[53] HA-Krupp FAH II P 103: A. Krupp to Procura on 19 May 1874.

[54] A. Krupp's reaction to these discoveries: 'Following the new revelations I am so worn out by sleeplessness and feverish anxiety. I could go mad.' HA-Krupp FAH II B 320: A. Krupp to Goose on 25 August 1874.

[55] The Seehandlung (state bank) provided a loan of 10 million thaler via a banking consortium, in which the most important banks in the Ruhr were involved. Krupp had to mortgage his entire firm as security. Although the loan was not originally connected with the discovery of the deficit, since Krupp had already asked for it beforehand, it nevertheless formed the foundation stone of the rescue package. Without the State's guarantee, the banks were no longer prepared to provide further means to finance Krupp's purchases, which was why he had already had recourse to the Prussian Seehandlung in 1873. As further developments at Krupp's showed in 1874, the banks' sceptical attitude had not been unfounded. Re the history of Krupp's loans, see Willi A. Boelcke, *Krupp und die Hohenzollern in Dokumenten, Krupp-Korrespondenz mit Kaisern, Kabinettschefs und Ministern 1850–1918*, Frankfurt am Main 1970, pp. 61–82.

thing.[56] His main idea was to operate the available Bessemer plant as intensively as possible, which would achieve 'immediate savings by the entire plant necessary for production and consequently offers the totality of our crucibles [converters U.W.], in the event of their being fully served by cupola furnaces, the possibility of producing a greater number of ingots, so that the overall costs etc. are spread over a much greater production'.[57]

Here, too, the primary motive behind the introduction of cupola furnaces was to increase the converters' productive capacity. Krupp expected to improve the profit situation above all by reducing the proportion of fixed costs and by increasing production. And, despite opposition from the Commission, he insisted on implementing this strategy, which as far as he was concerned obviously included cupola furnaces and ingot rolling mills. On the other hand, he gave way on other technological and organisational questions, such as the size of the castings, the lay-out of cranes, the rails saws and the sequence of alteration work without too much fuss. This was remarkable in so far as the Commission had been shown accounts, according to which the old reverberatory furnaces were more economic than cupola furnaces.[58] Krupp's only back-up in this affair was his English partner, Longsdon, who was above all the technical representative in his procuration and had links with British iron manufacturers. According to Krupp, the Commission's cost reckoning did not take sufficient account of the higher production capacity resulting from the use of cupola furnaces and, among other things, was not aware of the possible fixed costs degression or the technical economies of scale effected by operating converters more frequently.[59] In view of these objections and the positive reports about using cupola furnaces not only from Great Britain but also from the Dortmunder Union and the Osnabrücker Stahlwerk, the Commission decided after all to recommend carrying out a trial using one plant in Bessemerwerk III, which was yet to be built.[60]

At bottom, the Commission's recommendations were always in accordance with Eichhoff's assessments, he being the manager of the Bessemer works who also compiled the summarising memorandum in the form of a reply to questions put to the Commission. The cost of a complete Bessemer plant,

56 HA-Krupp WA IV 56: A. Krupp to Bessemer-Kommission on 3 September 1874.

57 HA-Krupp WA XI a 2,27: A. Krupp to Bessemer-Kommission on 19 September 1874.

58 HA-Krupp WA IV 56: Protokoll der Zweiten Konferenz der Bessemer-Kommission 7 October 1874, pp. 15–25. In these minutes the Bessemer Commission showed that its members were already sceptical about the cupola furnaces even before a cost comparison had been drawn up, which then came out on the side of reverberatory furnaces, much to their satisfaction. See note 66.

59 HA-Krupp WA IV 56: handwritten marginal comments by A. Krupp in the minutes of the second conference of the Bessemer Commission, p. 20.

60 HA-Krupp WA IV 56: Protokoll der Dritten Konferenz der Bessemer-Kommission 10 October 1874, p. 39.

consisting of two converters, two big cupola furnaces for the pig iron, two small Spiegel-furnaces and a steam-driven crane, were set by Eichhoff at 139,000 marks. Existing apparatuses to the value of 73,050 marks could be employed, so that a further 66,450 marks had to be found for the trial.[61] This new plant would enable a production rate of 18–20 charges in 24 hours, whereas previously the whole works, operating as it did then with 13 big and 2 small converters, could produce a maximum of 60 charges in 24 hours, which was only three times as much while using seven times as many converters.[62] At Krupp's, the two small converters were always accounted the same as one large 5-ton converter and, although the ratios were not viewed quite so crassly in the plant's replacement-cost standard valuation, they were still in favour of the new plant with cupola furnaces. The existing works' 11 reverberatory furnaces[63] and 14 converters had cost 567,000 marks, whereas the two new converters and three cupola furnaces belonging to them, and two furnaces for Spiegeleisen cost 96,750 marks, that is 27 per cent of the cost of the entire works.[64] Interest and depreciation were calculated at 13 per cent in the Bessemer works, which worked out, according to Eichhoff's reckoning, at a probable interest and depreciation of 0.28 marks per ton for the new plant.[65] In his opinion, however, this advantage of the faster new plant was more than compensated for by the reverberatory furnaces, which smelted the charges more economically. The end result was that steel produced according to the previous procedure was 6.49 marks cheaper per ton.[66]

Krupp did not agree with Eichhoff and the Commission's views, and he questioned the accuracy of their calculations. He insisted on changing to an alternative plan, drawn up by Longsdon according to the British model. The latter envisaged, not an additional trial operation, but the reconstruction of the existing works. Like Krupp, Longsdon was of the opinion that profits could only be permanently improved by cheap mass production: 'It is also desirable to more than double the works' whole trade, to achieve an annual turnover which is more in proportion to capital . . . Our overhead costs for present production amount to circa 9 per cent, our depreciation to 13 per cent, and this high percentage is solely due to our low production.'[67]

61 HA-Krupp WA IV 56: Beantwortung der von der Commission für Bessemerwerksangelegenheiten gestellten 27 Fragen, Eichhoff, pp. 123–55.

62 Ibid., pp. 136f.

63 Shortly beforehand Krupp had just replaced twenty-six old and small reverberatory furnaces with eleven bigger ones fitted with Siemens regenerators. HA-Krupp WA XV a 102: Bessemer-Werk I und II; XV 4,81;Bessemer-werk I und II; XV 4,33: 10,000 lb smelting furnaces in Bessemer-Werk I.

64 HA-Krupp WA IV 56: Beantwortung, p. 136.

65 Ibid., p. 137.

66 Ibid., p. 141.

67 HA-Krupp WA IV 56: A. Longsdon on 10 March 1875, p. 173. The following hypothetical calculations are all derived from this memo, pp. 176ff.

The production capacity of the existing Bessemer works stood at 90,000 tons maximum per annum. The conversion costs, i.e. the cost of the finished steel minus the price of the iron employed, came thereby to 27.34 marks per ton. Longsdon's proposal envisaged eight converters with the pertaining cupola furnaces, of which four were to be in constant operation and four under repair. In the case of a total of nine converters, when demand fell three could be in use and six under repair. In the Bessemer works, the second version was preferred because it was not believed that they could operate with fewer reserves. The possible level of production was estimated at 153,000 tons and 114,000 tons respectively per annum, with conversion costs of 20.80 marks and a workforce requirement of 477 or 448 men respectively. Raising production by 34 per cent would thus require only a 6.5 per cent increase in manpower. Thus the level of mechanisation of the planned steelworks was already so far advanced that wages practically assumed the nature of fixed costs. This meant that reducing or increasing the workforce was now no longer used as a means of adjusting to economic fluctuations.

The structural alterations required by this project, which included removing the big reverberatory furnaces and setting up a new crane system, were set at a cost of 270,000 marks, which was still 50 per cent less than the replacement cost standard value of old and less productive plant. Because the conversion costs were so low, the alteration costs were paid off within three or four months of full-scale operations. In any case, Longsdon considered that it would not be possible to sell all the achievable output, in view of the previous year's production of a good 50,000 tons. However, he did think that 90,000 tons could be achieved at the lower prices which were then possible; this was enough to justify the structural alterations. Under these conditions, the project was accepted and the piecemeal alteration of the two Bessemerwerke I and II was approved. The 'Commission for matters concerning Bessemer works' was re-formed into a 'Commission for alterations to Bessemer works and rails-rolling mills' (*Commission für die Änderungen im Bessemerwerk und Schienenwalzwerk*), which began working on 12 March 1875.[68] The first 'System' according to the new method, using three converters and two cupola furnaces, was put in operation in the same year on 2 September. The second system followed suit on 26 May 1876, and the third on 16 November 1878. The last one was, however, only used for a while to produce castings for steel tyres. The reverberatory furnaces were shut down when the second system came into operation. For the time being, the structural alterations to Bessemerwerk II were not implemented.[69]

68 HA-Krupp WA IV 56: Ite Sitzung der Commission für die Änderungen im Bessemerwerk und Schienenwalzwerk, 12 March 1875, pp. 182ff.

69 HA-Krupp WA IV 656: C. Reuter to Hoecker on 2 December 1881.

The hammering process experienced much the same fate. The Bessemer Commission had been sceptical about the rolling plant at Dowlais as well, and had come down in favour of retaining their own tried and tested plant.[70] Menelaus, the director of Dowlais, had sent Longsdon drawings of his cogging mill, which could roll 1,000 tons a week using a workforce of two men and two boys.[71] As with the converters, in this case too the Commission's opposition was overcome. In 1876 the new cogging mill came with two triple rolling lines, which was sufficient for the anticipated yearly production of *c.* 100,000 tons.[72]

These changes were concealed as far as possible from outside, because Krupp assumed that his customers objected to 'Schnellbetrieb' (rapid driving) as much as Eichhoff and the Bessemer Commission did in his own establishment. Before work on the alterations started he issued the following directive:

> Our earlier expensive manufacture has at least secured us our reputation which we must make the most of. It is necessary for every agent and everyone in the firm to speak in harmony about the manufacture and when explaining the possibilities of such low prices. We may not say that we are manufacturing in another way, since this would give rise to reservations about the quality. We must explain it in terms of the increase in successful products from a given quantity of material, which has been so significantly improved by using our excellent Spanish ore.[73]

Altering the steelworks for 'rapid driving' was a success from the start in the organisational and technical sense. Krupp's proviso, not to experiment but to copy tried and tested methods, bore fruit. No complaints arose about the efficiency of the new plant in functional terms and the 18 to 20 charges in 24 hours which had been envisaged at the start of the alteration work, were soon exceeded without a hitch. No loss in quality as a result of the accelerated pace of work was observed. Consequently, half a year on, further economies in the Bessemer works could already be achieved; 'that the converters should only be heated for the first charge from the cupola furnaces, since they retain the heat completely because the charges follow one another so quickly'.[74] Previously the converters had always cooled down so drastically during the long pauses between charges that they had to be re-heated to glowing point by a coke fire every time.

Now, however, given that production had attained a level of organisation which made continuous operation possible, the constant necessity of repairing converter bottoms must have been increasingly viewed as a bottleneck, even in the Krupps Bessemer works. Thus it was in January 1878 that the chief of the

70 HA-Krupp WA IV 56: Protokoll der Dritten Konferenz der Bessemer-Kommission 10 October 1874, p. 45; Beantwortung, pp. 148ff.

71 HA-Krupp WA IV 56: Menelaus to Longsdon on 28 July 1874, pp. 30ff.

72 HA-Krupp WA IV 737: Die Walzwerke der Gußstahlfabrik 1852–1910.

73 HA-Krupp WA IX a 172; A. Krupp on 10 February 1875.

74 HA-Krupp FAH III B 200: Narjes to F. A. Krupp on 31 March 1876.

Bessemer works, who three years earlier had still been pleading for the slower variant, turned impatiently to the management: 'can the building of an American crucible [he means converter] of the latest construction, which was decided on long ago, not be speeded up? As long as there is no end to this crucible calamity, we are not at all in a position to apply the third system to advantage, although it is increasingly necessary.'[75] A reply to this request can no longer be found in the internal correspondence. Once the Bessemer works had been reorganised, the procuration had no more problems with their technical operation. They then turned all the more keenly to questions of securing the market and expanding sales.

The reduction in production costs effected by introducing cupola furnaces was not restricted solely to savings in fuel and repairs, but depended essentially on exploiting the 'economies of scale' which were now possible, as far as could be managed. Krupp had indeed made it clear over and over again that he saw in this the actual significance of the structural alterations. With the new plant, each of which always had one converter in operation, 77,338 tons of steel were produced in the working year 1876/7, and as much as 89,600 tons of steel in the following working year 1877/8.[76] Since this stage of the building project included only two-thirds of Longsdon's 'slower' alternative, this meant that the productive capacity of 38,000 tons per plant, which was then accepted, had been achieved from the start and had already been clearly surpassed by the second year. Once the technical success of the alterations had thus become obvious, the inmates of Krupp's estimating office concentrated on determining its commercial success compared to previous practice as precisely as possible. The conversion costs provided the standard for this, as in every other case known to me. Since the processed pig iron constituted only a 'transitory item' for the steelworks, this value is far more revealing when it comes to assessing the economic viability of the plant than are the overall costs.[77] In the working

75 HA-Krupp WA IV 656: C. Reuter to Hoecker on 16 January 1878.

76 HA-Krupp WA VII f 484: C. Reuter to Hoecker on 16 October 1880.

77 In any case, one weakness of this measurement is that the varying iron losses during the conversion process, as when more iron was thrown out during the boiling period, are not indicated. This iron loss was called *Abbrand* (metal wastage) and was already included in the cost of the pig iron. To some extent this metal wastage was the necessary and desired result of the refining process, since the 5–8% impurities in the pig iron (carbon, silicon and manganese) were supposed to be removed from it, producing a corresponding weight loss. However, iron was also lost in the course of combustion and extrusion, reducing the quantity of steel achieved. The total metal wastage was consequently greater than the proportion of impurities extruded. During the 1870s it swung between 10 and 13 per cent. HE-Krupp, VW IV 976: Bessemer- und Schienenwalzwerk, Produktionskosten und Löhne 1873–82. As may be seen from the following analysis of enterprise costs, an alteration of only 1 per cent in the metal wastage had a greater effect on the books than, for instance, all the redemption payments. Since, however, the composition of the pig iron constantly varied and with it the proportion of impurities, it is scarcely possible to establish reliably a minimum level of metal wastage. Slightly higher

Table 2. *Analysis of the production costs for 1 ton of Bessemer steel at Krupp's in 1873/4*

	marks	
1. Pig iron	166.32	(= 76.18%)
2. Various scrap	10.61	(= 4.86%)
3. Wages and interest	11.83	(= 5.42%)
4. Fuel	7.07	(= 3.24%)
5. Fireproof materials	4.13	(= 1.89%)
6. Magazine materials	0.68	(= 0.31%)
7. Ingot moulds	2.53	(= 1.16%)
8. Repairs to furnace	2.88	(= 1.32%)
9. Iron implements, tools etc.	4.63	(= 2.12%)
10. Various	0.38	(= 0.17%)
11. Steam, water, gas, railway	6.20	(= 2.84%)
12. Redemption payments	1.09	(= 0.50%)
An average casting of 1,000 kg	218.39 (= 100%) marks	

year 1873/4, the last 'normal' year before the reorganisation, these had amounted to an average of 41.42 marks, with a total production of 71,102 tons. Table 2 provides a list of the assembled costs.[78]

In order to obtain the conversion costs, we have to subtract the cost of the pig iron and the various scrap metal, which could be smelted in the reverbatory furnaces. This leaves 41.42 marks. The 17.34 marks mentioned by Longsdon are not to be compared with this, since on the one hand they apply to 1875, when prices were already considerably lower, and on the other they started from an annual production of 90,000 tons. In order to eliminate these influences and to determine the actual effect of the alterations, the estimators at Krupp's re-worked the accounts for 1873/4 using 1877 prices. This demonstrated that 1,000 kg of steel produced by the old method would have cost 118.52 marks, with conversion costs of 34.79 marks.[79] The same procedure was applied to the operating costs accounts for the altered plant. The average

wastage did not necessarily mean higher iron losses, but could also relate to that particular pig iron's higher carbon, silicon or manganese content. Further variations were produced by the variable proportion of scrap, which was smelted together with the pig iron. On practical grounds, this led to leaving the different metal wastages out of calculations about conversion costs. Controlling the metal wastage depended on how conscientious the foreman was in the Bessemer works. He had to take care when directing the blast that as little as possible was wasted and that the blowing process was stopped at the right moment before the iron began to combust more intensively.

[78] HA-Krupp WA IV 976: Bessemer- und Schienenwalzwerke, Produktionskosten und Löhne, 1873–82, p. 1.

[79] Ibid., p. 31.

Table 3. *Analysis of the production costs for 1 ton of Bessemer steel at Krupp's in 1877 (according to 1873/4 prices)*

	marks
Pig iron and scrap	176.60 (= 88.20%)
Wages and interest	4.59 (= 2.29%)
Fuel	5.73 (= 2.86%)
Fireproof materials	2.62 (= 1.31%)
Magazine materials	0.19 (= 0.09%)
Iron implements and tools	0.13 (= 0.06%)
Ingot moulds	3.53 (= 1.76%)
Furnace repairs	0.72 (= 0.36%)
Various repairs	1.71 (= 0.85%)
Various	0.17 (= 0.08%)
Redemption payments	0.73 (= 0.36%)
Steam, water, gas, railway	3.51 (= 1.75%)
	200.23 (99.97%) marks

Table 4. *Comparison of the conversion costs for 1 ton of Bessemer steel at Krupp's in 1873/4 and 1877*

	1873/4 prices	1877 prices
Old plant (1873/4)	41.42 marks	34.79 marks
New plant (1877)	23.63 marks	18.48 marks

costs for the months of April, May and June 1877 for 1,000 kg of steel amounted to 102.82 marks, of which the conversion costs were 18.48 marks.[80] In terms of 1873/4 prices, this would have been 200.23 marks for 1,000 kg of steel and 23.63 marks conversion costs.[81] Table 3 compares the production costs for 1873/4 with the production costs of the altered plant in the months of April to June in terms of 1873/4 prices, so that a direct comparison with individual items may be drawn.[82] Table 4 compares the conversion costs of old and new plant with almost the same annual production.

When one compares the individual items in both these lists with the 1873/4 prices, it is apparent that the greatest savings were achieved in 'Wages and Interest'. This meant that producing about the same quantity of steel required not only fewer converters but also considerably fewer workers.

80 Ibid., p. 30. 81 Ibid., p. 27. 82 Ibid.

A costs list for the Bessemer works in 1878 reveals that an average production rate of 98 tons per man in 1873 had risen to 236 tons per man in 1878.[83] This corresponded to a 140 per cent increase and more or less exactly to the increase in productive capacity of the altered plant from 10 to 25 charges a day, as illustrated in Figure 7.

At this point, it should also be mentioned that the above work performance of 236 tons per man in 1878 adds up to a total workforce of *c.* 380 men to operate two plants. Longsdon had allowed for 448 men for three plants. If we keep to these proportions, then enlarging the plant from two to three would have already involved a proportionally smaller increase in manpower (plus 17.5 per cent), along the lines of Longsdon's notions about working four instead of three plants. Contrary to the puddling process, variations in the quantity produced did not entail a proportional increase or reduction in the workforce. Thus the organisation of the refining operation no longer depended on how well individual workers performed but on how well the process had been planned.

Likewise, the productivity of both capital and labour rose as a result of the changes to Krupp's Bessemer works. This also included the obvious disappearance of the item 'iron implements and tools', which was indeed closely associated with the number of workers and the degree of mechanisation. Franz Meyer's recollections of his years of employment in Krupp's Bessemer works between 1872 and 1890, tell the same story. He considered the most remarkable event of the 1870s to have been the increased intensity of work following the introduction of cupolas, which had led to rapid driving with sixteen instead of the earlier five charges.[84] The savings in the items 'fuel' and 'furnace repairs' demonstrate the greater efficiency of cupolas over reverbatory furnaces in this respect. On the other hand, the four collated items 'Steam, water, gas, railway' point once again to the degressive effect of rising output on cost increases. The same steam-generators, steam engines, water supply systems and, compared with earlier systems, even more railway connections could also be used more intensively in the course of continuous operations. This aspect of the continuous use of all the works apparatuses as a source of a visible fall in costs had caused a furore at the iron workers' discussion which took place right at the beginning of the 1870s. In the final analysis, it was this aspect which allowed Krupp to push the Bessemer Commission's and Eichhoff's opinions to one side, caught up as they were in detailed calculations.[85]

[83] PEE Tafelband, Anlage zu I.b.2 und II.15, Bessemer-Werke.

[84] HA-Krupp WA VIII 123: recollections by Franz Meyer on 5 December 1908.

[85] HA-Krupp WA IV 822: Geschichte der Kruppschen Bessemerwerke von Dr. Kahrs, p. 35. The new faster Bessemer process which he had fought against, was adopted in September 1875 and managed by C. Reuter.

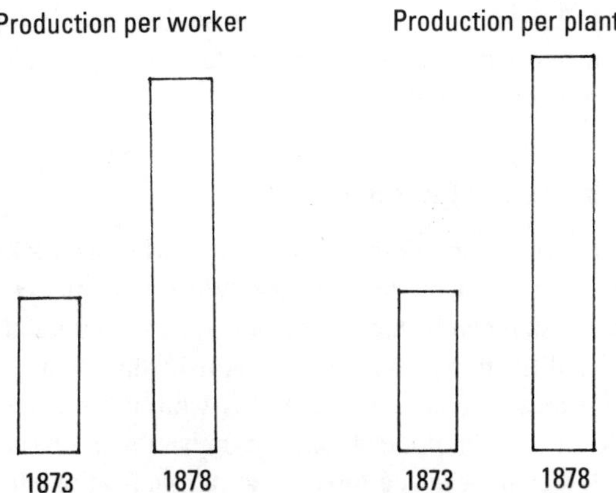

7 The rise of productivity in Krupp's Bessemer works

Very considerable savings were also achieved by simplifying the rolling mills. During the Eichhoff era Krupp had worked them at a level of expenditure untypical for that time, although this was not adequately reflected in the sales prices. While rolling costs in the rails mill were reduced only slightly from 23.22 marks per ton in 1873/4 to 18.39 marks per ton in 1877 (in 1877 prices),[86] huge savings were made by introducing cogging mills connecting the Bessemer works with the rail mills. Roughing costs there amounted to 5.41 marks per ton in the second quarter of 1877. The corresponding processing stage according to the old method of hammering had cost 46.76 marks per ton in 1873/4, in 1877 prices.[87] However, such huge expenditure was quite untypical at the time for this processing stage. The competition's hammering costs were considerably lower, the equivalent of six and eight marks less.[88]

The foregoing comparison of the cost of Bessemer rails, including redemption and interest payments, has revealed a reduction from 203.00 marks to 136.67 marks per ton (still in 1877 prices) as a result of simplifying and speeding up the manufacturing process.[89] The comparatively minor economies made in the rails mills from 23.22 marks to 18.39 marks rolling costs per ton

86 HA-Krupp WA IV 976: Bessemer- und Schienenwalzwerk, Produktionskosten und Löhne 1873–82, pp. 30f.

87 Ibid.

88 GHH-Archiv, 300 108/6: Vergleichende Selbstkostenberechnungen für Bessemer-Stahlschienen, Vorlage zur Aufsichtsratssitzung am 9 Januar 1879. For Rheinische Stahlwerke, see Thyssen-Archiv, RSW 420: Ausgaben Hammerwerk, X. Betriebsmonat, 1 March 1877–28 March 1877.

89 See note 86.

show that when the Bessemer process was introduced there, the mature technology and organisation of labour which were developed in the puddling works, were already being applied there.

3 Rapid driving versus fuel economy

The fundamental reconstruction of the Bessemer works, as took place at Krupp's, was not at all typical of the industry. We chose this case as our example above all because it can be used to demonstrate the cumulative effect of the individual modifications, which distinguish the first generation Bessemer works from those which came later. As well as this, the delay on Krupp's part led to the technological and organisational changes being implemented in one fell swoop, thus allowing the rationalisation strategies of the early seventies to be viewed in their entirety and in relationship to one another. Unlike those works which rose up later, the optimisation of the material input and the temporal duration of the production process coincided. At the same time, Krupp's adoption of considerable parts of the American plan documents the technical dimension of the organisational strategy behind 'rapid driving'.

As demonstrated by a comparison with the plans for Prussian Bessemer works, this idea was put into action only after the start of the depression in the 1870s, when otherwise unchanged converters and cranes were being set up. All works built before 1871 already had cupola furnaces set in raised positions behind the converters, but they nevertheless still retained the small deep casting pits, since they as yet clearly did not envisage any increase in their production capacity. It was only in 1874, when the third and fourth plants in the Rheinische Stahlwerke were ready, that they abandoned the casting pit, as Krupp was to do the following year. For the time being, casting onto the shopfloor was considered only as a facility and not as an essential condition for speeding up the conversion process. In any case, works which operated with an old pit kept thoroughly abreast of European developments in the organisation of production until the end of the 1870s. In most cases, the introduction of cupolas and improved bottoms created sufficient opportunities for condensing the production process.

The Bochumer Verein had already demonstrated very early on just which possibilities resided in a conventional plant with casting pit and cupola furnaces. According to a Krupp memorandum of 20 October 1873, the plant there, which was set up in 1870, could already achieve 24 charges in 24 hours.[90] The other works soon followed suit, as we have already seen. So for those

[90] HA-Krupp FAH II P 103: Nachlass Richard Eichhoff aus den Jahren 1857 und 1861–1874, Notizen pro 20 Oktober 1873.

Table 5. *Conversion costs in relation to quantity of steel produced: Krupp's, 1878*

June 1878	1,294 charges – 18.5 mk per ton
July 1878	1,514 charges – 15.3 mk per ton
Aug. 1878	1,537 charges – 15.4 mk per ton
Sep. 1878	1,482 charges – 15.8 mk per ton

works which were equipped with cupola furnaces from the start – and they were the majority – the wave of rationalisation in the 1870s was to a great extent limited to optimising the length of the production process. This was primarily an organisational problem, compared with which any differences in the works' technical equipment were of a minor order.

The most important quantities in assessing the Bessemer operation were, as already shown in the Krupp case study, the conversion costs. Just how successful the technical and organisational advances of the 1870s were may be deduced from the way these costs developed. Figures in published technical debates fell from about 40 marks per ton in 1868,[91] after a temporary inflationary peak in 1873, to about 15 marks per ton in 1879–80, according to figures quoted in the course of the debate about the introduction of the Thomas process.[92] According to Krupp data, which are available for individual months, it becomes clear that the conversion costs were closely connected to the quantity of production (see Table 5).[93]

The markedly higher conversion costs for the July production, which was about 15 per cent lower, stand out. They provided pressing evidence that the available production potential had to be fully exploited on grounds of cost. The same effect could be observed at GHH. In June 1879 the production in its Bessemer works was reduced to slightly more than half. The result was to increase the cost of its steel ingots by more than 6 marks per ton.[94] As it was,

[91] Edouard de Billy, Note sur l'invention du procédé Bessemer pour la fabrication de l'acier, in: *Annales des mines* 14 (1868), p. 43.

[92] *Zeitschrift des berg- und hüttenmännischen Vereins für Steiermark und Kärnten* 12 (1880), p. 249, p. 499 and, for the lowest value of 13.20 marks per ton, p. 505.

[93] HA-Krupp WA IV 976. Bessemer- und Schienenwalzwerk, Produktionskosten und Lohne 1873–82, pp. 53–64.

[94] Steel production fell from 3,379 tons in May to 1,715 tons in June. While the costs of pig iron hardly changed from 63.03 marks per ton in May and 63.15 marks per ton in June, the production costs for steel rose from 85.22 marks per ton in May to 91.62 marks per ton in June. Since the wastage in June was even a little less than it had been in May (10.29 per cent as opposed to 10.42 per cent), the rise in costs can be attributed solely to higher conversion costs due to under-capacity plant utilisation. They amounted to 15.62 marks per ton in May, and in June to 21.97 marks per ton; a rise of 40 per cent! GHH-Archiv, 300 108/6, Vorlagen zu den Aufsichtsratssitzungen am 9 Juli 1879 und 12 August 1876.

GHH and Krupp did not come off very well, compared with the competition. Krupp's procuration had already realised by early in 1876, much to Alfred Krupp's surprise, that, in spite of their great efforts at rationalisation, Dortmunder Union, Hoesch and Rheinische Stahlwerke were still better equipped.[95] We have comparable figures for the second and third quarters of 1878 for the conversion costs of neighbouring works. The picture provided by Table 6 is based on these figures.[96]

Unfortunately comparable data could not be found for British Bessemer works, thus preventing any straightforward comparison of conversion costs. We must, for the time being, be content with a few scattered figures. An article published in the Swedish *Jernkontorets Annaler* about the introduction of the Thomas process assessed conversion costs in Cumberland in 1879 at 11.50 marks per ton, basing itself on Snelus, the steelworks director of the West Cumberland Iron and Steel Co. Ltd.[97] This amount was clearly lower than those in the Ruhr and appears even more extraordinarily good value, when one considers that the West Cumberland Co.'s Bessemer works, which were known to be operating at a rather leisurely pace, had in the meantime worked up to a mere eighteen charges per day.[98] In the Ruhr they would have been well in the fore of the competition. If their conversion costs were still very low, this was due to a procedure which did not affect the actual conversion process. There, the liquid pig iron was conducted directly from the blast furnaces to the converters, thus saving in the overall costs of the cupola process. Snelus set the total anticipated savings at four marks per ton.[99] Thus, in the case of a conventional works with cupola furnaces, the conversion costs would have lain above fifteen marks per ton, which came closer to what one would expect on

95 HA-Krupp WA IV 1439: Meyer to Goose on 5 March 1876 and on 8 March 1876. In any case they managed to catch up with Rheinische Stahlwerke by the following year. Thyssen-Archiv, Rheinstahl 412: Betriebskostenrechnung Bessemerprozess, 1 März–28 März 1877.

96 HA-Krupp WA IV 967: Bessemer- und Schienenwalzwerk, Produktionskosten und Löhne 1872–83. GHH-Archiv, 300 108/6: Vergleichende Selbskostenberechnungen für Bessemer-Stahlschienen, Vorlage zur Aufsichtsratssitzung am 9 Januar 1979, giving figures for Phönix, Dortmunder Union and Osnabrücker Stahlwerk as well.

97 N. Lilienberg, Meddelanden från utlandet, in: *Jernkontorets Annaler, Ny serie, Tidskrift för Svenska Bergshandteringen* 1880, p. 261.

98 Steel production by the West Cumberland Iron Co. in the year ending 1880 according to *Iron* 17 November 1882 was about 70,000 tons. Given two plants with 6-ton and 8-ton converters this works out at a daily rate of about 18 charges per plant. In 1875, a year after Holley's removable bottoms were introduced to the works, its director was proud to be able to make 460 tons of steel in a week with one plant thanks to this innovation (= *c.* 10–12 charges a day). *JISI* 1875 part 1, pp. 204f. However, the sharpened competition forced the Bessemer works to use its plant more intensively, which, after making a loss for the first time in 1880, finally managed a good 30 charges per day in 1881/2. Speeding up operations in this way also turned out to be profitable, so that in September 1882, 8 per cent dividends could once again be paid out. J. Y. Lancaster and D. R. Wattleworth, *The Iron and Steel Industry of West Cumberland, a Historical Survey* (BSC Teesside Division), 1977, p. 51.

99 *JISI* 1876 part 1, p. 21.

Table 6. *Steel conversion costs in German works in 1878 (marks per ton)*

Krupp	15.30–15.80
GHH	17.04
Phönix	14.04
Dortmunder Union	12.36
Osnabrücker Stahlwerk	13.99

the basis of the not very great number of charges. The 'direct process', as conversion without re-smelting the pig iron in cupola furnaces was called, enabled even those works with a slower production rate to reduce their conversion costs to the same extent, or even more than by further speeding up the conversion process. This meant that towards the end of the 1870s there was, in addition to speeding up rapid driving, an alternative strategy for reducing costs in the Bessemer works. Working directly from the blast furnaces was first applied successfully in the 1860s to blast furnaces in Sweden and Austria, using their high-quality but extremely expensive charcoal-fired pig iron.[100] The first attempts using the usual pig iron did not, however, meet with the desired success. In Great Britain, Barrow had taken the lead in these attempts, as with the introduction of cupola furnaces, but had abandoned them soon after.[101] In Germany at around the same time in the 1860s, dispositions for the direct process had been set up Königshütte, but it too was soon abandoned. The same thing happened a few years later with the Dortmunder Union's trial works.[102]

It had become apparent that the quality of steel produced in this manner was very variable, since it was still not possible to maintain the composition of the pig iron inside the blast furnaces within sufficiently narrow bands. The chief difficulty lay apparently in fixing the silicon content, which could vary considerably from one tapping to the next.[103] These differences were always compensated for by processing in the cupola furnaces, since they smelted a chance mixture of pig irons from a variety of ores. Although it was in principle possible to economise on the cost of re-smelting in the cupola furnace, these

100 Wilhelm Hupfeld, Eisen und Stahl im Jahre 1874, in: *Zeitschrift des berg- und hüttenmännischen Vereins für Steiermark und Kärnten* 7 (1875), p. 294. Hupfeld mentions that the direct process, which was presented as novel in England, had already been carried out some time previously in Sweden and Austria.

101 The reason for abandoning the tests was, in the words of the works manager there 'that there was one most fatal objection to it, i.e. it was utterly impossible, when the iron was in the furnace, to know what its quality was'. *JISI* 1874 part 1, p. 357.

102 Hupfeld, *Eisen*, p. 294.

103 Snelus of the West Cumberland Iron Co. in a discussion held at the Iron and Steel Institute. *JISI* 1874 part 1, p. 361.

difficulties caused the works once again to abandon working directly from the blast furnaces at the beginning of the 1870s. Their prior task was to ensure the production of a Bessemer pig iron with a permanent and constant composition, which French and Belgian works were the first to achieve around 1875. They were followed shortly afterwards by the great British coastal works, Bolckow, Barrow, West Cumberland, Rhymney and Dowlais, which secured a leading position in steel production during the second half of the 1870s.[104]

In a report on ironmaking at the Paris Exhibition of 1878 it was already claimed that working from blast furnaces was becoming increasingly general.[105] This should also have been qualified by adding 'where this is possible', since it was by no means the case that all the very different steel-works had the buildings or space for introducing the direct process. In Germany, at any rate, the opportunities were very limited. In addition, towards the end of the 1870s, imported Bessemer pig iron was mostly cheaper than the German product. For this reason, Bochumer Verein, for instance, had steadily delayed putting its second blast furnace, which was intended for the direct process, into production. Its board, nevertheless, had already provided the means for working directly from the blast furnaces, should the occasion arise. However, even when the duty on pig iron was re-introduced this did not happen. Two years later it was decided to put the plant into operation using only cupola furnaces, since the direct process was said to be too slow and irregular.[106]

Krupp was very impressed by the news which Longsdon brought him from England about the successful implementation of the process and he proposed constructing blast furnaces in Essen, or, as a radical alternative, erecting a completely new works in Spain on his iron mines there.[107] Although neither of these two ideas came to fruition, they do nevertheless indicate which criteria were applied to assess a Bessemer works' chances of success: a coastal position; an ore base of its own and integration with a blast furnace works.

It is not possible to provide generally viable figures for the economies implemented by the direct process. Since the blast furnaces could not all deliver their charges according to the same tempo and rhythm as was required in the Bessemer works, all the works retained their cupola furnaces. This led to

[104] Jeans, *Steel*, pp. 418, 421ff., 435. BSC-Northern RRC, loc. no. 04898: minutes book 5, board meeting 18 October 1877. Glamorgan CRO, D/DGE 3: report year ending 31 March 1877, p. 3.

[105] Lecture by Kupelwieser according to the *Berg- und hüttenmännische Zeitung* 37 (1878), p. 340.

[106] Werkarchiv Krupp-Bochum, BV 12900 Nr. 12: Protokoll der Verwaltungsratssitzung vom 19 Mai 1880. See too BV 12900 Nr. 11: Protokoll der Verwaltungsratssitzung vom 16 März 1878.

[107] HA-Krupp WA IV 1439: Meyer and Goose on 8 March 1876; FAH II B 313: A. Krupp to A. Longsdon on 4 May 1876.

the frequent practice of mixing directly conveyed pig iron with cupola pig, in order to even out variations in quality. Thus it would surely be excessive simply to delete the items 'fuel' and 'furnace repairs' from the cost accounting. This is clearly what Snelus, of the West Cumberland Works, had in mind when he drafted his euphoric report about the direct process in Belgium and France, in which he computed the possible savings at ten shillings per ton.[108] With regard to his own practice, however, savings of just over four shillings per ton were mentioned.[109] In 1880 Windsor Richards reckoned with even more modest savings of at least two shillings per ton.[110] The cost-degression effect of the direct process thus failed by a long margin to play the same role as the speeding up of the refining process through cupola furnaces and improved bottoms. The hope expressed in *Iron*'s leading article of July 1876, that the British works would once again, by introducing the direct process, be able to compete with the American works' technological and organisational lead, on the costs side at least, was not wholly fulfilled.[111] However, as the example of the West Cumberland Steel Co. had shown with a production tempo, which lay at the lower end of the scale compared with the rest of Europe, they were certainly capable of matching the faster German works where conversion costs were concerned.

The unreliable composition of pig iron was still a problem, which was only solved technically when the pig-iron mixer was introduced at the beginning of the 1890s, which could load several charges and mix them. Until then a great variety coexisted, combinations of both direct process and cupola smelting. The direct process had no immediate effect on the operation of the Bessemer plant, since it was in principle a matter of indifference whether the liquid pig iron came from a blast or a cupola furnace. The main thing was that there should be enough to keep the plant fully employed.

4 The *modus operandi* of the rationalisation strategies

Hitherto we have concentrated mainly on the technical dimension of the wave of rationalisation which took place in the 1870s. However, the optimisation of the material input does not in itself suffice to explain the capacity effect achieved, although it was the necessary precondition for it. Firms did not engage in the acquisition of technology as an independent activity. This process only became effective in connection with a specific element in the organisation of production. The driving motive behind technical and organisational advance was the attempt to stabilise profits under the critical

108 *JISI* 1875 part 1, p. 204.
109 *JISI* 1876 part 1, p. 21.
110 *JISI* 1880 part 2, p. 567.
111 *Iron* 22 February 1876, p. 98.

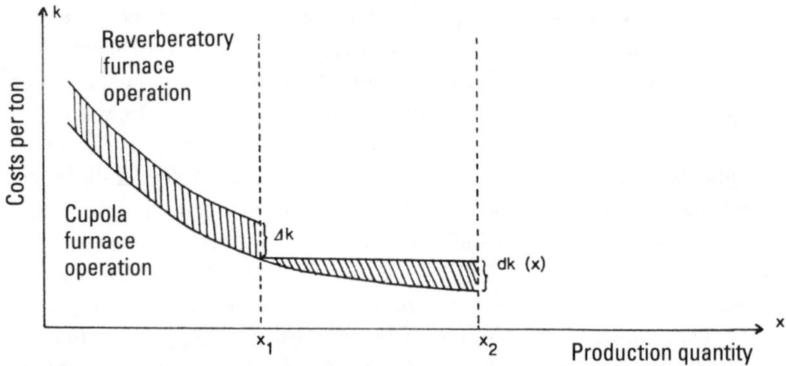

8 Changing from reverberatory furnace to cupola furnace: the effect on the cost function

deflationary conditions of the 1870s. At the same time, developments in the technological and organisational domain certainly did not always occur along parallel lines.

When the cupola furnaces were introduced in 1868 we saw that the possible rise in productivity which they entailed was not immediately reflected in the figures relating to the works' production capacity. The labour organisation of the conversion process was still much too uncertain about the novel apparatus and methods and was not yet competent to exploit its technical potential. In order to do so the sequence of the production process had to be reorganised: it had to be cut by curtailing all the standstills. Initially the cupola furnaces only brought the converter works a proportional reduction in the variable factor pig iron. Years later, the same thing applied to the direct process as well. It was only when the labour and material input were reorganised that a dynamic component emerged, which was attained by raising output while continuing to degress cost increases. Alfred Chandler called it 'economies of speed' in his survey of the rise of mass production in the nineteenth century.[112]

The different natures of these two methods of reducing costs, cheapening the pig iron input and speeding up the conversion process, are presented in terms of their effect on the cost function in Figure 8.

Replacing the reverberatory furnaces with cupola furnaces brought, independently of the quantity produced, a cost saving Δk, proportional to the reduced fuel consumption. Without making any changes to the organisation of production, this applied up to quantity x_1, the plant's previous capacity. With

[112] Alfred D. Chandler Jr., *The Visible Hand, The Managerial Revolution in American Business*, Cambridge, Mass. 1977, pp. 281ff.

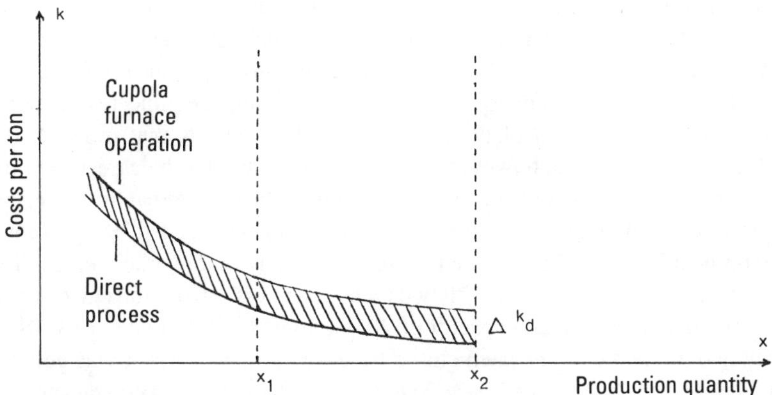

9 Changing from cupola furnace to the direct process: the effect on the cost function

the help of changes to the organisation of production in the sense of intensifying the production process, the falling costs (the degression effect of the fixed costs and economies of scale, which continued beyond quantity x_1) meant that the new capacity limit x_2 could be achieved. Fitting the cupola furnaces had in the meantime become a necessary but on its own not sufficient condition for cost saving dk(x) in excess of the old capacity limit of x_1.

The introduction of the direct process by further reducing the cost of the pig-iron input, once again effected a proportional economy in the cost per ton Δk_d. The capacity limit of the operation with cupola furnaces x_2 remained unaffected by this.

The introduction of the direct process differed from that of cupola furnaces in that it did not create any new latitude for further developments in the organisation of production. Consequently, it did not provide any additional motives for achieving further cost reductions by speeding up the conversion process since, as demonstrated in Figure 9, the full extent of the possible cost reduction Δk_d applied to the whole area between x_0 and x_2.

Realising the possible degression in costs by means of increased plant and workforce utilisation depended on how far it was possible to exhaust the production potential of the technical system, the limits of which had been extended by the cupola furnaces and were as yet unknown. This depended on, among other things, the ability of management and workforce to learn.[113] When

[113] However, we must at this point establish that this learning success was very different from that known as the Horndal-effect, since in the Bessemer works speeding up the refining process was the declared aim of the production organisers and not the result of year-long routines using

Müller was writing about the construction of the Osnabrücker Stahlwerk he recalled that the productive capacity of the plant as delivered had been twice as high as previously calculated;[114] concealed behind this was none other than an intensification of the production process by planning the time taken more efficiently. The delivered plant was certainly the same as that ordered. Its productive capacity was, however, only initially limited by the assessment of its organisational potential and not by technical factors. What is more, in Osnabrück too 'double' capacity was once again soon exceeded.[115] The same development has been documented in the Rheinische Stahlwerke's case. The works there were built in 1871 with two plants, each with two 5-ton converters, and estimated at an annual capacity of 20,000 tons. In 1877 this very works, which had in the meantime been closed because the few orders it was getting could be supplied by the firm's second and newer Bessemer works, was moved to Russia. At this time its annual capacity had already been assessed by the works management at 50,000 tons.[116] There are no clues as to which changes were made to the mothballed works to explain this 150 per cent capacity increase.

The new orientation towards rapid driving required first of all changing the traditional and familiar work process and the values subscribed to by the technical management. Krupp's quarrels with Eichhoff and the Bessemer Commission showed this clearly. Estimating the productive capacity of a works depended on the successive number of charges which was considered possible in organisational terms. Longsdon took this into account in his proposal, by offering two ways of organising technically identical plants. It was clear that their productive capacity depended on organisation and not on technology. The works manager decided in favour of the variant which made less demands on the organisation.

At Cockerill's in Belgium, whose Bessemer works had been newly built according to the 'American design' and was put into action in 1874, they were also aware of the fact that its technical potential was initially greater than their own organisational capabilities. As in the case of the new plant, these capabilities had first of all to be learnt, also according to the American model. Consequently, Cockerill's initial assessment for the works capacity was set at

unchanged technical equipment. Correspondingly, the growth in productivity in the Bessemer works was greater than the 2 per cent per annum in the case of Horndal's ironworks. See Nathan Rosenberg, Factors Affecting the Diffusion of Technology, in: Idem, *Perspectives on Technology*, Cambridge 1976, pp. 197f.

114 25,000 tons per annum instead of 12,500 tons per annum. Müller, *Georgs-Marien*, p. 71.

115 By the first half of 1874, the Osnabrücker Stahlwerk was already producing 14,500 finished goods with one plant, which corresponded to more than 30,000 tons of crude steel per annum. *Zeitschrift des VDI* 18 (1874), col. 704.

116 Thyssen-Archiv, RSW 126-00 2: Bericht des Vorstandes in der Generalversammlung am 29 November 1877.

c. 100 tons (approx. 14 charges) a day per plant. Once they had speeded up the removal of the steel ingots the daily rate was supposed to climb from 130 tons to 150 tons (approx. 18–21 charges).[117] In actual fact, this turned out to be 180 tons a day (approx. 25 charges), before it became necessary to implement new technical changes in 1881, to increase further the plants' productive capacity.[118]

The same phenomenon may be observed in more compact form at Rhymney. The works here were built in 1877 for a weekly production capacity of 500 tons. By the first year the plant was already regularly producing twice that and it soon achieved a weekly rate of 1,250 tons.[119]

We could supplement the four examples presented above with a few more showing clearly that the productive capacity of a works was not the exclusive property of its technical plant, although in the steel manufacturers' minds this was often assumed to be the case. The calculations relating to the capacity of British and German Bessemer works which are quoted below reflected not only the number of the plants but also what was recognised at the time as 'good business practice' in each case. In no case can one assume that high-capacity figures represented a proportional increase in the invested capital.

A technological barrier was reached again only when the rate of successive charges was limited by the low durability of the converter bottoms. The development of Holley's removable bottoms thus had a different significance to the introduction of cupola furnaces, since it served to remove a technical hindrance to an organisation of production, which had in the meantime become very efficient, so as to provide a greater margin for achieving the additional cost degression which was considered possible on the organisational level.

The cost factor is represented by a graph (see Figure 10) which shows how this gave rise to a further extension of capacity limit to x_3. Unlike the direct process, but similar to the intensification of operations following the implementation of cupola furnaces, the potential costs degression dk(x) could only be achieved if output were increased to the area within x_2 and x_3.

The works strategy of simultaneously increasing the productivity of labour and capital by intensifying the production process was developed out of the potential of the cupola furnace works. On this basis, then, adequate

117 *Iron* 26 September 1874.

118 *Stahl und Eisen* 3 (1883), p. 185. In *Stahl und Eisen* the capacity of the converters was given as six tons, whereas in *Iron* it was as much as seven tons. Taking these six tons as a basis would mean that they made twenty charges a day. Since it is unlikely that smaller converters were fitted later on, contrary to the trend in other works, it may be assumed that the converters were not filled so full in order to reduce metal loss. The significance of wastage and associated metal losses for the production costs has been mentioned in note 77. It is also very probable that they began by making fourteen charges a day with seven-ton inputs and later thirty charges a day with six-ton inputs.

119 Jeans, *Steel*, p. 424. *Iron* 4 December 1878, p. 748.

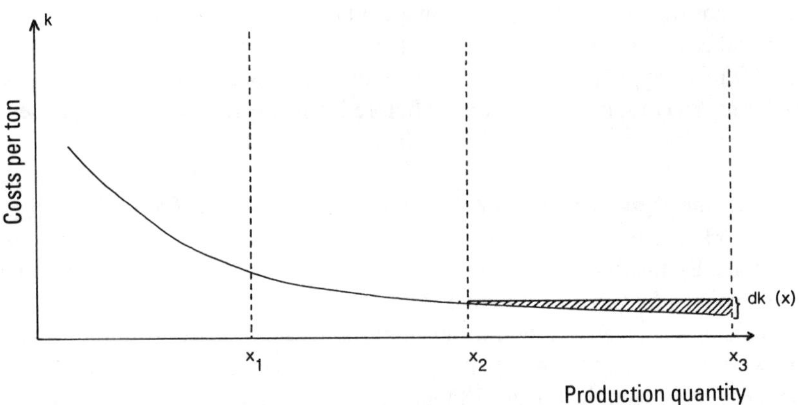

10 Introduction of Holley's removable bottoms: the effect on the cost function

development strategies could in their turn be formulated in the technical domain. The results were combined in the 'American design' of the Bessemer works, with Holley's removable bottoms and improved transportation systems.

Thus the technical alterations to the Bessemer plants during the seventies document at the same time a change in the organisation of production. Whereas in the first phase they were directed towards interpreting the technical potential, the inverse applied during the second phase. Now it was the newly acquired organisational potential which determined the development of the technical plant. The whole process is presented in Figure 11 in the form of a flow chart.

This flow chart implies that the German and British works were trying to maintain an existing organisation of production as long as possible, and were aiming in the first instance at reducing costs in other ways. That meant that they only worked at speeding up the conversion process until the possibilities of Column 1 were exhausted or until they became demonstrably unviable. The steel masters were basically conservative where the organisation of production was concerned and they were prepared to innovate only when this was required to stabilise satisfactory yields.

When one compares the practice in British and German Bessemer works, it is noticeable that towards the end of the seventies, the Rhineland-Westphalian works were using their plant on average more intensively. The highest production of a Bessemer installation with two plants in Great Britain lay in 1878 at 88,000 tons, using 8-ton converters.[120] This involved *c.* twenty charges

[120] *BITA* 1878, p. 33.

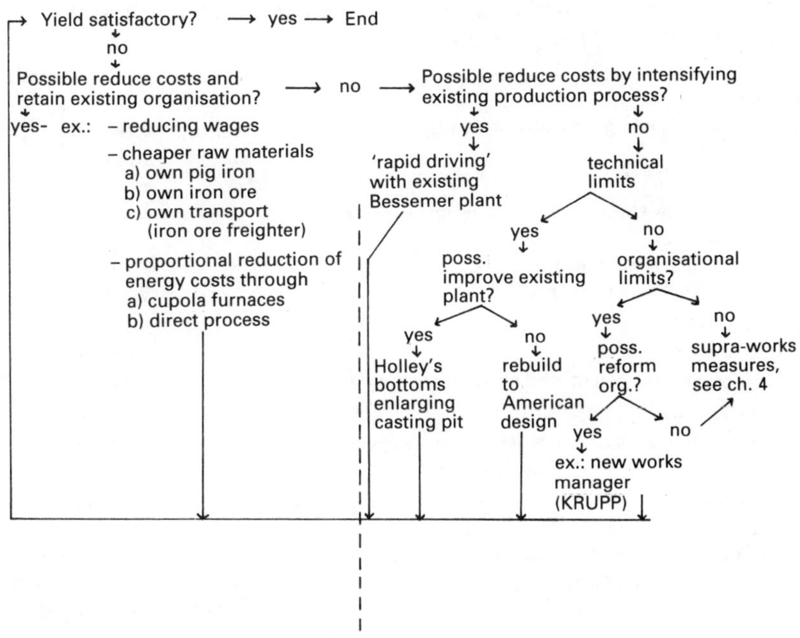

11 The problem-solving procedure employed during rationalisation of the Bessemer works in the 1870s

per plant over twenty-four hours. Krupp, Hoesch, Bochumer Verein, Osnabrücker Stahlwerk, GHH and Phönix all did better during the same period.[121] In terms of the works' technical equipment, on the other hand, a few British works, especially those which had only come into being since the mid-seventies, were in advance of the German. Some of them had already been constructed according to the American design, and the direct process was far more widespread in Great Britain than in Germany. However, their newly installed technical facilities were not utilised to the same degree as in the German works.[122] This meant that the British works had greater latitude for

121 Brauns, col. 93.

122 In the case of the British record-keeping for 1878, this would have meant that it could have produced at the same tempo of production as the Bochumer Verein the same amount of steel with one instead of two plants. Re Bochumer Verein, see above p. 71. It is suggested here that the British Bessemer works could still have achieved cost reductions had they applied the same production strategy as the German works, i.e. reducing the number of running plant per works while simultaneously intensifying operations with the remaining plant. The same conclusion is suggested by developments in South Wales at the start of the 1880s, see below p. 147. The costs were not minimised as far as they could have been in technical terms.

Table 7. *Comparative British and German production costs for Bessemer pig iron 1878–1879 (1 shilling = c. 1 mark)*

May 1878		
Bessemer pig iron (F.o.B.) Cumberland GB (sales price)	No. 1	64s.
	No. 2	62s. 6d.
	No. 3	61s. 6d.
Krupp (from its own Spanish iron ores, internal works price)		65.00 mk
Phönix (cost price)		70.87 mk
Dortmunder Union (cost price)		70.00 mk
Osnabrücker Stahlwerk (cost price)		70.74 mk
3rd quarter 1878		
GHH (cost price)		67.87 mk
English Bessemer pig iron (market quotations by GHH)		52s.
2nd semester 1878		
Krupp (from its own Spanish ores, internal works price)		65.00 mk
1st semester 1879		
Krupp (as above)		58.50 mk
Millom & Barrow, Cumberland (sales price 11.3)		49s.
Workington, Cumberland (cost price, in shillings and	4.1.79	52/3.95 – 56/4.51
pence, and decimals of pence)	15.3.79	51/9.39 – 50/9.84
	6.9.79	49/0.54 – 49/5.04
2nd semester 1879		
Consett, Cleveland (cost price)		44s. 9d.

achieving the required cost adjustments without, or with comparatively minor, modifications being made to the organisation of production. By far the greatest item in their cost-price accounting, which was furthermore not affected by the quantity produced in the steelworks, was the starting product, pig iron. It was precisely here that the British works had clear cost advantages, as indicated by the figures in Table 7.[123]

The cost comparison which Carl Lueg presented to the GHH board supervisor on 9 January 1879 revealed that English Bessemer pig iron (F.o.R.

[123] May 1878 Cumberland, according to *ICTR*, 24 May 1878, p. 585; 28 June 1878, p. 725. May 1878 Krupp in HA-Krupp WA I 859: Umfang und Selbstkosten der Schienenproduktion 1875–91, here 1. Halbjahr 1878. May 1878, Phönix, Dortmunder Union and Osnabrücker Stahlwerk in GHH-Archiv, 300 108/6: Vergleichende Selbstkostenberechnungen für Bessemer-Stahlschienen, Vorlage zur Aufsichtsratssitzung am 9 Januar 1879. III/1878 = Mai 1878 GHH Archiv, 2. Halbjahr 78 = Mai 1878 HA-Krupp, here 2. Halbjahr 1878. 1. Sem. 79 Krupp = May 78 HA-Krupp, here 1. Halbjahr 1879. 1. Halbjahr 79 Millom & Barrow in GHH-Archiv 300 108/6: Vorlage zur Aufsichtsratssitzung am 11 März 1879. 1. Halbjahr 79 Workington in BSC-Northern RRC, loc. no. 04465: Workington Iron & Steel Co., blast furnace cost & production book, Jan. 79–Dec. 80. 2nd half year 79 BSC-Northern RRC, loc. no. 04698: Consett, cost analysis book 1, average pig iron, July–December 79.

Table 8. *Freight charges for iron ores from Bilbao in October 1878 (whole freight), to:*

Dortmund Hoerde	18s.
Bochum	17s. 9d.
Ruhrort	16s.
Wales	9s. 3d.–9s. 6d.
Middlesbrough (Cleveland)	9s. 6d.–10s.
Newcastle	10s.
Cumberland	10s.

Ruhrort) was still 5.87 marks cheaper than GHH's Bessemer pig. Using English pig would make the production of one ton of steel cheaper by 7.45 marks.[124] The higher prices for Bessemer pig iron in the Ruhr were a result of the higher freight costs for low-phosphorus iron ore, compared with Great Britain (see Table 8). The British centres for the Bessemer rails industry were sited along the coast, where low-phosphorus iron ore could be transported at low cost primarily from Spain. Even the Cumberland works, which disposed of their own iron ore in their immediate neighbourhood, took in additional Spanish ore.[125] These geographical advantages were correspondingly reflected in their freight charges (see Table 8).[126]

Given that *c*. 2.1 tons of iron ore were required for every ton of pig iron,[127] the differential in transportation costs alone meant that the British works' costs were lower than the Ruhr's by between 12.50 marks and 18 marks per ton of pig iron. GHH had worked out at the same time that the British works enjoyed a price advantage of sixteen marks over conditions there.[128] Krupp too, who obtained his ore from Bilbao in his own ships, had similar freight costs. The sea freight from Bilbao to Rotterdam alone cost him between thirteen and fourteen marks per ton. This was clearly more than the passage to the British ports, although the distance was about the same, because Krupp's ships were unable to find a lucrative return freight from Rotterdam, whereas the British ships

124 GHH-Archiv, 300 108/6: Vergleichende Selbstkostenberechnungen für Bessemer-Stahlschienen, Vorlage zur Aufsichtsratssitzung am 9 Januar 1879.

125 The *ICTR* was already reporting this on 3 June 1874, p. 636. It can be concluded from a discussion held in the Iron and Steel Institute that Spanish ores were better suited to the direct process, on account of their higher manganese content and uniformity, than were the ores from Barrow. *JISI* 1875 part 1, p. 209. They were probably mixed with West-Coast ores, in order to facilitate working directly from the blast furnace.

126 Werkarchiv Krupp-Bochum, BV 0100 Nr. 1; Müller (Dortmunder Union) to Baare on 24 October 1878. Nearly identical figures can be found in Bundesarchiv, R 13 I, Nr. 343: Bl. 438, Vergleichende Zusammenstellung der Frachten bei Verwendung von fremdem Eisenstein.

127 BSC-Northern RRC, loc. no. 04698: Consett, cost analysis book 1, Bessemer pig iron 1879/80.

128 See note 123, May 1878, GHH.

could carry coal from Wales or Cleveland on their return journeys.[129] For the time being, however, the management considered the conversion costs in their Bessemer works very satisfactory. The Bessemer works worked best. If the works were not fully competitive on the international level, this was now only due to the price of pig iron.[130]

The price differential between English and Westphalian Bessemer pig iron was so great that what were in principle cost-cutting measures, such as producing one's own pig iron and adopting the direct process, effected only unsatisfactory savings or no savings at all compared with prices for imported iron. The production of home pig iron in the Ruhr could only be justified on economic grounds if the losses to the blast furnace plant were minimised or if otherwise unsaleable stocks were processed. In the latter event, it was mostly a matter of panic buying at top prices during the raw-materials shortage at the beginning of the seventies. In GHH's case, they decided in 1875 against temporarily stopping their occasionally unprofitable rails production, because it would have meant reducing the activity of the blast furnaces and large stocks of iron ore would have lain around without being processed. Cutting back operations would have resulted in redemption rates and overall costs for the whole works rising from 14 per cent to 31 per cent. 'The reduced operations could never carry that!' – was the response. Consequently GHH preferred to go on processing its own expensive pig iron, even though they ended by making some losses in their rails business.[131] In Phönix's case, they too were forced in 1874 and 1876 to reduce their iron-ore stocks as a very important measure.[132] It was Krupp who expressed the general desire at the start of the Bessemer Commission's work: 'May our production of Bessemer steel, even if it cannot bring any sizeable profits, at least turn our stocks to account by achieving a large turnover.'[133] It was because they were unable to process their contracted raw materials in time that Rheinische Stahlwerke ran into financial trouble in 1877. They had to pay 800,000 marks compensation for deferred deliveries of pig iron, which brought about the collapse of the credit house that had supported the firm.[134] In the same way, the collapse of Bochumer Gesellschaft für Stahlindustrie (formerly Daelen, Schreiber & Co.) was also closely bound

129 HA-Krupp WA IV 1643: M. Jebsen to C. Meyer on 11 March 1878.

130 HA-Krupp WA IV 656: Reuter to A. Longsdon on 6 July 1877.

131 GHH-Archiv, 300 108/2: Promemoria für den Fortbetrieb der einzelnen Werke und Branchen der Gutehoffnungshütte (by F. Kesten, 15 October 1875) und Erwiderung des Vorstandes der Gutehoffnungshütte auf das Promemoria des Herrn Kesten.

132 Mannesmann-Archiv, P1. 25.35: Protokoll der Sitzung des Administrationsrates am 9 Juni 1874; ditto am 11 Dezember 1876.

133 HA-Krupp FAH II P 94: A. Krupp to Procura on 27 June 1874.

134 Thyssen-Archiv, RSW 126-00 5: Generalversammlung vom 31 Oktober 1878, Bericht des Aufsichtsrates, p. 1.

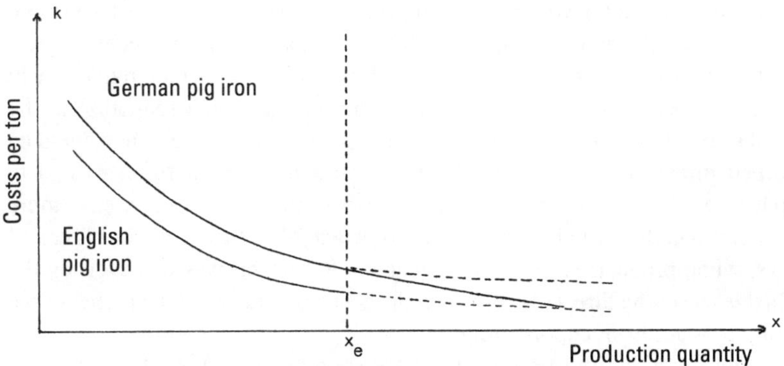

12 Cost functions for use of German and English pig iron in the Ruhr

up with their extensive and overpriced pig-iron transactions during the raw-materials shortage in 1872/3.[135]

For the conversion process, using one's own pig iron, whether from stocks or from one's own blast furnaces, represented only a further cost, which somehow had to be compensated for, if one wanted to appear on the world market with competitive prices. Given that their raw-materials costs could not be reduced to the levels enjoyed by the British competition, and that wages, which could in any case only be slightly adjusted because they formed only a small part of overall costs, were already driven as low as possible, the only remaining solution was to make greater than average use of production plant. That which was lost against British competition via col. 1 of the flow chart had to be regained as far as possible via col. 2.

A further motive for increasing production may, in the case of the vertically integrated works which had to use their own pig iron, have lain in the fact that their first costs would fall more sharply than they would in relation to the original cost function if they increased the proportion produced with cheaper imported pig iron (see Figure 12).

Up to quantity X_e, an expensive pig iron of their own is used. After X_e the cost curve draws near to the curve which would apply to those firms using cheaper imported pig iron exclusively.[136] In Great Britain, on the other hand,

135 Werkarchiv Krupp-Bochum, BV 16303 Nr. 4 a: Gesellschaft für Stahlindustrie, Protokoll der Verwaltungsratssitzung am 24 April 1874; ditto am 27 Mai 1874.

136 This effect is clearly revealed in the Bochumer Verein's cost accounting. The highest production costs were reached in those months when the proportion of their own pig iron was higher than average. Werkarchiv Krupp-Bochum, BV 01000 Nr. 1: Vergleichende

this motive could not have existed at all. Moreover the direct process was more widespread there, and although the full exploitation of the technological potential of the Bessemer works had to bring about a further reduction in conversion costs, it did not necessarily entail an equivalent reduction in the overall steel costs as well. If more pig iron was converted to steel than could be absorbed directly from the blast furnaces, then the cupola furnaces had to be put to work as well. The pig iron smelted in these was in any case more expensive than that which came directly from the blast furnaces. As Figure 13 shows, when production rose above quantity X_d, which was achieved by the exclusive use of the direct process, then the cost curve rejoined the higher curve relating to work from cupola furnaces.

It is not possible to distinguish from the source material available whether the cost curve for quantities greater than X_d rose, sank or ran parallel to the x-axis. However, it is certain that further cost savings in the conversion process were countered, in part at least, by the more expensive method of resmelting pig iron.[137]

Thus it is not inconceivable, in the case of the British coastal works, most of which operated directly from blast furnaces, that exploiting all the technical potential of their plant did not lead to an improvement in their yields. What we have here is the problem of the optimal lot-size of a plant, which, in our case, was obviously smaller in Great Britain with the direct process, than in Germany, where they operated almost exclusively from cupola furnaces. In the long term, however, because they could get cheaper raw materials they were not motivated to develop the organisation of production further, which eventually must have led to losses; this was already felt in comparison with the American works and also pointed towards the German steel industry. Massenez of the Hörder Verein wrote: 'Abroad, especially in England, the installations are also excellent. However, they do not operate as well in many of the works there as with us.'[138] Thielen of Phönix wrote: 'Where the method

Zusammenstellung der Selbstkosten diverser Fabrikate, July 1877–September 1878. Accordingly the Bochumer Verein had long avoided putting their newly built second blast furnace into operation. Although it was anticipated that working a second blast furnace would make their own pig iron cheaper by 4 marks per ton, it was long thought that this would not enable them to prevail over imports of British pig iron. Ibid. BV 12900 Nr. 11: Protokoll der Verwaltungsratssitzung am 13 September 1877.

137 The example of the West Cumberland Co. cited on p. 88 shows that by using the direct process, what were otherwise very high conversion costs with a moderate production tempo could be reduced to excellent values, compared with the competition. Lan stressed that extending production was not possible as straightforward with the direct process as with the cupola process, which was the only one to allow optimal use of the steam engine and blowing engine. According to *Berg- und hüttenmännische Zeitung* 42 (1883), p. 127. It was said of the Rhymney Bessemer plant, built to American plans with Holley's removable bottoms, that its production was limited by the two accompanying blast furnaces, whereas the plant itself certainly still had considerable reserves. Jeans, *Steel*, p. 425.

138 PEE, Massenez, p. 409.

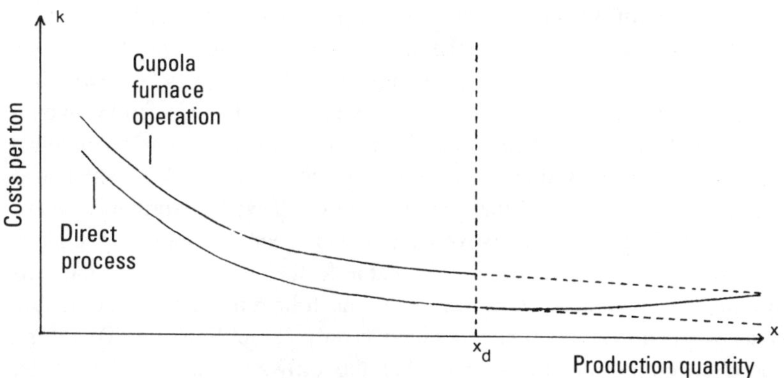

13 Abandoning the direct process: the effect on the cost function

of manufacture is concerned, the mechanical equipment of the English works is the very best of the age. Where order and accurate work are concerned, the Rhineland-Westphalian works are superior. If the English were to work with the same accuracy as we do, they would be much more dangerous for us.'[139]

Finally, we have this very decisive statement from Hoesch: 'our manufacture is based on a higher and more ideal point of view than the English'. Only in the case of the modern Bolckow, Vaughan & Co. works did he concede that it could compare in technological and organisational matters.[140] All this meant that the British works had reserves and could have reduced their production costs still further. For the time being, however, the competition in Britain was still not strong enough to encourage further rationalisation measures on the broader front. This came only when additional Bessemer works were built at the beginning of the 1880s, above all in Wales, where just such an optimisation of the temporal duration of production took place as had already been effected in the 1870s in Germany.

5 The export 'safety valve'

(a) In Great Britain

Expanding production would allow the first costs to be cut and was urgently required in view of the more intense competition, but the precondition was

139 Bundesarchiv, R 13 I, Nr. 171: Materialien für Enquête-Fragebogen. Englands Eisenindustrie (Herr Direktor Thielen), p. 5.

140 PEE Hoesch, p. 243.

finding a market for Bessemer steel which surpassed the earlier expectations of 1869–70. This meant primarily the rails market, on account of its more restricted application. In this respect, however, the domestic railways had hardly anything left to offer. The Bessemer works already had a firm grip on it and, in view of the general depression, the railway companies were not inclined to invest further in new lines. That left only the export market, which was predominantly controlled by the puddling works. It was, nevertheless, in an extremely desolate condition, as we have already seen. An inquiry published by the Iron and Steel Institute in 1878 came to the conclusion that the rails requirement of the whole world lay at around four million tons per annum, whereas production at this time amounted to only 2.7 million tons. Therefore, the poor economic situation meant that the railways had failed to order 1.3 million tons of rails,[141] with the consequence that many British puddling works, which had been wholly geared to the world market, collapsed. Hopes for a rapid recovery of iron-rails sales had not been realised; the market was pronounced dead, and for good.[142] Dead only with regard to British conditions, it must be stipulated, since, as Table 9[143] shows, in 1877 about 300,000 tons of iron rails were still being made. By 1878, the figure was only half that. Compared to 500,000 and 630,000 tons of steel rails in the same years, this was a particularly modest figure. The market was taken over by the Bessemer works, which were eagerly realising their opportunities for expanding their sales area.

It was high time that this export 'safety valve' for steel functioned, as was demonstrated by growing problems in this branch of industry. In Sheffield the heightened competition had already forced John Brown and Brown, Bayley & Dixon to cut back production in their Bessemer works by the end of 1874.[144] Whereas John Brown averted this threat, as did his neighbour Charles Cammell, by diversifying, Brown, Bayley & Dixon went bankrupt.[145] However, once the works had been fundamentally reorganised and the Bessemer plant modernised, B, B & D, which had in the meantime become the most rapid works in Great Britain, achieved a net profit of 18 per cent in

[141] C. Wood, Statistics respecting the production and depreciation of rails, and notes on the application of wrought iron and steel to permanent ways, with a description of a new kind of railway sleeper and clipchair, in: *JISI* 1878 part 1, p. 76.

[142] *Iron* 8 July 1876, p. 37. It was said in the annual report of the run-down Nantyglo and Blaina Ironworks that 'The highest authorities in South Wales have stated that the trade of Iron Rails has gone, never to return.' City of Wakefield Metropolitan District Archive, Goodchild loan MSS – Joseph Abbott MSS, No. 6: report year ending 31 August 1876.

[143] Table according to *BITA* 1879, p. 34 and p. 41.

[144] *Iron* 10 October 1874, p. 462; 5 December 1874, p. 718.

[145] Re diversification by J. Brown and C. Cammell, see above p. 48. For B, B & D see *Iron* 13 May 1875, p. 336; 4 June 1875, p. 681.

Table 9. *Production, consumption and export of rails: Great Britain, 1870–1879 (1,000 tons)*

	Total production	Proportion steel	Inland consumption	Exports	Proportion steel
1870	1,358		299	1,059	
1871	1,365		386	979	
1872	1,268		331	945	
1873	1,104		319	785	
1874	1,105		323	783	
1875	865		319	546	
1876	852	400	338	515	174
1877	817	508	319	498	235
1878	780	634	340	439	251
1879		520		464	328

1877.[146] After this brief flowering, the firm was wound up yet again at the turn of 1880/1.[147]

Yet another victim was the Mersey Iron & Steel Co., whose ten converters were only occasionally operated after 1877.[148] This firm had never been really healthy since it was founded in 1864. Of its £800,000 nominal capital, only £300,000 had actually been paid up. At its second start-up the nominal capital was reduced to £500,000, most of which was actually paid. However, its newly won stability did not last long. Eighteen months later, in February 1872,

146 *ICTR* 8 March 1878, p. 267. *Iron* 6 April 1878, p. 425.

147 *Iron* 31 December 1880, p. 489. B, B & D owed their brief boom period in the late 1870s basically to the introduction of 'American' methods, for instance the introduction of Holley's removable bottoms. They had thereby been able to raise their production tempo to rates of over twenty charges a day, which were by this time already normal levels in Germany. In 1878 the maximum rate stood at fourteen charges per shift (approx. twenty-eight per day). In order to surmount the structural disadvantage of their unfavourable situation re transport, B, B & D thus took the same course as that adopted by the works in Rhineland-Westphalia which were experiencing similar problems. Their success in speeding up the conversion process and the measures taken to this end were presented in a lecture by the works manager to the Iron and Steel Institute in March. C. B. Holland, On the Manufacture of Bessemer Steel and Steel Rails at the Works of Messrs. Brown, Bayley and Dixon (Limited), Sheffield, in: *Iron* 13 April 1878, pp. 456f. He provided detailed figures about the extension to the plant and in particular about Holley's removable bottoms, which aroused Krupp's special interest. Holley's removable bottoms were just about to be introduced to his works, as they were urgently demanded by his works manager in the Bessemer works. See above p. 81. This was why they had deviated from their usual practice and had prepared a special translation for use in his Bessemer works. HA-Krupp WA VIIb 103: Geschichtliche Angaben über die Fabrikation im Bessemerwerk 1875–86. The translation does not have a source reference, but follows Holland's lecture word for word.

148 W. M. Lord, The Development of the Bessemer Process in Lancashire, 1856–1900, in: *Transactions of the Newcomen Society* 25 (1945–7), pp. 163–80, here p. 173.

capital was reduced to £240,000 once again, which does not seem to have secured a net improvement. Nevertheless, the Mersey Iron & Steel Co. was still the only Bessemer works in Great Britain whose capital had been reduced during the depression in the 1870s – quite unlike conditions in Germany.[149] In June 1875 the Phoenix Bessemer Steel Co. was bankrupted by a London merchant. The works were bought by a 'betting agent' and were run successfully from 1876 onwards.[150] Ebbw Vale in South Wales also fell into great difficulties, which were only solved by bringing in Spanish ores.[151] Manchester Steel and Plant Co. and Carnforth Hematite Co. had also closed their Bessemer works by 1878 at the latest.[152] The Glasgow Bessemer Steel Co. had likewise been out of action for some time now.[153]

Two developments had drastically narrowed down the price gap between iron and Bessemer rails, which had previously been wide enough to retain the overseas export market for the former. One was the iron ore market which had settled down once Spanish ores could be obtained for the Bessemer process. The Carlist revolts had prevented any regular operations on the iron-ore fields around Bilbao from July 1873 to March 1876, which led to a shortage of low-phosphorus iron ore and kept prices high.[154] On the other hand, however, the rationalisation measures described above were now taking effect and helping to keep prices down, just as the ensuing heightened competition between Bessemer works was driving their profit margins down.

It is clear from Table 10[155] that the slight price differences during the second half of the decade must have brought the point at which it became worthwhile for the railway companies to buy Bessemer rails even nearer. This also means that it was now cost-effective to buy steel rails for the less-intensively used lines overseas. The decline of iron rails had now moved into its second stage. 'The enormous growth of the steel trade has, moreover, involved as its

149 Select Committee on Companies Acts of 1862 and 1867, in: P.P. 1877, VII, p. 375 and Appendix, pp. 175ff.

150 The individual stages of this procedure are well documented in: *Iron* 12 June 1875, p. 749; 19 June 1875, p. 781; 24 July 1875, p. 109; 7 August 1875, p. 172; 18 September 1875, p. 365; 30 October 1875, p. 554; 25 March 1876, p. 396 and *ICTR* 18 June 1875, p. 743; 6 August 1875, p. 956; 29 October 1875, p. 1319.

151 The losses incurred by Ebbw Vale were supposed to have totalled £670,000 by 1877. Burn, p. 29.

152 *BITA* 1878, p. 34.

153 Peter L. Payne, *Colvilles and the Scottish Steel Industry*, Oxford 1979, p. 23.

154 J. C. Kendall, Iron Ores of Spain, in: *Transactions of the Federated Institution of Mining Engineers* 3 (1891/2), p. 606.

155 Cols. 1 and 2 according to Carr and Taplin, p. 96 and *BITA* 1879, p. 41. The enterprise costs for Dowlais which have been produced for the purpose of comparison, applied in each case to the business year starting on 1 April and ending on 31 March of the following year. Dowlais was one of those works which survived the 1870s without making a loss, but did not qualify as the works with the most economical production, Glamorgan CRO, D/DG E 3: report for year ending 29 March 1884, appendix.

Table 10. *Average rail prices (ex works) and production costs in Great Britain, 1872–1879*

	Puddled rails £.s.d.	Bessemer rails £.s.d.	Production costs Bessemer rails Dowlais £.s.d.
1872	9. 6. 0	13.17. 6	9. 9. 7
1873	11. 4. 4	15.10. 0	10.10.11
1874	8. 9. 0	9.17. 6	9. 1. 4.5
1875	6.19. 6	8.17. 6	7. 9. 5.4
1876	6. 0.10	7. 2. 6	6. 8. 5
1877	5.13. 3	6. 7. 6	5.13. 7
1878	5. 2. 1	5.12. 0	4.17. 5.5
1879	4.18. 3	5. 2. 6	4. 9. 7.6

corollary a much more entire depression of our malleable iron trade than would otherwise have occurred.'[156]

The long-awaited upswing of Indian demand came too late for the puddling works. After its unhappy experience with American railroad bonds, British capital in search of investment opportunities moved towards safer government stock in the colonies.[157] *The Economist*'s *Investor's Manual* was already indicating at the end of 1874 that the colonies were compensating in part for the loss of foreign government stock.[158] By the end of 1876, Australian and Indian stock in particular were already being described as 'extensive'.[159] Most of the capital flow in these places went to finance railway construction, and so found its way back to heavy industry at home. In 1877 half of the exported railway material went to India alone.[160] German steelworks also attempted to get a share of this trade. Bochumer Verein and Phönix undercut (according to their own figures) the entire British competition for Indian or Australian rail tenders, but they were still not accorded the contract.[161] The 'invisible hand' of the colonial administration preserved this extended 'inner' market for British

156 *BITA* 1877, pp. 1–2. In the meantime, the American railway companies in the American West had also gone over to laying their lines with steel rails, whereby the hitherto biggest customer for iron rails was lost. H. Wedding, Das Eisenhüttenwesen der Vereinigten Staaten von Nordamerika, in: *ZBHSW* 24 (1876), pp. 328–50 and pp. 402–86, here pp. 457f.

157 S. B. Saul, *Studies in British Overseas Trade 1870–1914*, Liverpool 1960, p. 112. *ICTR* 25 May 1877, prospects of the iron trade, p. 573.

158 *Economist* 26 December 1874, p. 438.

159 *Economist* 30 December 1876, p. 442.

160 *BITA* 1877, p. 40.

161 PEE, Thielen, p. 19, and Baare, p. 809.

industry. So as not to lay themselves open to the accusation of contravening their own Free Trade dogma, the British authorities sheltered behind 'technical' directives, which served the same purpose. The exclusive use of British pig iron was made an essential requirement for government supplies for Great Britain and the colonies.[162] Furthermore, the British steel industry could profit from the fact that the construction of railways overseas using British capital was entrusted to British technicians who resorted, when in doubt, to familiar domestic material. On the other hand, the first 'tied loans' for railway construction came only in 1907 for the Chinese railways and indeed Burn and Saul consider this nineteenth-century 'bias' in favour of British railway materials to have been still very effective.[163]

By squeezing the wrought-iron works out of the export trade, the British Bessemer works had created a space large enough for marketing their growing production. 'It is only the displacement of one form of product by another', as the British Iron Trade Association commented laconically.[164] There were two ways out of the situation for the big iron rail works, if they did not want or did not have to close down completely. They could retain their puddling furnaces and alter their rolling mills and go over to manufacturing metal plates for shipbuilding. This course was particularly viable in Cleveland, where there was already a large shipbuilding industry. The Britannia Works, built only in 1871 for the American rails market and closed since 1876, were bought up by the Skerne Company in Darlington to produce metal plates for shipbuilding.[165] Consett, which had closed down its rails production in 1876, apart from the odd sporadic commission, decided in October 1876 to convert one of its rail mills to a plates mill.[166] By March the following year the success of this business led to the decision to carry out a further conversion of this type.[167] In order to secure their new business, the iron plate manufacturers of Cleveland quickly came to working agreements similar to cartels for limiting production quantities.[168]

162 Protokoll der 3. General-zugleich Wanderversammlung des Berg- und hüttenmännischen Vereins für Steiermark und Kärnten, abgehalten in Klagenfurt am 8. September 1877, in: *Zeitschrift des berg- und hüttenmännischen Vereins für Steiermark und Kärnten* 9 (1877), p. 374.

163 Burn, p. 262. Saul, *Studies*, p. 75. Re the first tied loans, see D. C. M. Platt, *Finance, Trade and Politics in British Foreign Policy, 1815–1914*, Oxford 1968, p. 21.

164 *BITA* 1877, p. 2.

165 *ICTR* 1 March 1878, p. 240.

166 BSC-Northern RRC, loc. no. 04687: minutes book 3, board meeting 3 October 1876; loc. no. D/Co 89: cost analysis book, pp. 157 and 178, rails. There this production of rails in 1876 is still set at 18,573 tons, compared with 27,642 tons in the preceding year. By 1877 it was only 3,042 tons.

167 BSC-Northern RRC, loc. no. 04687: minutes book 3, board meeting 9 January 1877 and 6 March 1877.

168 Ibid., board meeting 1 April 1879. The *ICTR* had already established on 25 January 1878 that such agreements did in fact exist.

The second course was the one preferred by Bolckow, Vaughan & Co., which was at that time still doing exceptionally well as the largest iron and steel producer in the world. The company had acquired the Lancashire Steel Co.'s Bessemer works near Manchester at the beginning of the 1870s, with the declared aim of simply using them as a stand-in until their own great works in Cleveland were built.[169] At the same time the company acquired extensive iron mines in northern Spain in order to secure its own supplies of low-phosphorus iron ore. Delayed by the Carlist risings, the Bessemer process could only be initiated in 1877 on the basis of its own ore in Eston/ Cleveland.[170] Due to its low raw material costs and to utilising the most modern Bessemer plants along American lines, Bolckow Vaughan soon had the lowest costs in the industry, thus giving rise to considerable concern.[171] The old works, acquired as a stop-gap, was sold to the Manchester, Sheffield and Lincolnshire Railway Co., which used them to produce their own rails.[172]

In view of its neighbour's success, Consett too considered erecting its own Bessemer works, since it also had its own supplies of low-phosphorus iron ore from North Spain.[173] These plans were, however, later abandoned in favour of

169 Carr and Taplin, p. 89. For a while they had also considered a plan for moving the steelworks from Gorton to Middlesbrough, but this was abandoned in favour of a bigger and more modern works. BSC-Northern RRC, loc. no. 04898: minutes book 4, board meeting 1 August 1874, 29 August 1874, 19 November 1874. In this latest meeting the decision was finally made to build a new works in Eston, near Middlesbrough, which would be able to produce 500 tons of rails a week. In accordance with notions at that time, two plants were set up, each with two 8-ton converters. As was only to be expected, within a few years its production capacity exceeded the quantity originally proposed many times over. By early in 1880 the works were making not 2,000 tons, but over 9,000 tons of rails per month. Ibid., minutes book 5, board meeeting 21 May 1880. Even taking into account the fact that the original estimate for its production capacity had been very cautious, the people at B, V & Co. could scarcely have reckoned with four times the quantity.

170 BSC-Northern R.R.C., loc. no. 04902: report of directors, year ending 31 December 1874 and year ending 31 December 1875.

171 Re the entry of B, V & Co. into the rails trade, Hupfeld wrote that the people there knew 'by economical and intensive operations, how to reduce production costs so that the above mentioned firm was already, after a brief period, able to penetrate the international rails trade, with prices which so shaped the export business that from then on it became a loss-making business for the entire competition.' Wilhelm Hupfeld, Der Einfluss der Roheisen-Entphosphorung auf die Weiterentwicklung der Bessemer-Industrie, in: *Zeitschrift des berg- und hüttenmännischen Vereins für Steiermark und Kärnten* 11 (1879), p. 250. By loss-making export business, Hupfeld certainly meant the Continental European Bessemer works, since he knew more about their circumstances. As a glance at conditions in Dowlais shows, this works was still keeping up well in the late seventies (see Table 22, p. 199). In any case, Hupfeld was not alone in his estimation of the relative strengths of B, V & Co. See *ICTR* 8 November 1878, p. 529, where the first working year for the Bessemer works in Eston is commented on in a similar fashion.

172 J. K. Almond, Steel Production in North East England before 1880, in: C. A. Hempstead (ed.), *Cleveland Iron and Steel, Background and 19th Century History* (The British Steel Corporation) 1979, p. 171. *Iron* 22 January 1876, p. 109.

173 BSC-Northern RRC, loc. no. 04687: minutes book 3, board meeting 7 November 1876. Extensive details about this are to be found in Chapter 5, s. 2(e).

an open-hearth works for manufacturing metal plates, which had in the meantime become so successful that Consett completely dropped all further rail production. Since the existing works was by then fully employed with iron plates and pig iron, it had become necessary to build an additional works with all the infrastructure, such as streets, worker dwellings and so on. Although manufacturing Bessemer rails was indeed considered profitable, the firm shrank back from spending the £300,000 required for building the necessary blast-furnace capacity plus infrastructure.[174] The Darlington Co., also in Cleveland, had been forced to close its works in October 1878 due to lack of orders. In 1873 it could still pay 30 per cent dividends and still a little in 1874, after which nothing.[175] Early in 1878 the deliberations there took shape in the form of a Bessemer works according to the Bolckow Vaughan model, which, however, only came into blast in 1880.[176]

Even worse were the conditions for iron rails in South Wales, since there was no active shipbuilding industry there, which might have made a shift to iron plates possible. A few works tried to keep their puddling furnaces above water by manufacturing bar-iron for the tin-plate industry, although in so doing they came up against established competition in a branch of industry which was already very sensitive to business trends.[177] As well as this, local supplies of iron ore were becoming increasingly exhausted, so that not only was the coal often of second-class quality, but their own raw-material supplies were no longer forthcoming. On the other hand, the lack of raw materials spurred on the changeover to importing low-phosphorus iron ore from Spain, and thereby the changeover to the Bessemer process, which was effective when the means for a successful entry into this business were available, although the amount of investment required had grown considerably in the meantime.

Along with Cumberland with its local low-phosphorus ores and Cleveland, South Wales was at that time the most cost-effective site in the world for the Bessemer process, since the works in all three cases lay along the coast, close to cheap transport, and disposed of one of the two raw materials in their immediate neighbourhood. This was particularly significant for overseas exports, which were now becoming increasingly important. Nantyglo & Blaina and Blaenavon were further wrought-iron works which failed financially when the market in iron rails collapsed.[178] Cyfarthfa had closed its works in time and

174 Ibid., minutes book 4, Messrs. Williams & Jenkins Steel Report, board meeting 6 March 1877.

175 *ICTR* 1 November 1878, p. 501.

176 Almond, *Steel*, p. 174. *ICTR* 29 March 1878, p. 354; 6 June 1879, p. 610.

177 Carr and Taplin, p. 81.

178 *ICTR* 20 December 1878, p. 698, 10 April 1875, p. 462. City of Wakefield Metropolitan District Archives, Goodchild loan MSS – Joseph Abbott MSS: Nantyglo & Blaina, no. 7, Reply by ex-general manager to Hugh Mason, 1877.

was seeking in vain for a buyer.[179] All three considered building a Bessemer works. Blaenavon finally managed to do so in 1880, with Cyfarthfa following suit in 1882.[180] The late changeover to the Bessemer process occurred relatively smoothly only in the case of Tredegar and Rhymney.[181] The model works in this region was Dowlais, whose production of Bessemer rails was still as profitable as ever, all the more so given that exports had moved into the foreground from 1877, as well as which they were finally able to process their own Spanish ores. One result at least was that the Cumberland works lost their previous competitive edge over the Welsh works.[182]

All the works mentioned up to now (Rhymney, Blaenavon, Tredegar, Cyfarthfa, Nantyglo & Blaina, Britannia, Darlington, Consett, Dowlais and Bolckow Vaughan) had started out as big wrought-iron works and belonged in 1873 to the exclusive club of the two dozen biggest in Great Britain.[183] They had been strongly involved in rails exports from the start. When the price gap between Bessemer and iron rails was closed, their hopes of recovering their sales were dashed and they were finally forced to change either the range of their products or their production process. A third course lay in mechanical puddling, to which the branch still attached great hopes at the beginning of the decade, but which proved unviable.

By taking over the export market, the thrust of the British Bessemer industry tipped towards the coastal works, which took advantage of their great advantage in the matter of freight. The expansion of this potential market for British steelworks was obviously large enough to make the addition of further works along the coast look promising. The same attraction did not apply inland. The home market was already fully exhausted, and bankruptcies had occurred there as well. In any case, the British railway companies continued to prefer steel from the Sheffield works, whose rails were often preferred to those from the coastal works on the home market, being considered better quality.[184] This

179 National Library of Wales, Cyfarthfa MSS, box 8: Joseph Maynard to Robert Crawshay on 24 March 1874; Robert Crawshay to Joseph Maynard on 25 March 1874; Robert Crawshay to John Waterhouse on 21 July 1874.

180 City of Wakefield Metropolitan District Archives, Goodchild loan MSS – Joseph Abbott MSS: Nantyglo & Blaina, no. 6 report and balance sheet, year ending 31 August 1876. *ICTR* 14 December 1877, p. 666; 1 November 1878, p. 501. Re adopting Bessemer process, see Gwent CRO, Blaenavon MSS, loc. no. D. 480.1 and 2: minute book 1, board meeting 27 January 1880. Re Cyfarthfa, see *Iron* 6 June 1884, p. 499.

181 *ICTR* 9 February 1877, p. 157; 1 June 1877, p. 609; 8 February 1878, p. 158. Jeans, *Steel*, p. 421.

182 Glamorgan CRO, D/DG E 2: statement of profits – 1864 to 1894. After the 'net profit on steel' for 1875 had amounted to just £500 and in 1876 to just £1,600, it rose in 1877 to £28,500, in 1878 to £25,100 and in 1879 to as much as £35,700.

183 Calculated according to the number of puddling furnaces in *Mineral Statistics*, 1873, pp. 109–14.

184 Gwent CRO, Blaenavon MSS, loc. no. D. 480. 1 and 2: minute book 1, board meeting 8 June 1880.

only served to emphasise still further the coastal works' orientation towards the export market.

(b) In Germany

Unlike in Great Britain, the domestic market for railway rails in Germany did not prove at all stable. As Table 11[185] shows, it was not possible to refer to a firm market base here after 1875.

Two developments in Germany had led to the rails business lagging behind expectations during the second half of the decade. On the one hand, there was the decidedly pro-cyclical behaviour of the liberal twosome Camphausen and Achenbach in the Prussian Finance and Trade ministry. During the economic boom of 1873 they had raised the tariffs on railways, thus increasing the heavy industry's material costs.[186] Furthermore, a report by a commission of the Prussian House of Delegates established a 'junctim' between raising the railway tariffs and further credits for extending the railway network.[187] Though the credits were granted, in view of the rapidly deteriorating economic situation they were not implemented. For the time being, those contracts which were believed secure were not being placed. Meyer, a member of Krupp's procuration, reported that discussions about the Slump were being held with representatives of heavy industry in Berlin, in which all agreed that 'Achenbach should be persuaded with all the means available finally to bring the 180 million, which has been granted for railway purposes, to the people.'[188] However, their efforts did not have much prospect of success from the start. Achenbach was not in the least interested in subsidising the needy steelworks. 'Only over the bodies of works incapable of surviving is he ready to issue his commissions.'[189]

185 The consumption of rails was calculated from production + imports – exports. Production according to *Reichsstatistik* 1871/p. 56, 1872/p. 172, 1873/p. 96, 1874/p. 98, 1875/pp. 85, 89, 91, 93, 1876/pp. 85, 89, 91, 93, 1877/pp. 115, 116, 126, 1878/pp. 117, 118, 128, 1879/pp. 121, 122, 132. Imports and exports according to *Statistisches Jahrbuch für das Deutsche Reich*, 2. issue 1881, p. 86.

186 Leese, Die Erhöhung des Gütertarifs der deutschen Eisenbahnen im Jahre 1874, in: *Jahrbuch für Gesetzgebung, Verwaltung und Volkswirtschaft* 1893, pp. 206f.

187 A loan of 50 million thaler was made, by raising the tariff, to depend on the real prospect of 'permanent and appropriate interest being paid on the capital invested in the railways'. Bericht der Kommission des Abgeordnetenhauses zur Vorberatung des Gesamtentwurfs, betreffend die Aufnahme einer Anleihe von 50 600 000 Thalern. According to Leese, p. 207.

188 HA-Krupp WA IX d 517: Meyer to Goose on 16 December 1875.

189 Ibid., an earlier letter from the firm of F. Krupp to Meyer already stated: 'According to information received here, the railway managements that operate under royal administration intend to avoid placing orders for the current year, provided this in no way jeopardised the safety of the service they provide.' HA-Krupp WA III 14: Friedrich Krupp to C. Meyer on 26 March 1875. The state railways thus did not differ at all from the private railways in their

Table 11. *Consumption and export of rails: Germany, 1871–1879 (1,000 tons) (incl. fish-plates)*

Year	Consumption	Export
1871	412	42
1872	441	71
1873	546	71
1874	524	85
1875	354	122
1876	247	133
1877	262	225
1878	290	207
1879	230	164

The second damper to the demand for rails was provided in 1875 by Bismarck's nationalisation project for the railways, which he launched to improve the State's income. After a tactically motivated and unsuccessful preliminary thrust in the direction of a Reich railway, this primarily involved the takeover of the viable private railways by the Prussian State.[190] The private railway managements reacted by throttling expenditure on new and replacement parts, in order to strengthen their position during settlement procedures by raising book profits as high as possible, and by squeezing out maximum returns for their shareholders in the time left to them.[191] In Prussia, this produced a particularly heavy slump in rails deliveries in the Rhineland-Westphalian works' 'inland' market after 1875 (see Table 12).[192]

Both developments thereby stressed the necessity of setting up an export 'safety valve' for the Bessemer works, which had already arisen following the rationalisation measures. In Germany there had been no rail exports worth mentioning until the seventies, which meant that there was no possibility here of robbing the local wrought-iron works of their export market a second time round as well, as in Great Britain, and in this way of realising the possible cost

pro-cyclical behaviour. The same theme in Bundesarchiv, R 13 I, Nr. 1: Bueck to Richter on 25 November 1873. This also contains complaints about the reluctance of the private railways to realise their concessions.

190 Hans Mottek, *Die Ursachen der preussischen Eisenbahn-Verstaatlichung des Jahres 1879 und die Vorbedingungen ihres Erfolges*, Diss. Berlin 1950, p. 79. There can, on the other hand, have been hardly any talk of rescuing 'necessitous railway bonds', as claimed by Böhme, p. 476. Railway bonds, unlike those for the iron industry for instance, were very much still paying dividends. See Eduard Wagon, *Die finanzielle Entwicklung deutscher Aktiengesellschaften*, 1870–1900, Jena 1903, p. 212.

191 Mottek, pp. 79f.

192 Bundesarchiv, R 13 I, Nr. 172: Schienenlieferungen an die preussischen Staats- und Privatbahnen, in: General-Versammlung, 22 October 1880.

Table 12. *Rail deliveries to the Prussian State and private railways, 1871–1879*

1871	165,000 tons
1872	155,000 tons
1873	175,000 tons
1874	115,000 tons
1875	202,000 tons
1876	115,000 tons
1877	50,000 tons
1878	70,000 tons
1879	51,000 tons

reductions. The export 'safety valve' could not be taken over; it had first of all to be created. And this was achieved, as Table 13[193] shows, to the disadvantage of Great Britain and Belgium, who had previously led the market and who, unlike Germany, had to take on losses during the 1870s.

The changeover to the Bessemer process brought in its wake the new orientation to the export market, which soon represented the bulk of the Rhineland-Westphalian works' sales. However, the fact that in Germany, unlike in Great Britain, not a single new Bessemer works was started up during the second half of the decade shows that this newly won market abroad did not appear nearly as promising there; quite the contrary. The Bochumer Stahlwerk, Steinhauser Hütte and Aachener Verein abandoned their Bessemer works in 1875.[194] Rheinische Stahlwerke were happy to be able to sell a complete Bessemer works with two plants to Warsaw.[195] Poensgen & Giesbers, Gienanth and the Alsace works engaged in only sporadic production during the second half of the seventies.[196] The Maxhütte in Bavaria and the Königin-

193 For figures in Germany, see note 185. For figures for Great Britain see note 143. Figures for Belgium according to Sering, p. 231.

194 Brauns, p. 92. Rabius, p. 38. Werkarchiv Krupp-Bochum, BV 16303 Nr. 4a: Gesellschaft für Stahlindustrie, Protokoll der Aufsichtsratssitzung am 29 Januar 1875.

195 This older of the two works was moved wholesale to Praga (Warsaw). The Rheinische Stahlwerke received in return shares to the value of 120,000 roubles (1 rouble = 2.10 marks approx.) from the new company. Their original demand for 200,000 roubles could not be maintained. Thyssen-Archiv, RSW 126-00 2: Bericht des Vorstandes in der Generalversammlung am 29 November 1877; RSW 123-00 1 2: Protokoll der Aufsichtsratssitzung am 6 Februar 1878. Their participation turned out after a very short time to be extremely profitable. The enterprise was encouraged by premiums from the Russian state and protected by high tariffs, and achieved fabulous profits. Rheinische Stahlwerke drew dividends of 10 per cent (1879) and, once initial running-in problems had been overcome, 43 per cent (1880), 50 per cent (1881) and 40 per cent (1882). RSW 126-00 9: Bericht des Aufsichtsrats der Rheinischen Stahlwerke für die Ordentliche Generalversammlung vom 7 Oktober 1881; RSW 126-00 10: Bericht . . . vom 26 September 1882; RSW 126-00 11: Bericht . . . vom 17 Oktober 1883.

196 Brauns, col. 92f.

Table 13. *Rail exports 1871–1878 (1,000 tons)*

Year	Great Britain	Germany	Belgium
1871	979	42	84
1872	945	71	
1873	785	71	73
1874	783	85	
1875	546	122	59
1876	515	133	
1877	498	225	45
1878	439	207	

Maria-Hütte in Saxony were protected from domestic and foreign competition and thus remained largely exempt from the ravages of the world-wide economic slump. They had no competition in either country and could rely firmly on their governments never infringing their monopoly with the State railways by accepting any offer from Prussia, however alluring. In spite of their privileged position, there was no question of using their plants to full capacity, the consequences of which were, nevertheless, less drastic than in Prussia because of the absence of competition on the spot and the practice of allotting contracts without putting them out to tender beforehand.[197] With the exception of Krupp and occasionally of the Rheinische Stahlwerke, every works had cut operations to one single plant.[198]

This one plant, which could not be subdivided any further, was nevertheless worked as intensively as possible, in order to enjoy the 'economies of scale' relating to the single plant. For this reason, further works cutbacks were not effected by slowing down the conversion operations in the last plant still being worked, but by occasionally stopping the night shift, as Phönix did between 1876–9 and Bochumer Verein between 1877–9. In both cases curtailing the work periods did not lead to a fall in production but, on the contrary, to a further production increase or at least to maintaining output.[199] This meant that

197 Königin-Marien-Hütte was awarded, without submitting a tender, the Saxony state railway's requirements at higher prices than in Prussia, after which it was still allowed to tender for the Prussian state railways, much to the annoyance of the Ruhr works. HA-Krupp FAH II P 106: Schriftstücke aus dem Nachlass Erhardts, Meyer to Erhardt on 26 November 1877; WA IV 1155; Firma Fried. Krupp to Maybach (Reichseisenbahnamt) on 6 May 1878. In the latter, Krupp complains that Königin-Marien-Hütte would be able to exploit its protected internal market in Saxony and offer dumping prices in Prussia. Re agreement with the Bavarian Maxhütte, see pp. 143f.

198 Brauns, col. 91f.

199 Mannesmann-Archiv, Pl. 25.11.1: Ordentliche Generalversammlung am 9 November 1876, Bericht der Direction; also Generalversammlung am 27 November 1877, 28 November 1878, 18 November 1879. Bochumer Verein, see above p. 71. In 1874/5, the Phönix Bessemer

limiting the allotted time was more than compensated for by disproportionately intensifying the production process during the remaining time.

Since the restricted sales meant that the degression effect of the fixed costs could not be fully exploited, the management found itself increasingly obliged to develop the organisation of production further in order to effect a reduction in production costs, in the domain of variable costs, by realising the technical 'economies of scale', which had been gained. Working all the installed plants would, by reducing the proportion of fixed costs, surely have had the same effect on the enterprise, and more simply; however, the prerequisite for this was a proportional increase in demand. But this was precisely what was lacking. Even when operations were cut back by closing about half of all the installed plants in 1877, demand did not improve because the rest were being worked more intensively. 'The duty incumbent on every factory establishment, to bring the means of manufacture in operation into full use, paralysed the effectiveness of works reductions which had been undertaken almost everywhere, with the result that uninitiated persons unacquainted with the trade found occasion in the naked production figures in the statistics to accuse the iron industry quite unjustifiably of fraudulent expansion.'[200] This is the complaint which was levelled against the steelworks' position at the Bochumer Verein's AGM that same year. The fact that the works' cutbacks coincided with their exports offensive phase demonstrates that this was not nearly enough to keep the existing works busy. The profit situation was not satisfactory either, as was demonstrated by the total absence of dividend payments.[201]

In the second half of the decade the Rhineland-Westphalian Bessemer works were forced to realise that the markets in the countries bordering theirs, which had previously been readily open to them, had in the meantime become almost inaccessible. The severe depression had produced a trend towards heavier protectionism in the commercial policies of the European states. Protecting home industry from foreign competition now took precedence over promoting international trade. On the Continent this had come to mean primarily protection against German industry; France, Austria and, more strongly after 1877, Russia, all protected their own Bessemer industries (which were less efficient) from German competition by means of customs duties,

works was still fully employed and producing 14,000 tons. By the business year 1878/9, on the other hand, operations had been reduced to one shift a day for eight months; during which 22,000 tons were produced. Mannesmann-Archiv, Pl. 25.11.1: Ordentliche Generalversammlung am 24 November 1875, Bericht der Direction; also Generalversammlung am 18 November 1879.

200 Werkarchiv Krupp-Bochum, BV 12600 Nr. 1: General Direktor Louis Baare auf der Generalversammlung am 17 Oktober 1877, p. 5.

201 Wagon, p. 212.

premiums and acceptance specifications.[202] It is somewhat ironic that the Rheinische Stahlwerke sold their closed Bessemer works to Warsaw and had a share in the company there, drawing good profits, whereas Konigshütte in Upper Silesia was complaining to the *Bundesrat* that they had been excluded from their rails market in Poland because factories had been established there which were favoured by the Russian state.[203]

During the mid-seventies discussions in the Union of German Iron and Steel Industrialists (*Verein Deutscher Eisen- und Stahlindustrieller*; VDESI) and its lobby in Berlin were dominated by demands for 'reciprocity' in the economic clauses. The survey conducted by the Iron Inquiry Commission of 1878 revealed that representatives of the Bessemer works were still preening themselves on the opportunities that would result from open borders to the foreign lands around them. Thielen of Phönix AG said: 'If Russia were to reduce its tariffs by a certain amount, then we would be able to destroy their entire industry. The same applied to conditions in France and Austria; the Bohemian industry would be definitely destroyed by Silesia.'[204] To quote Hoesch again:

> here I must make a very specific statement: if the tariffs were to come down at all, then we are totally ready to fight, meaning that we feel we are on an equal footing . . . What we would lose in the coastal regions, we would definitely win back inland. We would win the whole of northern France, we would be wholly on a par with Belgium, we would conquer Switzerland for ourselves, occupy Russia victoriously and take complete possession of Austria for ourselves; therefore by perfect reciprocity I see only a very powerful and robust advance of our steel industry in Germany . . . Once we achieve reciprocity we could immediately enlarge our works.[205]

202 Joachim Mai, *Das deutsche Kapital in Russland 1850–1914*, Berlin 1970, pp. 118ff. Emil Zweig, *Die russische Handelspolitik seit 1877*, Leipzig 1906, p. 173. Sering, p. 203. For France, this was not only a matter of protecting their internal market, but also of opening up similar opportunities for dumping exports, as were already available to the German steel industry. The French government paid export-premiums (*acquits-à-caution*) on steel exports, which played an important part in this. M. Rogé, an iron industrialist from Pont-à-Mousson put the French steel industry's problems with exports as follows: 'France cannot seriously consider exporting to Germany. It is not to help them go to Germany that our exporters must be given assistance, which could never be in sufficient quantity to be effective, it is to go to Italy, Austria, Switzerland, Spain, countries where Germany's compensatory rights are not effective. On the contrary, it is to be able to fight in these various countries, above all with the Germans, that we need these *acquits-à-caution*.' And, elsewhere, 'With regard to future negotiations we think that too much attention is being paid to the competition which England can furnish and not enough to that of Germany. This competition is capable of becoming terrible, notably in the case of metallurgy.' Archives Nationales, Paris, C 3223: Réponse à la circulaire de M. le ministre du 24 mars 1877 relative au projet du tarif général des douanes, mémoire de M. Rogé, maître de forges à Pont-à-Mousson et membre correspondant de la Chambre de commerce de Nancy.

203 PEE, Richter, p. 746.

204 PEE, Thielen, p. 20.

205 PEE, Hoesch, p. 245. This series could be continued by adding similar quotations from representatives of Krupp, Königs-Laura-Hütte etc. See PEE, pp. 81, 746. 811.

However, since these countries were, and with good reason, careful not to grant German heavy industry any trading facilities, they could only export to such countries as had no steel industry of their own to protect. Massenez from Hörder Verein defined the term 'abroad' before the Iron Inquiry Commission as follows:

When we refer to abroad here, we do not thereby mean the industrial countries England and Belgium, but those countries which have no developed iron industry of their own, such as Italy, Spain and Russia. The Belgians sell in Belgium more expensively than they do abroad, as do the English in England. We are all trying to cast abroad what we produce in excess of consumption in our own country.[206]

Alfred Krupp was particularly far-seeing in this matter and formulated a plan for giving the Emperor of China a narrow-gauge railway, including engines and models one-quarter and one-eighth the normal size, to incite him to build a railway.[207] The Bochumer Verein also once briefly ventilated the possibility of building a railway in China in connection with plans for setting up an artillery factory there.[208] However, for the moment these were but future dreams and they had no commercio-geographical advantages against British competition in those countries, which Massenez had described as relevant foreign markets, unlike the markets in the bordering countries, which they had still hoped to secure during the foundation of the Reich. On the contrary, southern Europe, as demonstrated by freight charges for ore travelling in the opposite direction, was cheaper to reach via the sea route from England. Attempts by the Oberbergamt (mining authority) in Dortmund to procure, at the instigation of the Prussian Trade Minister, an agreement between the North Sea shipping companies and the steelworks in its district, whereby the latter would be compensated for their disadvantageous freight costs in the Italian export trade, compared with British and Belgian works, came to nothing.[209]

206 PEE, Massenez, p. 401.

207 HA-Krupp WA IX a 172: A. Krupp, Niederschrift am 13 Februar 1875.

208 Werkarchiv Krupp-Bochum, BV 12900 Nr. 11: Protokoll der Verwaltungsratssitzung am 19 September 1876.

209 Werkarchiv Krupp-Bochum, BV 00100 Nr. 10: letter from Oberbergamt Dortmund to Bochumer Verein on 18 September 1878. The letter bears an ironical comment on its margin: 'That will help!'

4 Surmounting the Slump: collective strategies

1 Dumping, tariffs and cartel formation in Germany

In order to prevail over the British Bessemer works' advantages with regard to raw materials and freight, the German works were obliged to make some considerable price concessions in their export markets compared with their home market. Until 1 January 1877 the latter was still being protected by a customs duty of 20 marks per ton of steel rails, over and above freight costs. This is why, starting in 1874, different prices were set for the internal and external markets, as shown in Table 14.[1]

Once such a lavish price differential beyond the scope of the different freight costs got through to the consumer, it was immediately assumed that suppliers were restricting competition on the home market. Successful dumping, as practised during the 1870s by the German Bessemer works, can, in the event of overcapacity, only function on the basis of cartels or cartel-like agreements.[2] Such agreements concerning iron rails had existed between the Rhineland-Westphalian works since the 1850s.[3] In the case of Bessemer rails, firm

[1] Table taken from Däbritz, pp. 160f. These figures were typical of the Ruhr and not just of the Bochumer Verein. Compare the almost identical figures in HA-Krupp, WA VIIf941: Produktion, Löhne, Selbstkosten 1867–1895/96 and FAH III B 194: Verwaltung allgemein 1875–1901 – Kosten und Rentabilität. Likewise the sales prices in the remaining works, which were exchanged in order to set up the Iron Inquiry (*Eisen-Enquête*) in Werkarchiv Krupp Bochum, BV 01000 Nr. 1: Ottermann (Dortmunder Union) to Baare on 25 November 1878, Goose (Krupp) to Oberbergamt Dortmund on 10 November 1878, Lueg (GHH) to Baare on 25 November 1878, Haarmann (Osnabrücker Stahlwerk) to Baare on 27 November 1878 and Servaes (Phönix) to Oberbergamt Dortmund on 9 November 1878.

[2] This refers in general to Josef Löffelholz, *Repetitorium der Betriebswirtschaftslehre*, Wiesbaden 1975 (5), pp. 552f.

[3] GHH-Archiv, 200 201/12: Verträge mit Schienengemeinschaft 1865/72. HA-Krupp, WA IV 1240: Schienenvereinigung 1878, Goose to A. Krupp on 23 April 1878. See too Fritz Blaich, *Kartell- und Monopolkritik im kaiserlichen Deutschland. Das Problem der Marktmacht im deutschen Reichstag zwischen 1879 and 1914*, Düsseldorf 1973, p. 54. Robert Liefmann, *Die Unternehmerverbände (Konventionen, Kartelle). Ihr Wesen und Bedeutung*, Freiburg 1897, p. 139. Rolf Sonnemann, *Die Auswirkungen des Schutzzolls auf die Monopolisierung der deutschen Eisen- und Stahlindustrie 1879–1892*, Berlin 1960, p. 61. Hans Gideon Heymann, *Die gemischten Werke im deutschen Grosseisengewerbe, ein Beitrag zur Frage der Konzentration der Industrie*, Stuttgart and Berlin 1904, p. 242.

arrangements had existed since 1868 at the latest between the Bochumer Verein, the Hörder Verein and Krupp.[4] At the time, however, these three works did not face serious competition in Prussia. Orders were divided in a ratio of half for Krupp and a quarter each for Bochum and Hörde until December 1869. As was already usual with these kinds of products, whenever tenders were submitted each works in turn would be 'protected' by the others, which would give higher estimates.[5] In October 1869, Krupp demanded a new ratio of ³⁄₅ for Krupp and ¹⁄₅ each for Bochum and Hörde, based on his steel works' high productive capacity. Krupp was able to produce twice as many steel rails as Bochum and Hörde together and within four weeks he would be in a position to produce as much as three times the quantity.[6] Hörde and Bochum gave way and on 11 December 1869 they agreed, primarily at Bochum's insistence, to the terms demanded by Krupp.[7] Matters proceeded on this footing until 1873.[8] Before the competition's Bessemer rails works in the Ruhr had started up, the sole purpose of their agreement had been to keep prices in Prussia as high as possible. It could not, however, pan out into open price-rigging because of the danger that the Prussian authorities would resort to buying in England, as they had done once before in the case of the iron rails agreement.[9] In 1871 came the additional task of fighting off competition from the new works.[10] However, the rapidly climbing sales prices during the boom, which persisted until 1873, soon robbed their purpose of its rationale. Instead, they held a series of secret conferences to ensure that all the works would keep to an agreed procedure in the event of price rises.[11]

The economic about-turn in the autumn of 1873 immediately re-introduced the customary quotas, with 'protected' prices for tenders.[12] In the meantime, however, a considerable change had taken place with the emergence of the new Bessemer works that had arisen during the years 1869 to 1873. The big Rhineland-Westphalian works now combined: Dortmunder Union, Hoesch, Steinhauser Hütte, Hörder Verein, Bochumer Verein, Phönix, GHH,

4 HA-Krupp, WA II 5: C. Meyer to Firma Krupp on 29 July 1868.

5 Ibid.: C. Meyer to Firma Krupp on 11 August 1868.

6 HA-Krupp, WA III 10: Hagemann to C. Meyer on 9 October 1869, on 14 October1869 and on 19 October 1869.

7 Ibid.: Hagemann to C. Meyer on 11 December 1869 and on 14 December 1869; WA IV 1006: Hagemann to Clausen on 13 December 1869.

8 HA-Krupp, WA III 12: Hagemann to C. Meyer on 11 December 1871 and on 27 December 1871. Similarly, the entire contents of the correspondence with Pieper, the representative of Firma Krupp, in WA III 162: letters from Pieper and Hagemann to various persons 1869–73.

9 This cautionary example, which had to be taken into account when setting prices, was pointed out by C. Meyer, the Krupp representative in Berlin, in HA-Krupp, WA II 5: C. Meyer to Firma Krupp on 17 October 1869.

10 HA-Krupp, WA IV 1003: Hagemann to Clausen on 23 September 1871.

11 HA-Krupp, WA III 13: Firma Krupp to C. Meyer on 5 October 1872.

12 Ibid.: Firma Krupp to C. Meyer on 3 November 1873.

Table 14. *Prices for Bessemer rails at the Bochumer Verein (marks per ton), 1874–1878*

	July 1874	July 1875	Oct. 1876	Oct. 1877	Oct. 1878
Internal	252	216	165	156	145
Foreign	216	184	138	138	115

Osnabrücker Stahlwerk and Rheinische Stahlwerke formed a cartel under the unofficial name of *Vereinigte Stahlwerke*. Krupp kept his distance at first but then in August 1874 came to an agreement with the *Vereinigte Stahlwerke* whereby he was allowed to win 25 per cent and they 75 per cent of all internal rails tenders.[13] This did not prevent the agreement from being subsequently circumvented in the case of individual deals.[14] It did not command a great deal of respect from every participant and was endangered right from the start by Krupp and Bochum, the two most efficient works.[15]

The final break came in 1875 and was implemented by Krupp, who was fully conscious of the rationalisation measures that he had just successfully carried out. One month after his first rebuilt Bessemer plant with cupola furnaces had been set in action on 15 April 1875, he primed his Berlin representative as follows: 'The unproductive rails association is on its deathbed and the first nail for its coffin has been forged over here.'[16] From then on, Krupp only cooperated with the rails association over a few deals, otherwise engaging in a price war with it. The association filed complaints about Krupp's low prices, which were conveyed by Russell, the head of the Dortmunder Union and the Diskonto-Gesellschaft's (a major bank) right-hand man in the Ruhr and one of the few men with whom Krupp was on talking terms; his mission, however, bore no fruit.[17] On the basis of his calculation of profits, Krupp was determined at all costs to establish an annual production of 30,000 tons per plant. This objective was viewed with some concern by the competition, which had shortly

13 HA-Krupp, WA IV 249: Schienenvereinigung 1874/5, agreement between C. Hagemann (Krupp) and Franz Nettke and Julius Riesenberg (Vereinigte Stahlwerke) on 21 August 1874.

14 Examples of this in HA-Krupp, WA IV 1462: C. Meyer to Goose on 30 November 1874, 4 December 1874 and on 23 December 1874.

15 Already, three months after the signing, they were saying at Krupp's that 'We must arrange to get us out of it'. HA-Krupp, WA IV 1461: Wiegand to Meyer on 1 December 1874. Similar hankerings were expressed at this time by the Bochumer Verein, which only changed its mind when the banks withdrew their credit. Werkarchiv Krupp-Bochum, BV 12900 Nr. 11: Protokoll der Verwaltungsratssitzung am 29 August 1874.

16 HA-Krupp, WA IV 249: Firma Krupp to C. Meyer on 15 April 1875.

17 References to Russell's position as a go-between in HA-Krupp, WA IV 249: Schnabel to C. Meyer on 10 July 1875. Meyer was based in Berlin, where he was in contact with the Diskonto-Gesellschaft.

beforehand combined in forming the Verein Deutscher Eisen- und Stahlindustrieller (Union of German Iron and Steel Industrialists). Krupp distanced himself in the main from all communal undertakings, from the rails agreement as from the VDESI which was very concerned about his behaviour:

Where conditions of delivery are concerned, I must point out to you that all our efforts are in vain if the Krupp factory won't join us. At the moment it is making quite outrageous attempts to draw where feasible the little business that there is in Germany to itself, so as to be exceptionally well employed. It offers long terms of payment and extraordinary price reductions and so will also enter into all possible terms of delivery, however bad, and is not inclined to enter into an agreement aimed at abolishing bad terms of delivery.[18]

The situation only became less tense for the competition when Krupp had secured enough orders to keep his works fully employed until the summer of 1876.[19] It was only when they had achieved this consolidated position that the Essen group was prepared to be involved in negotiations at all: 'We can always engage in such negotiations; once a way has been found to secure us a notable part of German orders at better prices, then we are all the better placed to seek out our focal point abroad!'[20] The cartel was thus quite clearly viewed as a way of facilitating exports.

The reason why the creation of a new railmakers' association in 1876 has until now scarcely been considered in the relevant literature and why its background has not been examined more closely anywhere is mainly due to the fact that the overwhelming importance of the rails market for the steel industry, which was then still in its infancy, had not been recognised. Even the latest comprehensive monograph about the iron and steel industry of the Ruhr by Wilfried Feldenkirchen simply mentioned it in passing as a not very successful specialised cartel, basically only interesting as the precursor of the Stahlwerksverband (Steelworks Syndicate).[21] In fact, however, controlling the rails market by means of a cartel in the 1870s meant that a good three-quarters of the steelworks' entire inland sales was managed collectively. This meant that there was practically no further internal competition so that, contrary to Feldenkirchen's claims, one can indeed refer to 'an effective syndicate-like organisation of this branch of industry' during the later 1870s and early 1880s.[22]

The motor force behind attempts at forming a cartel in 1876 was Louis Baare

[18] Werkarchiv Krupp-Bochum, BV 83000 Nr. 1: (Bochumer Verein) Baare to Rentzsch (Secretary of the VDESI) on 24 April 1875.

[19] HA-Krupp, WA III 34: Firma Krupp to H. Haass on 5 July 1875, 21 July 1875 and 14 April 1876; WV IV 656: Schwab to A. Longsdon on 31 May 1875.

[20] HA-Krupp, WA 1439: C. Meyer to Goose on 13 December 1875.

[21] Feldenkirchen, *Eisen- und Stahlindustrie*, p. 120.

[22] Ibid., p. 121.

of the Bochumer Verein.[23] Although the Bochumer Verein had the most potent Bessemer works in Germany after Krupp and, as we have already seen, had achieved a top position regarding its organisation of production, it suffered greatly from the liabilities it had amassed by raising huge loans during the economic boom.[24] This was expressed in its cost accounts by an exceptionally high redemption rate for pig iron, an average of 18.3 per cent during the working year 1877–8 for instance.[25] Consequently the low costs resulting from the rationalisation of the Bessemer works by using its own pig iron were cancelled out yet again. It was clear that the Bochumer Verein had taken on too much at the time by building its blast furnace plant, and would soon bitterly rue the loans it had been obliged to raise. Subsequently this became a motive for creating reserves more intensively and for paying out less generous dividends.[26] On account of its mistaken business policy, the Bochumer Verein could not use its excessive production capacity to engage in an overt price war with the competition, as Krupp was doing. Whereas Krupp was saved in 1874 by a then unprecedented loan from the Prussian Seehandlung, the Berlin banks withdrew their support from the Bochumer Verein by drastically curtailing their credit levels. By the beginning of 1876, bank insiders were already able to report that Baare 'has been informed by an influential source that they cannot and will not view further price reductions with equanimity! Bochum is in deep trouble with its bankers!'[27]

As the strongest among the weak, Louis Baare thus assumed the lead in the Rhineland-Westphalian steelworks' dispute with Krupp. For a brief moment it

23 According to a communiqué from Russell to C. Meyer, reproduced in HA-Krupp, WA IV 1439: C. Meyer to Goose on 5 March 1876.

24 It had become necessary to take on loans, since the decision to implement a planned capital increase had been made too late, after the stockmarket collapse of May 1873, which was why it did not bring the anticipated results. The enterprise was greatly in debt to the banks, which considerably reduced its freedom of movement. This was subsequently bitterly regretted and was a reason for increasing its reserves once the crisis had been surmounted. Wilfried Feldenkirchen, Kapitalbeschaffung in der Eisen- und Stahlindustrie des Ruhrgebietes 1879–1914, in: *Zeitschrift für Unternehmensgeschichte* 24 (1979), p. 46, see too note 26.

25 Werkarchiv Krupp-Bochum, BV 01000 Nr. 1: Vergleichende Zusammenstellung der Selbstkosten diverser Fabrikate.

26 This is what was said on behalf of the board at the general meeting on 26 October 1878, after it had been admitted that the yearly results would justify paying a dividend of 2–3 per cent: 'But, looking back at the results of the mistakes made three years ago, the distribution of a dividend of 2 per cent amounting to 300,000 marks constitutes at the moment a completely inadmissible weakening of the financial situation. At the time, the big Berlin financial institutions were gradually reducing their credits of 800,000 thaler [*c.* 2.4 million marks U.W.], which they had formerly eagerly proffered, to 200,000 thalers.' The result of this was that the dividend payment of 300,000 marks, which was then determined on, could only have been produced with the greatest difficulty. Werkarchiv Krupp-Bochum, BV 12600 Nr. 1: General-Versammlung to 26 October 1878.

27 HA-Krupp, WA 1439: C. Meyer to Goose on 13 March 1878. Emphasis in original text. Meyer was referring to information supplied by Hansemann and Russell, the directors of the Disconto-Gesellschaft.

still looked as if the Dortmunder Union wanted to abscond. Not because it was doing particularly good business – the opposite was the case – but because it disposed of unlimited recourse to the Disconto-Gesellschaft, being its creation. 'Actually at this moment the Union is the Disconto-Gesellschaft. The D.G. could not care less whether it works at a loss; it wants to and will work, that is what Hansemann and Russell have most earnestly assured me!' reported Meyer, the banks' contact in Krupp's procuration, following a conversation with the two about possible market agreements.[28] The Union did not have to worry about redemption payments. In view of its strength, Baare offered the Dortmunder Union the same share in the cartel as the Bochumer Verein, that is a sixth each. Krupp was to get a fifth.[29]

In this connection he also weighed up the possibility of coming to an agreement with the British, French, Belgians and Austrians about an approved procedure in warranty cases, which could perhaps be extended to cover further agreements. The preconditions for this were already present, since the British had founded an association similar to the VDESI, the British Iron Trade Association. The Belgians would always be happy to come to an agreement and, for Austria and France, this could possibly be grounds for forming their own association.[30] Thus were laid the foundations of the international cartel constellation in the 1880s. As yet only the British works had no reason for associating with the weaker competition on the Continent.

Much more urgent than these future projects was the task of getting Krupp, once Dortmunder Union had given its consent, to join the national cartel, which could not function without him. On account of their 'special relationship' to particular offices of state, Krupp and his procuration were of the opinion they would do better on their own. 'We get more secretly than we would be allotted, given the extremely small demand, by the rails association!'[31] In addition, the firm had a particularly hot iron in the fire right at the end of 1876, when the negotiations about forming a combination entered their decisive phase. Krupp had offered Maybach, the president of the Imperial Railway Office the chair in his procuration and had clearly seen that he was interested. Naturally the affair was kept strictly secret, but it did not advance quickly enough to introduce yet another twist into the question of a cartel.[32] In the meantime the other works had declared, in the event of Krupp refusing any longer, that 'they would have to begin the knife fight which would lead to their mutual destruction'.[33] Faced

[28] Ibid.

[29] Werkarchiv Krupp-Bochum, BV 54601 Nr. 1: Baare to Müller (Dortmunder Union) on 29 February 1876.

[30] Ibid.

[31] HA-Krupp, WA IV 1439: C. Meyer to Goose on 25 January 1876.

[32] HA-Krupp, WA IX d 517: C. Meyer to Goose on 7 December 1876; FAH II P 114: A. Krupp to Baumann on 21 January 1877.

[33] This is what Louis Baare told the Iron Inquiry Commission retrospectively.

with the closed ranks of the entire Rhineland-Westphalian competition, Krupp got the wind up and joined them in November 1876.[34] His attempt to win Maybach for his procuration also ended in failure, as did his attempt in the summer of 1877 to win Johannes von Miquel, the speaker in financial and political matters for the National Liberal faction in the Reichstag.[35]

The banks seem to have played a decisive part in the formation of the cartel. Bochum was obviously acting under pressure from its creditors, meaning the Schaffhausen Bankverein. However, the casting vote was provided by the Disconto-Gesellschaft since Russell combined membership of its Board with leadership of the Dortmunder Union. On account of its almost unlimited financial powers through the Disconto-Gesellschaft, the Union was assessed by Krupp as his only 'dangerous opponent' and he weighed up the pros and cons of working with it[36] given that Hansemann and Russell were still assuring him in March that 'the rest of the competition will soon give up'.[37] Hansemann must have known that because he also sat in on Phönix's board,[38] which was in debt to his Disconto-Gesellschaft and the Schaffhausen Bankverein to the tune of 4 million marks. A few months previously, he had reached an understanding with the latter which protected Phönix from insolvency.[39] Apart from which, the Disconto-Gesellschaft had just withdrawn 'its previously almost unlimited credit' from the Bochumer Verein. The enterprise avoided bankruptcy entirely thanks to the determined support of the Cologne, Schaffhausen and Oppenheim banks.[40]

Because the big banks' archives for the period seem to have been lost, it is no longer possible to determine the extent to which the Bochumer Verein's

34 HA-Krupp, WA IV 1641: Wiegand to C. Meyer on 23 November 1876.

35 HA-Krupp, WA IX d 517: C. Meyer to Goose on 5 June 1877.

36 HA-Krupp, WA IV 1439: C. Meyer to Goose on 9 March 1876.

37 Ibid. C. Meyer to Goose on 13 March 1876.

38 Mannesmann-Archiv, P1. 25.11.1: Ordentliche Generalversammlung am 19 Oktober 1869. Hansemann was replaced in April 1880, when Phönix was once again standing on its own feet, by a director of the Disconto-Gesellschaft. P1. 25.35: Protokoll der Administrationsratssitzung am 2 April 1880.

39 Mannesmann-Archiv, P1. 25.35: Protokoll der Administrationsratssitzung am 19 November 1875.

40 The Schaffhausen Bankverein was soon after also involved in the second biggest rescue package in this sector, this time for Rheinische Stahlwerke. They had been forced to declare their insolvency at a gathering of their creditors in December 1877, after Suermondt, the Aachen cloth merchant and majority shareholder, had withdrawn his bank warranties. This led to the immediate collapse of its credit structure, whereupon the firm was immediately reconstructed by its creditors, among whom was the Schaffhausen Bank Union, at an extraordinary general meeting on 5 February 1877. Thyssen-Archiv, RSW, 126 00-2: General-Versammlung am 29 November 1877; 126 00-4; ausserordentliche Generalversammlung am 5 Februar 1878; 126 00-5: General-Versammlung am 31 Oktober 1878. This general meeting was almost exclusively concerned with the collapse and reconstruction of the firm. Re the position of its creditors, see too Mannesmann-Archiv, P1. 25.35: (Phönix) Protokoll der Administrationsratssitzung am 18 Dezember 1878.

unexpected rescue and the apparently contrary policy of its supporting banks caused the Disconto-Gesellschaft to abandon its plans to decimate the steel industry. In any case, this change of heart proved decisive for Krupp and the formation of the rails cartel, since it was when the Dortmunder Union swung into line with the cartel camp and he was threatened with a 'knife fight' that he was laid low and left with no other option but joining. However, Alfred Krupp did angrily refuse the chairmanship which was immediately offered him in order to enable them to boot out Louis Baare, who had often been rather too loud and was not much liked.[41] The post finally went to Servaes of Phönix – and thereby *de facto* to the Disconto-Gesellschaft.[42]

The allotment of quotas, unlike the old agreement with the Vereinigte Stahlwerke (United Steelworks), had been altered very much to Krupp's disadvantage. In the meantime only the Steinhauser Hütte whose closed Bessemer plant was finally sold wholesale to Angleur in Belgium had given up.[43] Aachener Verein had used this opportunity to make a successful claim, since it had in the meantime managed to bring its faulty blowing engine, which had condemned the enterprise to closure, into full operation. It did not join up immediately, however, waiting until 1879 when the works finally resumed larger-scale operations to do so. The final offer to Krupp thus looked as follows:

Krupp	18 per cent
Bochumer Verein, Dortmunder Union	12.5 per cent each
Rheinische Stahlwerke	9.9 per cent
Osnabrücker Stahlwerk, Hoerder Verein, GHH	8.9 per cent each
Hoesch	8.5 per cent
Phönix	7.4 per cent
Aachener Verein	4.5 per cent[44]

Since the Aachener Verein did not initially lay claim to any orders and the Osnabrücker Stahlwerk was only included at the end of 1877, the other works actually ended up with rather larger shares.[45] When Konigshütte in Upper Silesia joined in October 1878 and was granted a 9.5 per cent share, it became necessary to make yet another allotment of quotas, which, however, adhered

41 HA-Krupp, WA IV 1462: C. Meyer to Goose on 17 November 1876.

42 Werkarchiv Krupp-Bochum, BV 54302 Nr. 1: Besondere Fälle des Verkaufs (Schienengemeinschaft transactions from 76) (Bochumer Verein).

43 The works there were successfully operated according to the Thomas process and were not considered to be at all worn. *ICTR* 19 March 1880, p. 327.

44 HA-Krupp, WA IV 1641: Wiegand to C. Meyer on 11 November 1876.

45 HA-Krupp, WA IX d 517: C. Meyer to Goose on 9 December 1877. Meyer felt that the Schienengesellschaft (Rails Association) was endangered by this increase. Detailed accounts for the period from 18 November 1876 to 30 September 1878 can be found in Werkarchiv Krupp-Bochum, BV 45302 Nr. 1: (Bochumer Verein) Besondere Fälle des Verkaufs (Schienengemeinschaft transactions from 76).

routinely to the old ratios.[46] Alfred Krupp was still the only dissenting element in the cartel, but he was successfully restrained by his procuration.[47] They were aware that, in the event of a price war, the other works would not go 'kaputt' as quickly as Krupp believed. Until that happened, the firm would have lost 'much, yes very much', money, which was better not risked at all.[48]

In November 1876, the Prussian Bessemer rails cartel, which had been called into being with such pains, now seemed in terms of profit to be only ephemeral, since on 1 January 1877 protective tariffs on steel rails were lifted. Their firm hopes that the tariffs would be extended were dashed at the last hour: 'In the trade ministry a total swing to our disadvantage had been effected! Camphausen has secured the upper hand', ran the tearful message from Berlin. All that was left was Achenbach's 'consoling assurance' that 'even if abroad is cheaper, we will still have first refusal! Very consoling!!!'[49] The entire cartel structure now rested on these informal statements. But it stayed up! The Prussian State Railways continued, even in a tariff-free age, to grant the Bessemer works higher prices than those demanded by the mainly British competition. This voluntary supplement was very capable of exceeding even the old customs rate of 20 marks per ton.[50] Although the strict free-trader Camphausen had indeed prevailed in formal terms, in terms of content, he had already been foiled. It was now only a matter of time before the protective tariffs were reintroduced, and of finding the right moment to do so.[51] The Prussian authorities did not now proceed any differently to what the Saxons and Bavarians had been doing for ages, in that they subsidised the works in their own land. In any case, the Prussian trade minister urgently requested the

46 HA-Krupp, WA IV 1057: Conferenz-Berichte der Stahl-Schienen-Gemeinschaft, Conference on 9 October 1878.

47 Although Krupp had temporarily dissociated himself from the Rails Association on 1 July 1878, this was basically only in order to be able to improve his quota slightly. HA-Krupp, WA IV 1642: Meyer to Goose on 3 February 1878. GHH-Archiv, 300 112/16: GHH to Servaes on 3 November 1879. Thyssen-Archiv, RSW, 123 00 1 5: Protokoll der Administrtionsratssitzung am 17 April 1878.

48 HA-Krupp, FAH II B 310: Goose to C. Meyer on 27 April 1878. Emphasis in the original text. It continues thus: 'when we can make the rails for 100 marks, then the matter of whether we want to abandon the Association can be raised again'. In fact, the cost accounting for steel rails in that month lay at 119.80 marks per ton. HA-Krupp, WA I 859: Schienenproduktion, April 1878.

49 HA-Krupp, WA IX d 517: C. Meyer to Goose on 3 November 1876. Compare also Böhme, p. 436.

50 On the one hand, this is already indicated by the rails prices shown in Table 14, p. 120. On the other hand, this was mentioned by the steel industrialists at the Iron Inquiry Commission itself. PEE, Thielen (Phönix), p. 17; Meyer (Krupp), p. 78; Baare (Bochumer Verein), p. 806; Hoesch, p. 239.

51 See especially the extensive description of Bismarck's rancour at the 'demolition' of Delbruck and Camphausen, in Böhme, pp. 385–9, and pp. 410–16.

steel industrialists not to spread the news that he was granting subsidies abroad. That applied with particular force to the indiscreet Baare.[52]

Where the Reich was concerned, these subsidy procedures could not be implemented in a straightforward manner, since the Reich Railways in Alsace-Lorraine were also responsible to those federal states which had no industry of their own and consequently were not particularly interested in how they thrived.[53] In individual cases this problem of inviting tenders was solved so that the works in the cartel were informed by the Reich Railways Office about the lowest British offers, and they could then adjust their own prices accordingly.[54]

With most of the private railways too, the higher internal prices during the tariff-free period could be pushed through, if they did not rise above reasonable limits.[55] Since the heavy industry provided the private railways in the West of the Reich with the bulk of their freight, they were extremely interested in having prosperous steelworks. Given the considerable distance between the sources of coal and of iron ore, the production of every ton of steel was tied up with long-distance railway freight. At the Bochumer Verein, for instance, the proportion of freight costs to overall costs was put at 36 per cent in 1878.[56] Since the higher internal prices were based on a mixed calculation involving the exported proportion as well as the works' entire freight requirement, the price mark-up for rails, which were anyway being ordered in very small quantities due to the threat of nationalisation, was in the end still thoroughly worthwhile for the railways. Both Baare's and Meyer's (Krupp) remarks before the Iron Inquiry Commission pointed in this direction.[57]

Where the Bessemer works were concerned, then, the tariff-free period existed only on paper. Although they nevertheless continued their propaganda campaign in favour of protective tariffs with greater vehemence in the central circuit of the north-western group of the VDESI, it was to secure legal confirmation of an informal situation. The alpha and omega of their success was still their ability to export, which could only profit from customs, as Richter, President of the VDESI and manager of Königshütte in Upper Silesia, explained: 'In my opinion, the reintroduction of the tariffs would simply make

52 Werkarchiv Krupp-Bochum, BV 00100 Nr. 10: Achenbach to Baare on 29 June 1877. HA-Krupp, WA IV 1462: C. Meyer to Goose on 25 January 1878: 'Baare would have done better to have held his tongue', is what he wrote about the latter's self-serving publicity work concerning the Railmakers' Association's affairs.

53 HA-Krupp, FAH II P 114: A. Krupp to Procura and to 'Fritz' on 19 February 1878.

54 HA-Krupp, WA II 19: C. Meyer's draft of a letter by the Dortmunder Union and Firma Krupp to the Reich Chancellery.

55 Excessive prices caused the private railways, like the Prussian state railways, to threaten to buy from the English. This meant that prices could not be set in a completely arbitrary manner. They still had to maintain a certain, informally determined relationship to prices on the export market. HA-Krupp, WA IV 1462: C. Meyer to Goose on 29 December 1887.

56 PEE, Baare, p. 780.

57 PEE, Meyer, p. 80; Baare, p. 806.

us better exporters, because we would be able to sell even more cheaply on the foreign markets, if we could achieve higher prices internally.'[58]

2 Dumping and cost accounting

In spite of the authorities' insistence on discretion, the different prices for exports and the internal market did not, of course, remain concealed from the foreigners for long, least of all since a few countries had already had experience in this kind of business.[59] For this reason by 1874 complaints were already appearing in the trade journals about the Germans dumping rails abroad and favouring German works in their own country in spite of cheaper offers by the British and Belgian competition.[60] The only consoling aspect was that the German steelworks were shortly expected to make a loss on their exports.[61] In fact, this is what seems to have happened. According to the maxim that 'to work cheaply one must produce much and be able to sell it',[62] the German Bessemer works also took on orders for exports, which did not pay enough to cover the calculated production costs. For Hoesch in 1878 'The profits came simply from the home market, which was not large. The foreign market operated at a loss. We had to sell on the foreign market as well in order to achieve a satisfactory production.'[63] As it was, considerations about employment politics were not what lay concealed behind this 'satisfactory production', although the VDESI tried to make its demands for protective tariffs 'attractive' to the public by using the motto 'Protecting the Nation's Work'.[64] The production had to be 'satisfactory' to reach a level of capacity utilisation of the plant at which the production costs would fall far enough to give rise to a notable 'Beneficium' on the very restricted internal market. On

58 PEE, Richter, p. 746.

59 Examples for France and Austria in Hupfeld, *Eisen*, p. 162. *ICTR* 16 February 1877, p. 182. Werkarchiv Krupp-Bochum, BV 54601 Nr. 1: Baare to Servaes (Rails Association) on 10 June 1876.

60 To take one example from innumerable others: *ICTR* 10 June 1874, p. 670; 6 December 1878, p. 639; 21 February 1897, p. 210.

61 Hupfeld, *Eisen*, p. 163. *ICTR* 1 November 1878, p. 496 and 28 November 1978, p. 539: 'Herr Krupp must lose largely by the order he has got', is what they were saying in Bristol about one of Krupp's rails transactions there. This was certainly the case, since Krupp was delivering his steel 15 marks a ton cheaper than the British works, whose prices in 1879 lay just under 100 marks a ton (ibid.). If one compares this with Krupp's production costs for steel rails at this time, he must have been 'losing' at least 20 marks per ton. HA-Krupp, WA I 859: Schienenproduktion 1879.

62 Hupfeld, *Einfluss*, p. 257.

63 PEE, Hoesch, p. 239.

64 The argument about employment used by the advocates of protective tariffs is reported in greater detail in Karl W. Hardach, *Die Bedeutung wirtschaftlicher Faktoren bei der Wiedereinführung der Eisen- und Getreidezölle in Deutschland 1879*, Berlin 1967, pp. 64–72. See, too, Ivo Nikolai Lambi, *Free Trade and Protection in Germany 1868–1879*, Wiesbaden 1963, pp. 128ff.

its own, the internal market absorbed only about a third of production.[65] At the Bochumer Verein they drew the same conclusion as Hoesch from this, 'That export business must of necessity be contracted even at too cheap prices, because business at home is too small and *on its own* is not lucrative'.[66] Or at Hörder Verein: 'it is only because the German rails rolling works secure at least half of their business abroad that they are in a position to be able to sell in Germany without harming the prices, which the former are paid by the German railways'.[67] The Dortmunder Union's financial report for 1877/8 contained this observation:

> Were one to disregard such an export, which entails a direct loss for each individual transaction, then one would have to reduce the production figure so significantly that the losses arising indirectly from this would be far greater than the direct loss in the case of individual transactions abroad: the efficiency of the works operating at such increased costs would be seriously questioned, if they did not become impossible.[68]

Similar arguments were employed at GHH and Phönix.[69]

In the meantime, therefore, the price differential between the internal and external market meant far more to the German Bessemer works than merely a means of increasing their returns by exploiting their monopoly in their own country. The foreign, namely British, competition had driven down the price of steel rails on the international markets so far that, protective tariffs and cartel notwithstanding, it was now no longer possible to set a price for rails on the German market which, given the low level of capacity utilisation, would have shown a profit for all the works in the cartel through internal sales alone. Under these conditions some works would have had to 'pack it in' in spite of the cartel. The high costs of blasting only a few charges a day were well known to the firms from the days before the rationalisation drive, even if this difference was not measured in deflated prices, as at Krupp's. This meant that they did not have to rely on guesswork. When they took on export contracts at 'losing prices', it was in the conviction that the reduction in their average production costs by making more intense use of the plant would work out better than the reduction in average yields effected by the dumping prices.

This produced the impression that individual export transactions, which

65 PEE, Hoesch, p. 239.

66 Werkarchiv Krupp-Bochum, BV 12900 Nr. 11: Protokoll der Verwaltungsratssitzung am 20 Juli 1878. Emphasis in original text.

67 PEE, Massenez (Hörder Verein). p. 402.

68 Bundesarchiv, R 13 I, Nr. 170: Geschäftsbericht der Dortmunder Union für das Geschäftsjahr 1877/8, pp. 9f.

69 GHH-Archiv, 320 100/1: Generalversammlung am 21 November 1874; 320 100/2: Generalversammlung am 24 November 1876; 300 108/3: Bericht des Vorstandes für das Geschäftsjahr 1875/6. Mannesmann-Archiv P1. 25.11.1: (Phönix) Ordentliche Generalversammlung am 24 November 1875; P1. 25.35: Sitzung des Administrationsrates am 11 Dezember 1876.

were closed at prices below average costs were 'losing transactions'. In the same way, contracting such 'losing transactions' was the absolute precondition for being able to operate at cost at all. The development of economic understanding limped along behind the actual problems of costs and sales. This entailed problems not only in naming the phenomena, but it also made it more difficult to work out a successful price-differential policy. There was absolutely no clarity about which sales levels lay within the area of profit. They were only sure, as the figures quoted above show, that they lay far beyond the quotas for the internal market determined by the cartel. The export trade at 'losing prices' was not an object of precise cost assessment but rather a field of bold entrepreneurial decisions, the effects of which would only be established months later. Even at Krupp's, with his comparatively sophisticated accounting methods, sales below the total production costs were permitted only with the consent of the entire procuration 'as when a thousand thaler were to be given away'.[70] During the mid-seventies Phönix only sold below cost price abroad when faced with acute liquidity problems.[71] In 1878 its management preferred to cut back operations in the Bessemer works and reckoned themselves lucky not to have to fight as the competition did for export contracts at 'losing prices'.[72]

There were no generally accepted rules for determining prices for export transactions. Every new contract was pondered afresh, since there was great uncertainty about the possible lower limit for prices, which were negotiable down to just above 'destructive loss'.[73] This meant that the enterprises were far indeed from the lapidary pronouncement of modern economic textbooks: 'The marginal costs set the lower price limit.'[74] The 'marginal revolution' which Jevons, Walras and Marshall brought to economic theory, and which provided the theoretical basis for this dictum, had not yet happened.[75] In Germany, the

70 HA-Krupp, FAH II B 320: Krupp to Procura on 2 December 1874. It is clear from the account of the production costs at Krupp's that they were calculated using unbroken-down averages. Since high production costs were always accompanied by a low rate of production, this meant that the half-yearly averages produced by the accounts office were always set too high, and the firm's own ability to compete was thus unwittingly reckoned worse than it actually was. WA I 895: Schienenproduktion.

71 Mannesmann-Archiv, P1. 25.11.1: Ordentliche Generalversammlung am 24 November 1875. Bericht der Direktion.

72 Ibid.: Ordentliche Generalversammlung am 28 November 1878. Bericht der Direktion.

73 The wording itself shows that there were no firm criteria. Another circumlocution was: 'as long as no considerable loss is incurred' – scarcely a precise business directive. Werkarchiv Krupp-Bochum, BV 12900 Nr. 11: Protokoll der Verwaltungsratssitzung am 13 September 1877.

74 Erich Gutenberg, *Grundlagen der Betriebswirtschaftslehre*, 2: *Der Absatz*, 10th edn Berlin 1967, p. 354.

75 Although Jevons' *Theory of Political Economy*, the first to apply the notion of marginal profits, had already appeared in 1871, it had at the time no effect on cost-accounting in the works. Joseph A. Schumpeter, *Geschichte der ökonomischen Analyse*, Göttingen 1965, p. 1009.

methods of calculating costs for setting price differentials, which were to be built on it, were only developed by Schmalenbach and Schmidt in the course of the 1920s.[76]

The premise for the dictum quoted above and for Schmalenbach and Schmidt's work on degressive costs (i.e. piece costs which fall as output increases), which here concern us, is as follows. Instead of calculating the average production costs, as was usual, and on this basis setting prices with varying profit premiums, with this method of calculation the variable and fixed-cost elements in the differentiated prices were considered separately. In setting prices, the most important role was now played by the variable costs. The decision as to whether to close an export transaction below full production costs was determined according to whether the yield from this sale would at least cover the variable costs (materials, energy, labour). Were this the case, then the enterprise suffered no 'losing business', no financial disadvantage, since the additional costs for implementing this contract (= approx. marginal costs) were covered by this yield. Each contract which brought in more than the variable costs arising from it further contributed to covering the fixed costs which had previously not been taken into account. In individual cases, the extent of their contribution to the fixed costs depended on the state of the competition in the relevant part of the market. In our case, the contribution was very great in domestic transactions, and very low in export transactions. This method of calculation meant that the enterprise's fixed costs were covered by different premium levels on top of the variable costs; these levels were adjusted according to what that part of the market could carry at the time. In this way price formation was a much more flexible affair than it is when the average production costs have been established. The firm made a profit when the sum of its premiums exceeded its entire fixed costs.

Although the steel industrialists of the 1870s calculated their average costs every month with the greatest precision, they lacked the concept of marginal costs for determining the lower price limit for exports. They did have a certain reference point for their price range in that they collated their production costs both inclusive and exclusive of overhead costs, and thus at least knew exactly what proportion of the overall costs they formed. This proportion lay, with very little variation, between 8 and 10 per cent.[77] Since the allowance for

[76] Eugen Schmalenbach, *Selbstkostenrechnung und Preispolitik*, Leipzig 1926 (3). F. Schmidt, *Kalkulation und Preispolitik*, Berlin 1930. An assessment of these two fundamental works can be found in Gutenberg, pp. 350–6.

[77] The cost accounting for Bessemer steel and rails still showed only the participating overhead costs of the Bessemer works, and the rolling mill respectively, which could produce the impression that they amounted to only 2 or 3 per cent. In these cost accountings, however, greatest attention should be paid to the development of the overhead costs within the narrow framework of this part of the concern. When forming export prices, on the other hand, the total overhead costs associated with making steel rails were taken into account, including those for

depreciation was also entered separately and only added afterwards, there was still quite considerable scope for price formation below total production costs (including overhead costs and redemption payments). The Bochumer Verein managed to maintain this sort of enterprise cost-accounting for a fairly long period, from the time of intensive exports dumping. They show that the average results for the business year 1877/8 give the following enterprise costs for steel rails:

plus redemption payments plus overhead costs	minus r.p. plus o.c.	minus r.p. minus o.c.
1.474 marks/ton	136.32 m/t	123.92 m/t[78]

In the Ruhr during the same period, prices for rails exports vacillated between 125 and 139 marks per ton,[79] thus covering the entire price range. Only domestic prices, which generally lay about 155 marks per ton,[80] still produced a notable premium above the total production costs. The low levels of production costs minus redemption payments and minus overhead costs did represent to some extent a first approximation to marginal costs, in that the fixed overhead costs and the redemption payments were not taken into account when setting the lower price limit. Dortmunder Union went a step further. According to a review of GHH in May 1878, it no longer took wages into account when calculating the production costs.[81] They had apparently also come to be considered as fixed costs.

I am not aware of any case where the low levels of production costs minus overhead costs and redemption payments were undercut. They were apparently regarded as the absolute bottom price limit. The 'losing prices' in the export trade thus fairly certainly still lay above the actual marginal costs, since the steadily rising rates of production and utilisation of individual Bessemer plant in the 1870s meant that ever more technical economies of scale could be realised. The variable costs were thus steadily falling without this being taken into account when calculating prices. Although managers were very well aware

making pig iron and for storage, for instance. Werkarchiv Krupp-Bochum, BV 010 00 Nr. 1 (Bochumer Verein) Vergleichende Zusammenstellung der Selbstkosten verschiedener Fabrikate. GHH-Archiv, 20021/8: Entwurf einer Fabrikbuchführung 1871. The principles developed here were expressed in the submissions to the board; 300 108/6: Vorlagen zu den Aufsichtsratssitzung PEE, Hoesch, p. 232. At Krupp's, only redemption payments were shown separately, but the percentage share of the overhead costs was also known. HA-Krupp, WA IV 56: A. Longsdon on 10 March 1875. See above too, p. 78.

78 Werkarchiv Krupp-Bochum, BV 010 00 Nr. 1: Vergleichende Zusammenstellung der Selbstkosten verschiedener Fabrikate.

79 See note 1.

80 Ibid. Werkarchiv Krupp-Bochum, BV 12901 Nr. 6: (Bochumer Verein) Nachweisung des Durchschnittspreises der am 1 April 1877 noch in Auftrag stehenden Schienen.

81 GHH-Archiv, 300 108/6: Vorlage zu der Aufsichtsratssitzung am 9 Januar 1879.

that 'both the operational and the overhead costs are reduced when the work quantum increases',[82] when estimating the price range for exports they took only the fixed-costs degression into account. Calculating operational costs in relation to level of plant-capacity utilisation, as is normal nowadays, was not undertaken, apart from very rough attempts by Krupp. Given increasing capacity utilisation and, as we have seen, the further falling production costs associated with it, managers rather tended towards a pessimistic assessment of how their own production costs were developing. From a present-day point of view, it is fairly certain from this careful calculation that the hopes built up by the firms on their 'loss-making export' trade, namely that it would improve their overall profit situation, would have been fulfilled. In the event, the German steelworks did manage to stabilise their finances, which had been so over-stretched in 1873/4, precisely in the years 1878/9, when they were most intensively engaged in exports dumping.

3 The temporary recovery of the rails market in 1879–1882

Although dumping their exports contributed substantially to the survival of the German Bessemer works, in that they made possible increased capacity utilisation combined with lower production costs, the German works' structural disadvantage, compared with the British works, could thereby only be attenuated, and not compensated. The higher costs of procuring low-phosphorus iron ores were, solely on account of their less favourable transport route, in the same order as the entire conversion costs.[83] On top of this came the higher costs of delivering the finished rails on account of the great distance between the German locations and the nearest ports. Under these conditions, their future prospects at the end of the unusually long-lasting depression in the 1870s were not exactly rosy. Even the anticipated tariffs could not bring the Bessemer works any material improvement, since these had in real terms been granted long ago. Alfred Krupp presented the situation in a long care-ridden letter to his son and procuration in February of 1878:

> All works, which do not work as cheaply as the competition and which have nothing to lose in the long term, will go down and even if we can still keep level with the competition's capability then we must reckon in advance on the prices getting even worse and on us earning nothing by exporting, when Menelaus [the director of Dowlais, U.W.] is still earning £1 per ton.[84]

It was a matter of great concern that the British still had reserves, in connection with both price formation and with enterprise costs, were they to adopt the

[82] PEE, Massenez (Hörder Verein), p. 402.
[83] See above, p. 100.
[84] HA-Krupp, FAH II P 114: A. Krupp to Procura and Fritz on 19 February 1878.

same practices as the German works. The latter would, in the final analysis, then be inferior and would have to 'pack it in'. The whole trade feared that the British Iron Trade Association could, just as the VDESI had done, become the catalyst for a cartel and drive the Germans out of the market by following their example on the Continent and dumping their exports themselves.[85] Given that the trade was very well acquainted with the means to this end, they saw no reason why the biggest British works should not permit themselves the expenditure of 'a couple of hundred thousand pounds' in order to wipe out the rising German competition, once and for all.[86]

Consequently the news in February 1879 that Wilson, Cammell & Co. had contracted to deliver rails for less than £4 7s. per ton ex works, caused a panic.[87] Bolckow Vaughan, which was generally accepted to have the lowest production costs in the world, was still demanding £4 15s. per ton at the time ex works.[88] The conclusion they came to was that Cammell's price, which was also for the first time lower than the price of puddled rails, could not possibly include any profit.[89] The rails went to the North Eastern Railway, previously a domain of the Cleveland Works, whose district it served.[90] This seemed to announce an internal British price war, which would certainly have repercussions on the export market.

Wilson, Cammell & Co. were in a similar situation in Great Britain to that of the German works *vis-à-vis* British competition. Their works lay far inland near Sheffield, and thus had to overcome disadvantageous freight costs, compared to the coastal works. Their operational strategy seems to have been fairly comparable, in that they did all they could to secure enough business, even when the prices they negotiated did not always cover all the costs. In 1879 the Sheffield area was the only district in Great Britain to suffer a reversal in

85 HA-Krupp, WA IX d 154: A. Krupp to Bismarck on 21 October 1875. In it, A. Krupp referred to a further motive for the English, in that taking such a step would enable them to reduce their production costs still further. It is quite clear that he thought that the German cartel's conditions and methods of calculation would be adopted.

86 Such were Thielen's, the director of Phönix's, very real fears in 1878. PEE, Thielen, p. 15.

87 *ICTR* 7 February 1879, p. 155. This report was so sensational that it spread quickly even without the help of the press. Werkarchiv Krupp-Bochum, BV 001 00 Nr. 10; Müller & Cie. (Düsseldorf) to Bochumer Verein on 5 February 1879. Bundesarchiv, R 13 I, Nr. 342: C. Meyer (Krupp) to Renzsch (Secretary of the VDESI) on 8 February 1879.

88 *ICTR* 7 February 1879, p. 155.

89 *ICTR* 14 February 1879, p. 185.

90 See note 87. Such behaviour on the part of the North Eastern Railway can only have been meant as a warning shot for Bolckow, Vaughan & Co., to induce the firm to reduce its local sales prices. Under normal circumstances the North Eastern Railway purchased its rails solely in Cleveland, since they did well out of freight agreements with the steelworks there and thus had a direct interest in its prosperity. For more information about this, see p. 260. The North Eastern Railway, by taking this unique step, was behaving just like the Prussian State Railway, which had also fired such 'warning shots' in order to drive down the prices demanded at home. See p. 119.

Bessemer steel production, which fell very clearly by 28 per cent.[91] It might have been worse, had they not been simultaneously diversifying away from steel rails. Production of Bessemer rails alone fell by 58 per cent against 1878.[92] Bessemer works which had not diversified in time and were still mainly existing by producing rails, ended by lagging behind the coastal works in Cumberland, South Wales and Cleveland. Brown, Bayley & Dixon were heading for bankruptcy again; Wilson, Cammell & Co. started preparations to shift their entire works to the coast.[93] Unlike in Germany, in Great Britain until this period it was only a minority of works which suffered from the heightened competition. This meant that for the time being there were still no grounds for getting the British steelworks to combine, in spite of the great anxiety felt on this score in Germany.

Combinations of this kind were not contrary to the nature of British iron industrialists, as was demonstrated by the price and quantity agreements which functioned in the case of puddled and pig iron, as well as by the way the works stood together during wage disputes with the workforce.[94] The Cleveland Ironmasters' Association, to which Bolckow Vaughan also belonged, was in no way inferior to the Prussian Bessemer works' Rails Association.[95] If the British Bessemer industry had, at this time, not yet formed their feared cartel, it was only because the majority of the firms could still make satisfactory profits in circumstances of free competition, and consequently did not need to relinquish their operational autonomy to a cartel.

The fact that, despite the unmistakable signs of weakness in the Sheffield district, it did not immediately come to this was due to the unexpected revival of the American demand for rails in the summer of 1879. For the Bessemer works, the end of the depression came in the form of a fresh American railroad boom, creating a demand which once again surpassed by far the productive capacity of the American rails works. Prices for all sorts of iron quickly recovered and in Germany this led to much false lustre accruing to the tariffs which had just been reintroduced on iron and steel, to which were attributed, by adroit use of propaganda, the good earnings at the end of the depression.

On the other hand, the other great event in the steel industry in 1879 was received with very mixed feelings. On 4 April 1879 the first charge to be

91 *BITA* 1879, p. 37.

92 Ibid.

93 See p. 48.

94 Unlike price and quantity agreements, they were much more likely to achieve a common method of dealing with pay disputes on a supra-regional level. Jeffrey Harvey Porter, *Industrial Conciliation and Arbitration 1860–1914*, Ph.D. thesis, Leeds 1968 pp. 125 and 181.

95 Re the CIA see below p. 167, esp. note 50. Unfortunately, there is as yet no history of the CIA, although this organisation's extensive documentary legacy is kept in the BSC Archive in London, and could be used to correct our picture of the unhampered workings of free competition in Great Britain in the nineteenth century.

blasted according to the Thomas process took place in the Bolckow Vaughan works in the presence of select British ironmasters. This meant that phosphoric pig iron could now also be used for making steel in a converter. The ironmasters of the Ruhr had also been working for years specifically to find the solution to this metallurgical problem, since they wanted to free themselves from their heavy dependence on foreign ore. The fact that it was precisely the industrial leader who also had the cheapest phosphoric iron ore on his doorstep who made the breakthrough, awoke all kinds of dreadful fears for the near future.

For the economic boom in the steel market which had just begun, however, the Thomas process came too late. The surviving Bessemer works, even the less favourably situated ones, were still in a position, after six lean years, once again to reap sizeable profits.

The first indications of the American demand for rails, which had already been given up for dead, reached South Wales in August 1879.[96] What was wanted – and that was the second sensation – was iron rails, although in the USA itself, these had been widely superseded by Bessemer rails during the seventies.[97] The reason for this lay in the very different duties on iron and steel rails which, in the case of exports to the USA, reintroduced the price differential, which had in the meantime disappeared in Europe.[98] However, in the USA itself, once the productive capacity of the Bessemer works there had been completely exhausted, the puddling works experienced an unexpected renaissance in the course of 1879, which was bound to meet the subsequent rise in demand.[99] According to a report by the *German Business Archive* (*Deutsches Handelsarchiv*) from New York, the fact that 70 per cent of the new railroad lay west of the Mississippi, where rail use was not expected to be very great for the time being, also played a role. 'Moreover', the report ran, 'one should not disregard the American practice, whereby, when laying railways, they build as cheaply as absolutely possible and only later on, when the railway has become profitable, are improvements introduced.'[100]

For South Wales, and Cleveland slightly later, this meant that iron rails works which had been unworked for years could once again be started up. The *Iron and Trade Review* reported emotionally that former great names, like

96 *ICTR* 8 August 1879, p. 118.

97 The production of iron rails had fallen from 900,000 tons (1872) to 300,000 tons (1878), while the production of Bessemer rails had risen from 190,000 tons (1874) to 650,000 tons (1879). Bundesarchiv, R 13 I, Nr. 170: fols. 209–12, Der Eisenbedarf der Vereinigten Staaten von Amerika und seine Deckung (report from New York, dated 6 March 1880, in *Deutsches Handelsarchiv*, Nr. 3, 16 April 1880), p. 5.

98 The entry duty payable on iron rails was £3 5s.4d. per ton, and on steel rails £5 16s.8d. per ton. *BITA* 1882, p. 154.

99 See note 97.

100 Ibid., p. 6.

Cyfarthfa and Britannia would once again appear on the market and produce great quantities of iron rails.[101] However, this anachronistic iron rails trade remained limited entirely to exports to the USA, amounting to over 70 per cent of the same.[102]

Much more important and reassuring for the Bessemer works was the economic upturn over a wide front in 1880, which brought with it a stronger demand both inland and on all the important foreign markets. The Board of Trade stated in its mid-year report in the summer of 1880 that even after subtracting exports to the USA a considerable increase could still be reported and that this, contrary to the speculative USA trade, was 'bona-fide business', which could be relied on in the near future as well.[103] Nevertheless, the fresh American rails boom also provided the economic turning point for the Bessemer works. Although rails sales in 1879 had clearly dwindled, the entire production of Bessemer steel had increased compared with 1878. This was not because new ways of using Bessemer steel had at last been discovered but once again because the American customs tariffs distinguished between steel ingots and steel rails and subjected only the latter to the higher prohibitive charge. This led to the active export of steel ingots which could be milled into rails in the USA using the ironworks' redundant rolling capacity there.[104] This explains, for instance, the increased steel production in Lancashire, where the Barrow Hematite Steel Co. had earlier already exported in strength to the USA alongside the simultaneous reduction in rails production in the district.

The prices for steel rails, which by the end of 1879 had already recovered from their lowest rate of £4 15s. to reach £6, swung back to £6 after a speculative high level of up to £10 in the early summer of 1880.[105] The fact that even the biggest beneficiaries of the American demand for iron rails were not convinced that this business would last is demonstrated by the construction of Bessemer works at Cyfarthfa, Rhymney and Tredegar in South Wales.[106] The market for steel rails was based on broader foundations. The railway companies both at home and abroad used the good economic situation to re-lay

101 *ICTR* 17 October 1879, p. 387; 7 November 1879, p. 456; 28 November 1879, p. 539; 23 January 1880, pp. 95f.

102 Board of Trade figures for February and March 1880, according to *ICTR* 19 March 1880, p. 320; 9 April 1880, p. 404.

103 In *ICTR* 9 July 1880, p. 43. The same tone was adopted in *ICTR* 2 April 1880, p. 376; 20 August 1880, p. 218.

104 *BITA* 1879, p. 36. *ICTR* 5 December 1879, p. 79.

105 *BITA* 1880, p. 33.

106 Cyfarthfa was once again producing 1,000 tons of iron rails a week. *Iron*, 23 April 1880, p. 299. Decision to build a new Bessemer works there in: *ICTR* 15 July 1881, p. 83. Report about big profits at Tredegar with iron rails in: *ICTR* 31 December 1880, p. 780. Construction of the Bessemer works there in: *ICTR* 15 July 1881, p. 83. Re Rhymney, see above p. 111.

Table 15. *Production, export and consumption of steel rails in Great Britain, 1879–1882*

Year	Production	Exports	Internal consumption
1879	519,718 t.	328,425 t.	191,293 t.
1880	739,910 t.	464,401 t.	275,509 t.
1881	1,023,740 t.	594,419 t.	429,321 t.
1882	1,235,785 t.	733,919 t.	501,866 t.

Table 16. *Production of steel rails in Great Britain by district, 1879–1882*

District	1879	1880	1881	1882
Sheffield	82,774 t.	151,174 t.	245,469 t.	310,000 t.
(change)	(–58.5%)	(+82.6%)	(+62.4%)	(+26.3%)
South Wales	189,503 t.	258,202 t.	305,043 t.	367,944 t.
(change)	(+10.3%)	(+36.4%)	(+18.0%)	(+20.6%)
Lancashire	73,870 t.	116,431 t.	135,543 t.	140,708 t.
(change)	(–34.4%)	(+57.6%)	(+16.4%)	(+3.8%)
Cumberland	103,969 t.	114,096 t.	121,093 t.	150,693 t.
(change)	(+35.0%)	(+9.7%)	(+6.1%)	(+24.4%)
Cleveland	69,255 t.	92,559 t.	216,004 t.	265,842 t.
(change)	(+14.6%)	(+33.7%)	(+133.4%)	(+23.1%)

their tracks with steel rails.[107] Internal consumption in Great Britain reached unheard-of heights, as illustrated in Table 15.[108]

If one also considers that a considerable part of these exports, *c.* 250,000 tons, went to the colonies, this gave the British rails producers a huge 'home' market.[109]

The expansion of production during the upturn could be fully met by the works which had been made redundant during the previous slump in sales. This also explains the remarkably resilient rise in steel production in Sheffield, as illustrated in Table 16.[110] It was there that most of the Bessemer plant stood, as before, and with the rise in the price of rails and especially the increased domestic sales, they could once again be worked at a profit.[111] No new works

107 *ICTR* 12 October 1883, p. 454.

108 Table compiled according to *BITA* 1882, p. 40. The figures for exports and internal consumption in 1882 have been corrected according to *BITA* 1889, p. 62.

109 Figures according to BT 1880, p. 92; 1881, p. 92; 1882, p. 92.

110 Table according to *BITA* 1880, p. 28 and *BITA* 1882, p. 34.

111 *BITA* 1882, p. 36.

were added there, unlike at the coastal sites, nor, in the light of their bad experiences during the seventies, were any planned. It was clear that only extraordinary economic conditions as at the start of the 1880s would make the production of rails profitable in this district.

In Germany the years 1880–2 brought an improvement in the Bessemer works' sales opportunities (see Table 17).[112] On the one hand, prices on the export markets increased due to the greater demand from the USA and to increased internal consumption in Great Britain. On the other hand, the German railways were now covering their replacement needs, which had accumulated during the economic slump while they were waiting to be nationalised.

The last year of the slump, 1879, had, however, brought a reversal in the production of steel rails in Germany, which had turned out, at just 10 per cent, to be much weaker than in Great Britain.[113] The threat was certainly felt much more subjectively in Rhineland-Westphalia, where they had already completely exhausted their repertoire for fighting the Slump within the industry. The relief they felt and the paeans of praise which were sung for Bismarck and his economic policy were consequently all the greater.[114] In Prussia, the tariffs and nationalisation of the railways had secured in law the opportunities for further dumping of exports and had stabilised internal demand for rails. These two measures produced scarcely any direct improvement in profits or increase in sales. The actual advantage would have lain much more in an improved ability to estimate future market conditions, since two important imponderables had been removed.

The decline in the production of steel rails in 1879 was, as in Great Britain, cushioned at first by the fact that towards the end of the year steel ingots could be exported in great quantities to the USA. According to a report by the *ICTR*, Krupp, the Bochumer Verein and the Hörder Verein alone were supposed to have sold 50,000 tons before the end of the year.[115] The export market for rails grew at first even faster than the internal market and in 1880/1 it reached its highest peak ever in the nineteenth century.[116] Internal sales were a whole year behind in following this trend, without, for all that, achieving the values of the early 1870s.

[112] Calculated from *Statistisches Jahrbuch für das Deutsche Reich, hrsg. vom Kaiserlichen Statistischen Amt*, Berlin 1907, part 1, pp. 265ff. and part 2, p. 93.

[113] Calculated from ibid., part 1, p. 266.

[114] Primarily on the part of the Union of German Iron and Steel Industrialists at their general meeting, which conformed to the customs tariffs under the auspices of good overall economic conditions. Bundesarchiv R 13 I, Nr. 172: general meeting on 20 November 1879 in Berlin and general meeting on 22 October 1880 in Berlin

[115] *ICTR* 5 December 1879, p. 559.

[116] All the same, at 230,204 tons in 1880, and 250,709 tons in 1881, German rails exports were still lagging well behind British figures. *Jahrbuch für das Deutsche Reich* (1907), part 2, p. 93.

Table 17. *Production, export and consumption of rails: Germany, 1880–1882*

Year	Production (1,000 t.)	Proportion puddled rails	Exports (1,000 t.)	Consumption (1,000 t.)
1880	481	12%	229	252
1881	560	7%	249	311
1882	564	6%	185	379

Since the price of rails on the export markets as dictated by Great Britain had risen to a good £6 per ton and was once again level with the entire production costs and more besides, and the cartel within the country could keep prices above 140 marks (£7), the rails trade was now showing respectable profits again.[117] The result of this was that plant which had been closed was put once again into operation, in so far as this was still possible. From time to time Krupp had all five of his plants in both his Bessemer works on the go, capable of producing almost 18,000 tons a month, which he occasionally did.[118] Even the Bochumer Gesellschaft für Stahlindustrie started up again, in order to work off its enormous bank debts. The Essen Creditanstalt, on which the works was totally dependent, and in whose rooms the directors' board meetings were held for simplicity's sake, clearly saw their chance of getting at least some of their money back.[119]

As might be expected in a rosy economic climate, keeping faith with the cartel was not particularly palatable. Alfred Krupp was prevented only with difficulty by his procuration from resigning from the cartel at the earliest possible date. He was afraid that the government would remove the tariffs again if the Railmakers' Association continued to exist and keep prices unnecessarily high. In addition, he wanted to have as little as possible to do with the competition, feeling that he had been lowered by the agreement to the level of 'cheats and swindlers'.[120] In any case, a bid by Krupp for the dissolution of the cartel would be good for the firm's image with the authorities, even if it did nothing else. The firm of Krupp was not to be involved

[117] Bundesarchiv, R 13 I, Nr. 170, fols. 280–7: VDESI – Bericht des Geschäftsführers an die Generalversammlung am 14 Dezember 1882, p. 7. The production costs for steel rails in the Ruhr swung between 103 marks/ton (Krupp) and 122 marks/ton (GHH). Krupp-Archiv, WA 1 859: Schienenproduktion. GHH-Archiv, 300 108/7 and 8. Monatsberichte und Bilanzen der Oberhausener Werke pro 1879/80 und pro 1880/81.

[118] HA-Krupp, WA VII f 484: Reuter to Hoecker on 10 August 1882, Bessemerwerksproduktion.

[119] Werkarchiv Krupp-Bochum, BV 16303 Nr. 4b: Gesellschaft für Stahlindustrie, Protokoll der Aufsichtsratssitzung am 12 April 1880, 9 November 1880, 18 März 1881, 29 September 1881, 24 Oktober 1882.

[120] HA-Krupp, FAH II B 356: A. Krupp to Procura on 22 July 1880.

in any harm done to the interests of customers by the tariff, and should resign from the cartel.[121]

However, Alfred Krupp could no longer impose his will on his procuration. It had become obvious that he had gradually lost his grasp of the enterprise as a whole and his comprehension of operational strategy, and his interventions in the management of the firm were becoming increasingly erratic and selective. He was no longer capable of taking far-seeing forceful action such as his extremely successful reorganisation of the Bessemer works in 1874/5. He was thus obliged to look on with some bitterness when his arch-rival, Louis Baare of Bochumer Verein, disbanded the cartel early in 1881. He complained bitterly to his procuration about their missed opportunity:

So the union has been dissolved. After all, that was indeed to be assumed, but better for us if we had done it, for now Baare will write to all the managers and be a good boy and so snatch the orders from beneath our noses, out of good will for him. He will enjoy its fruits, which I so wanted to keep for us, if he now says that he would have considered it unjust to persist with a combination which was no longer justified since the introduction of the tariff.[122]

He recommended as a countermeasure reducing prices immediately to 125 marks per ton of rails on the German market. This would still leave a profit of 20 marks a ton, as well as bringing in enough orders.[123]

Unlike Krupp, however, the Bochumer Verein wanted to terminate the cartel not in order to return to free competition, but in order to get the quotas re-allotted by restricting activity generally and, more especially, to regulate their competitive relations with the new basic Bessemer works on the left bank of the Rhine and with the Gesellschaft für Stahlindustrie. To get a foot in the door of the market, these works had submitted tenders up to 15 per cent below the cartel price and had thus upset its whole price structure.[124] At the Bochumer Verein, it was also felt, and with good reason, that their quota of the German market no longer related to the different increases in the works' productive capacities, and consequently needed to be adjusted.[125] Since the other works' interest in continuing with the Railmakers' Association was unaffected,[126] the cartel arose again on 30 October 1881, only a month after it had been dissolved

[121] Ibid.: A. Krupp to Jencke on 10 August 1880.

[122] HA-Krupp, FAH II P 114: A. Krupp to Procura on 2 April 1881.

[123] Ibid.

[124] Werkarchiv Krupp-Bochum, BV 12900 Nr. 12: Protokoll der Verwaltungsratssitzung am 3 März 1881.

[125] Ibid.: Protokolle der Verwaltungsratssitzungen am 25 März 1881, 28 April 1881 und 20 Juni 1881.

[126] Thyssen-Archiv, RSW 123-00 334: Protokoll der Aufsichtsratssitzungen am 20 April 1881. Werkarchiv Krupp-Bochum, BV 12900 Nr. 12: Protokoll der Aufsichtsratssitzungen am 28 April 1881 and 20 Juni 1881. In it, it was said that this attitude on the part of the other works boded very well for the Bochumer Verein's future share.

by the Bochumer Verein.[127] The control centre for negotiations seems once again to have been the Disconto-Gesellschaft. The Bochumer Gesellschaft für Stahlindustrie, which was chasing after orders, did not have any dealings with the president of the Railmakers' Association, Servaes of Phönix, but dealt directly with the director of the Dortmunder Union, which belonged to the bank, and who himself was to some extent one and the same as the bank's board of directors.[128]

With the exception of the Düdelinger Eisenhüttenverein (Dudelange) all the steel rails producers joined the new association. Their final quota distribution is given in Table 18.[129]

Although Maxhütte in Bavaria was not formally accepted into the cartel, it did reach an agreement with it whereby it would retain its monopoly in Bavaria so long as it did not submit any tenders in the rest of the Reich. By so doing, the association was only confirming a situation which had already long been in existence.[130]

4 The last solution: an international cartel

As had happened ten years previously, the collapse of the American economic boom gave the signal this time too for the start of a world-wide depression. While the steelworks were covered for a long time to come by their orders, the remarkable drop in enquiries was evident by late summer of 1882. The approaching recession promised to become extremely unpalatable, specifically in Great Britain, since it was here that a few new high-performance acid Bessemer works had been added in South Wales (Tredegar, Rhymney, Cyfarthfa and Blaenavon) and two big basic Bessemer works in Cleveland (Bolckow Vaughan and North Eastern Steel Co.), all of them fitted out for the production of rails. In view of this situation the dozen biggest steel railmakers met up in the autumn of 1882 to settle between themselves the entire production of steel, right down to the last insignificant bits, of around 1.2 million tons, and to reach an understanding similar to the Rhineland-Westphalian rails

127 Werkarchiv Krupp-Bochum, BV 16303 Nr. 15: Gesellschaft für Stahlindustrie, Otterman (Dortmunder Union) and Koehler on 31 October 1881.

128 Ibid. and Otterman to Gesellschaft für Stahlindustrie on 22 July 1881, 15 October 1881, Koehler to Dortmunder Union on 18 October 1881. The agreement concluced on 30 October 1881 was subscribed to, in the first instance, only by Königshütte and Laurahütte, Gebrüder Stumm and de Wendel & Cie., as well as the Rhineland-Westphalian works. The remaining works joined it in succession. For the list of works which joined the cartel, see ibid.

129 Werkarchiv Krupp-Bochum, BV 16303 Nr. 15: Gesellschaft für Stahlindustrie, Nachtrag zum Vertrag zwischen Gesellschaft für Stahlindustrie und der Schienengemeinschaft.

130 See note 128. Däbritz, pp. 233ff. Friedenshütte was the only works which produced rails from open-hearth steel. The Ilseder Hütte's rolling mill in Peine was absent, since the terms of its licence for the Thomas process did not allow it to produce railway rails. See p. 181.

agreement.[131] However, these negotiations did not immediately result in unity and, given that steel prices rose immediately after their meeting, it did not promote the formation of a cartel either.[132] The project was put on ice, but nevertheless acquired new relevance a year later, when the favourable economic conditions for rails had definitely ended.[133] Production of steel rails fell off in 1883, in spite of an enlarged productive capacity, and only a few new orders were placed.[134] The internal market had already been drastically reduced by 1883; 1884 promised a similar reduction in deliveries to the colonies, which in 1883 alone had still been showing a massive increase.[135]

A clear sign of this development was provided by the complete collapse of railmaking in Sheffield, where only 46,326 tons were produced in 1884, that is, 15 per cent of the quantity produced in 1882.[136] This meant that the marginal tenderers practically disappeared from the market. Cammell and Samuel Fox & Co. were the only ones there which managed to continue producing rails.[137] The fact that the latter firm was almost the sole owner of the Stockbridge Railway and thus had a more or less personal use for its rails, may possibly have played a role here.[138] The 8,000 tons which it produced in 1884 was less than 40 per cent of its meagre total production, which ensured that, when the British market was shared out, S. Fox was given last place with a share clearly under 2 per cent.[139]

Since the autumn of 1883 preparations for closing individual plants had been started in Great Britain. The very fact that Bolckow Vaughan, which had been able to dictate prices on the world market at the peak of the previous

131 This is extensively reported in: *ICTR* 15 September 1882, p. 270; 13 October 1882, p. 398. The *ICTR* was close to the heavy industrialists in Cleveland who were most deeply involved in the cartel, and supported every attempt at forming a cartel, since it thought that this would ensure the industry's survival, whereas *Iron* was without exception extremely critical of these agreements.

132 *ICTR* 22 September 1882, pp. 334 and 336.

133 By April 1883, *Iron* had already established that the rails boom was over and that production capacities were too great for future requirements. *Iron* 20 April 1883, p. 332. Ditto *ICTR* 11 May 1883, p. 548; 12 October 1883, p. 454; 26 October 1883, p. 517. In *ICTR* 9 November 1883, p. 578 it was already being said rather too pointedly that steel could now only be sold in any shape other than rails.

134 *BITA* 1883, p. 34.

135 BT 1883, p. 92 and 1884, p. 30. Compare too *BITA* 1889, p. 62.

136 *BITA* 1884, p. 31.

137 *ICTR* 14 November 1884, p. 632, in which are also mentioned Steel, Peech & Tozer (formerly Phoenix). At this time, however, rails were only being produced sporadically here. *ICTR* 11 January 1884, p. 84. A year before the *ICTR* had already pointed out on 1 June 1883, p. 644, that the production of rails in Sheffield was no longer paying and would have to be abandoned in favour of other steel products.

138 BSC-East-Midlands RRC, loc. no. 42129: minutes of 14th ord. general meeting, 10 August 1884.

139 BSC-East-Midlands RRC, loc. no. 42128: minute book no. 2, report by Mr Sharp, 10 July 1889.

Table 18. *Allocation of quotas in the German Railmakers' Association, 1881*

Friedrich Krupp	152,000 parts.	
Bochumer Verein	98,840 parts	
Dortmunder Union	91,857 parts	
Gutehoffnungshütte	58,987 parts	
Hörder Verein	58,987 parts	
Stahlwerke Osnabrück	58,987 parts	
Rheinische Stahlwerke	58,987 parts	
Eisen- und Stahlwerke Hoesch	57,598 parts	
Phönix	56,107 parts	
Aachener Hüttenverein Rote Erde	56,107 parts	
		748,456 parts
Rhineland-Westphalian group		
De Wendel & Cie, Hayingen	42,593 parts	
Stumm Gebr., Neunkirchen	31,945 parts	
		74,538 parts
Saarland-Mosel group		
Vereinigte Königs- und Laura Hütte	59,583 parts	
Königin-Marien-Hütte, Cainsdorf	49,652 parts	
Friedenshütte, Morgenroth	35,000 parts	
Stahlindustrie, Bochum	32,771 parts	
Group of individual works		177,006 parts
Total		1,000,000 parts

Slump, now had to abandon part of its rails manufacture and to discharge 400 men, tells us much about the way competitive conditions had altered.[140] In the meantime, the new Bessemer works in South Wales had taken over Bolckow Vaughan's position. They had been constructed according to the latest 'American' aspects and worked by the direct process and, where the organisation of labour was concerned, they were also heavily orientated to the 'rapid' American example, more heavily in any case than Bolckow Vaughan, whose head, Windsor E. Richards, also spoke sceptically on his return from America about the possibility of introducing similar forced working methods in Great Britain.

He felt that the biggest impediment was 'that with such hurried work, which we term "driving", we could not fulfil the conditions of the exacting specifications of English and Continental engineers, and so, requiring more time, we are compelled to do with four converters and four sets of men what the

[140] *ICTR* 26 October 1883, p. 517.

Americans do with two converters and three sets of men.'[141] Richards was here referring to the introduction of the eight-hour day with three shifts in the American steel industry, and he was clearly not very convinced of its value.

The eight-hour day was introduced first in the Edgar Thompson Works, when it was realised that the workers were not up to the speeded-up conversion process which they were planning to introduce. The threat of an increase in defective products and accidents was looming. Given that wages made up a very small proportion of the overall costs in the meantime almost any increase in wages seemed thoroughly viable, in view of the expected general reduction in production costs by further speeding-up operations. Captain William Jones, the manager of the Edgar Thompson Works, tells us that:

> therefore it was decided to put on the turns, reducing the hours of labor to 8. This has proven to be of immense advantage both to the company and the workmen, the latter now earning more in 8 hours than the former could in 12 hours; while the men can work harder constantly for 8 hours, they have 16 hours for rest.[142]

According to Jones, this concentrated, faster work did not have a negative effect on the quality of the steel. On the contrary, the shorter working hours and the continuous, uninterrupted process meant that everything ran excellently. Indeed, everything had to run excellently, since the tempo could not otherwise be maintained at all. This meant that the production tempo achieved in the steelworks ranked as an indicator of the steel's uniform quality.

Alexander Holley had countered his British critics in a similar manner: 'The statement sometimes made in England that the rapid production in America impairs quality of product is but a cover for inadequate plant. Steel is obviously no better because five hours instead of one are consumed in setting a vessel bottom, or because it may take twice as long in an English works to handle materials and product.'[143] As far as I know, no one in Great Britain contradicted him. Even if Richards's argument about quality was correct in the case of the USA, it would still not explain why other works in Germany, Belgium and South Wales were also very capable of satisfying the railways' quality requirements when using a faster process. Edward P. Martin, the General Manager at Blaenavon, who like Windsor E. Richards had also been sent to America by his directors to study the organisation of production there, came to different conclusions.[144] Although he too considered American rails inferior he thought

[141] Windsor E. Richards before the Cleveland Institution of Engineers, according to *ICTR* 27 January 1882, p. 101.

[142] *American Iron and Steel Association Bulletin*, 15, 18 May 1881, p. 121. The quotation here is taken from William T. Hogan, *Economic History of the Iron and Steel Industry in the United States*, Lexington, Mass. 1971, p. 220.

[143] Paper read before the American Society of Mechanical Enginers, 4 November 1880, in: *Iron*, 24 December 1880, p. 474.

[144] Gwent CRO, loc. no. D. 480. 1 & 2: minute book no. 1, board meeting 24 November 1881.

he could learn something from their production technique and organisation, and straightaway procured the necessary means to do so.[145] Shortly afterwards, he was made an offer by Dowlais, the leading steel manufacturer in the region, which he accepted.[146]

In South Wales in 1882/3, daily charges in the number of 30–40 were already the order of the day, and that using converters with capacities of 7 to 10 tons, which enabled weekly production rates of over 2,000 tons per plant. The Phoenix of Steel, Tozer and Hampton near Sheffield, which in 1879 had already reached a very rapid conversion process in British terms, was now producing 2,000 tons a week from each of its two new plants. The two new plants, each with two 10-ton converters, had been set up in 1880 with the explicit aim of achieving a total weekly production of 4,000 tons with the lowest possible production costs.[147]

In technical and organisational terms, then, these works were the best available at the time, on top of which they had a few minor cost advantages over Cleveland, which gave them, all in all, a slight lead. The rates for freighting Spanish ore to Wales were slightly less, as were the wages for the unskilled Irish labourers employed on the blast furnaces, steel works and mills.[148] The *ICTR* summed it up as follows: 'It is now generally conceded that the Welsh works are able to turn out the best steel rails at as low a figure as any district.'[149] The changeover from iron to steel rails was thereby successfully concluded in Wales, as well as catching up with Bolckow Vaughan's initial lead. The latter's previous over-mighty position came under pressure from another side too. The two big Bessemer works in the Sheffield district, Charles Cammell & Co. and Wilson, Cammell & Co., merged.[150] Simultaneously, the new firm proceeded to acquire the Derwent Iron Co., on the West Cumberland coast, in order to expand both its own raw-material base, and to win a good coastal site for exporting.[151] The Bessemer plant and rolling lines belonging to what had been Wilson, Cammell & Co., were moved from Dronfield to the

[145] Ibid.: minute book no. 1, board meeting 11 January 1882.

[146] Ibid.: minute book no. 1, board meeting 19 April 1882. He had thereby reached the top position in his profession in South Wales – an acknowledgement of his great organisational abilities. His successor was able to announce for the first time a weekly production of more than 2,000 tons, using plant which had been fitted out according to Martin's plans, and applying the organisation of production which he took over from him. Gwent CRO, loc. no. D.480. 1 & 2: minute book no. 1, board meeting 16 May 1883.

[147] *ICTR* 4 May 1883, p. 516; 31 December 1880, p. 779.

[148] *ICTR* 27 January 1882, p. 110; 26 October 1883, p. 517.

[149] *ICTR* 27 October 1882, p. 458. The principal reason for this was said there to be the introduction of modern mechanical equipment which was extensively employed. Qualified workers were now, as earlier in the case of the puddlers, no longer required.

[150] *Iron* 31 March 1882, p. 250. *ICTR* 31 March 1882, p. 364; 11 August 1882, p. 120. In addition Cammell bought further supplies of low-phosphorus iron ore in Cumberland, in order to increase his supplies of raw materials for the Bessemer process. *ICTR* 16 March 1883, p. 310.

[151] *ICTR* 11 August 1882, p. 120.

coast to take over the firm's rails business there. The move in the summer of 1883 meant that one of the biggest railmakers in the country was out of action for six months, and its return to the market augured no joy for the competition. The works was well furnished with new orders for their start-up, and began operations in October that year.[152]

The downwards trend in prices for Bessemer rails continued apace, as was to be expected. Towards the end of the year, between £4 10s. and £4 12s.6d. were still being achieved; by the start of 1884 rails could already be had in South Wales for £4 a ton.[153] Given production costs of rather more than £3 12s. per ton for steel ingots alone,[154] these must have been 'losing prices', which were only accepted to keep the works going. It is quite certain that no steel rails could be milled for 8s., quite apart from the steel wastage lost in the process.

In such a situation, the British railmakers looked enviously at Germany. Many of them had already praised the nationalisation of the railways as exemplary, as well as recommending it for their own country,[155] so the advantages of establishing a cartel in their home market must now have seemed newly obvious. Thanks to the 1883 cartel, the average price for all steel rails from German works was still 134.25 marks per ton.[156] This high average price, with its clear profit margin,[147] could be maintained in spite of their price-dumping on the 'neutral' export markets. In 1883 German exports had shrunk only slightly compared with the previous year and the Rhineland-Westphalian works could well afford the falling prices because they were still being kept sufficiently busy by the home railways.[158] Nevertheless, as before in the seventies, their export transactions could often only be concluded at 'losing prices'.[159] For the British steelworks this meant that, since their own market had no effective protection, forming a cartel would only make sense if they could get the German and Belgian works to agree to keep out of the British market or, better still, to regulate the entire sales market. Baare's dream of the seventies, that competitive relations could be wholly determined by means of an international agreement, seemed about to be realised. From now on, the once over-mighty British steel industry's ability to adapt to the market, had

152 *ICTR* 19 October 1883, p. 482.

153 *ICTR* 1 February 1884, p. 137.

154 These were the production costs at Dowlais. Glamorgan CRO, D/Dg E 3: report on balance sheet for year ending 31 March 1884, appendix p. 28.

155 *ICTR* 2 April 1880, p. 370 and p. 375.

156 *Reichstatistik 1883*, p. 132.

157 The production costs for steel rails in the Ruhr swung between 101 and 105 marks per ton (Krupp) and 120 and 124 marks per ton (GHH). HA-Krupp, WA I 859: Schienenproduktion. GHH-Archiv, 300 108/11: Berichte für den Aufsichtsrat, 1883/4.

158 *Berg- und hüttenmännische Zeitung* 42 (1883), *Montanproductenmarkt*, 9 November 1883, p. 538.

159 Werkarchiv Krupp-Bochum, BV 12900 No. 14: (Bochumer Verein) Protokoll der Verwaltungsratssitzung am 4 Oktober 1883 und am 29 November 1883.

come to an end and induced it to seize the initiative as its role on the international scene was diminishing. In December 1883 the British steelmakers invited their German and Belgian colleagues to London to discuss the formation of an international rails cartel.[160]

On the Continent by this time, the steelworks' accord had already spread far beyond their national cartel. The German Rails Association had made an agreement with the Austrian steelworks, who had their own cartel, which came into effect at the start of 1883. This agreement officially confirmed a situation which had long ago been established by the tendering system applied by the German and Austrian authorities, in that the German works engaged not to tender in Austria-Hungary, and the Austrian works not in Germany. The partners in this Rails Association on the Austrian side were: Österreichische Alpine Montangesellschaft, Erzherzoglische Teschensche Vertretung, Ternitzer Walzwerk und Bessemer Stahlfabrication-Actien Gesellschaft, Prager Eisen Industrie Gesellschaft, Wittkowitzer Eisenwerke Direction and Teplitzer Walzwerk und Bessemerhütte. The leader in Austria was the Prager Eisen Industrie Gesellschaft.[161] The first negotiations for this agreement had already been held in 1879 but, probably because conditions in Germany were unclear due to the addition of new works there, they were only finalised after the new German national cartel had been finalised.[162]

The 'German–Belgian Railmakers' Convention', which existed from the end of 1882 at the latest, regulated conditions with their western neighbour, where the State authorities had for some time ensured that their home industry enjoyed a *de facto* monopoly. The terms of this agreement are not known, but a letter by Servaes, the Director of the Railmakers' Association, suggests that it must in effect have been similar to the German–Austrian agreement.[163]

This agreement by the German, Belgian and Austrian steel producers meant that their unofficial State protectionism was safeguarded and confirmed on the industrial level. It meant that they stopped submitting tenders at dumping prices, which had been hopeless from the start, and thus removed an important prop in the argument used by Liberal critics of the State's subsidies policy. The steel industrialists on the Continent could not be accused of strengthening, even only indirectly, Free-Trade and Liberal currents, and the authorities' problems

160 GHH-Archiv, 300 108/11: Berichte für den Aufsichtsrat, 8 December 1883.

161 Werkarchiv Krupp-Bochum, BV 16303 Nr. 15: Nachtrag zum Vertrag zwischen Gesellschaft für Stahlindustrie und der Schienengemeinschaft, Vertrag vom 3 Januar 1883 und 27 November 1882 am 15 Januar 1883 an Gesellschaft für Stahlindustrie geschickt: Anlage zum Zirkular der Schienengemeinschaft vom 5 Oktober 1882.

162 Werkarchiv Krupp-Bochum, BV 12900 Nr. 12: Protokoll der Verwaltungsratssitzung am 8 Dezember 1879.

163 Werkarchiv Krupp-Bochum, BV 16303 Nr. 15: Servaes to Gesellschaft für Stahlindustrie on 1 December 1882.

about the legitimacy of covering the cartel's high prices, were unnecessary and harmful to the political general-weather situation.

The British initiative about forming a cartel for the international market thus encountered well-prepared partners. It is clear from the description of how the accord came into being by Josiah Timmis Smith of Barrow Co., the first president of the British cartel that, unlike the continental groups, the primary aim was to regulate exports.[164]

After some time the makers in England, all except one firm, agreed to join the association, and it was decided to endeavour to associate the Belgians and the Germans with us as being the only two countries that exported rails. The outcome, after taking the figures of three years of the exports from the three countries, was that Great Britain kept 66 per cent of the entire export trade. Belgium had 7 per cent and Germany 27 per cent. We have since modified the division a very little, and given Germany 1 or 2 per cent more and Belgium ½ per cent . . . The next thing that we had to do, having agreed upon what proportion each country was to have of the orders of the world, was to agree amongst ourselves how we should divide those orders, and we thereupon assessed the capabilities of each work, each company representing a certain number of parts out of 100 parts. The effect of this has been . . . [that] we have maintained a price of £4 13s. a ton at the works.

And to the question: 'What would be the position of a man opening a new firm?', he replied:

the position of a man opening a new firm would be, that if he would not join the union we should have to put our price up to the point that would prevent other people coming into it. The aim by which we regulate our price is to minimise competition as much as we can.[165]

The similarity to the German Railmakers' Association and how it came into being is obvious. The internal allocation of quotas in Great Britain is shown in Table 19.[166]

The negotiations with the Steel Company of Scotland, the only steel producer working according to the open-hearth process, were still in process when the contract was concluded. This meant that the quotas were altered a bit when the SCS joined.[167] The SCS's share is not known but, given this work's limited production of rails, it must have been around the lowest limit. Both cartels, the national as well as the international one, were contracted to run for a period of

164 The severe market slump of 1883 had led to the formation in 1885 of the RC on the Depression in Trade and Industry. Royal Commission on the Depression in Trade and Industry, minutes of evidence and reports, in *Parliamentary Papers 1886*: C.6421, 4715, 4715-I, 4797, 4893.

165 RC Depression, evidence Smith, p. 60.

166 BSC-Northern RRC, loc. no. 04898–04900: minute book 7, Railmakers' Association – Deed of Association – 1884. Also in BSC-East-Midlands RRC, loc. no. 042128: S. Fox & Co., minute book no. 2, loose sheet.

167 Ibid.

Table 19. *Allocation of quotas within the British Group of the International Railmakers' Association, 1884*

		Percentage
1	The Blaenavon Company Ltd	6,476
2	Crawshay Brothers Cyfarthfa	4,084
3	The Dowlais Iron Company (Guest & Co.)	8,850
4	The Ebbw Vale Steel Iron and Coal Company Ltd	8,237
5	The Rhymney Iron Company Ltd	7,148
6	The Tredegar Iron and Coal Company Ltd	4,084
7	Bolckow, Vaughan & Co. Ltd	14,295
8	The Darlington Iron and Steel Company Ltd	3,063
9	The North Eastern Steel Company Ltd	4,629
10	The Barrow Hematite Steel Company Ltd	11,232
11	Charles Cammell & Company Ltd	14,295
12	The Mossbay Hematite Iron and Steel Co. Ltd	5,786
13	The West Cumberland Iron and Steel Co. Ltd	5,957
14	Samuel Fox & Company Ltd	1,873

two and a half years from the beginning of 1884. They called themselves the 'Railmakers' Association', that is to say, the 'International Railmakers' Association' (IRMA) and their registered office was in London.

On the Belgian side the members of the IRMA were:[168] Société John Cockerill, Société de Thy le Château, Société de la Fabrique de Fer d'Ougrée, Société des Aciéries de l'Angleur, and La Louvière. Of the members of the German Railmakers' Association, thirteen works dealing in exports joined the IRMA, which meant basically the Prussian works. In any case, J. T. Smith complained that two German works had not joined the cartel, whereas in Great Britain only one had refused and in Belgium all the works had cooperated.[169]

The Hörder Verein, Osnabrücker Stahlwerk, Königs-Laura-Hütte and the Saxon Königin-Marien-Hütte did not join it immediately.[170] However, Hörder Verein came in soon after, although it occasionally made a lot of trouble within the IRMA,[171] so much so that the Rheinische Stahlwerke, which were always very set on stable conditions within the cartel, were even prepared to reduce their share of the German rail-sleepers cartel, to which many railmakers also belonged, in favour of the Hörder Verein merely in order not to put the

168 Däbritz, p. 235.

169 RC Depression, evidence Smith, p. 60. He was, of course, referring only to exporting works, and thus not to Maxhütte in Bavaria or to Gienanth, for instance.

170 GHH-Archiv, 300 108/11: Berichte für den Aufsichtsrat, Lueg to Haniel on 31 January 1884.

171 Werkarchiv Krupp-Bochum, BV 12900 Nr. 14: (Bochumer Verein) Protokoll der Verwaltungsratssitzung am 26 Juni 1884.

international rails agreement at risk.[172] The Bochumer Verein was even demanding an end to all internal cartels, until such a time had been reached when all the rails manufacturers had joined the IRMA.[173]

The Osnabrücker Stahlwerke, on the other hand, went on refusing to join. By 1881, anyway, the firm's export business, which to a great extent involved deliveries to Russia, was falling off. The firm could no longer match the 'give-away prices' on the world market, and had to a great extent abandoned its share of foreign rails tenders. H. Müller, the works co-founder, wrote retrospectively about the sales policy of that year that: 'Only a few large foreign contracts at loss-making prices were taken on, if the lack of other work made it necessary to keep the works going.'[174] Necessity drove him in 1884 to conclude a big rails contract with the Danish Railway at 84 marks per ton, which was way below the IRMA's cartel price. This sporadic export business certainly posed no threat to the IRMA's existence, given that the Osnabrücker Stahlwerk had to suffer a considerable cutback in production starting in 1884.[175] It is not at all sure which of the two remaining works, Königs-Laura-Hütte and Königin-Marien-Hütte, ended by joining the IRMA and which one stayed out. Nevertheless, like the Osnabrücker Stahlwerk, both belonged to those works whose strength lay above all in their regional sales and whose share of German exports was insignificant. The outsider among the British works was probably Brown, Bayley & Dixon, which had been reactivated yet again after going bankrupt. However, since there is no positive statement to this effect we cannot be sure of this.[176]

The internal markets of the three participating countries were reserved for their national cartels, much like the agreement which already existed between the Belgian, Austrian and German works. The British colonies constituted a special case. The market there was reserved exclusively for the British Railmakers' Association, and was treated in this respect like their internal market. The quantities exported to the colonies were, however, charged to the British quota within the IRMA.[177]

The IRMA was given a further, albeit not decisive, boost in the summer of

[172] Thyssen-Archiv, RSW 123 005 62: Protokoll der Aufsichtsratssitzung am 18 Februar 1885.

[173] Werkarchiv Krupp-Bochum, BV 12900 Nr. 14: Protokoll der Aufsichtsratssitzung am 28 Februar 1884.

[174] Müller, *Georgs-Marien*, p. 82.

[175] Ibid., p. 140. The slowdown in production from business year 1882/3 to 1883/4 amounted to almost 30 per cent.

[176] Indicative of this is that, of the other potential rails producers, Steel, Tozer & Hampton's Phoenix works had abandoned its rails production when the cartel was formed, and the Patent, Shaft & Axletree Co. had already stopped making rails in 1883. Added to which, Sheffield, the home of B, B & D, was the only district in which railmaking was conducted on an irregular basis during the cartel period and did not follow the general trend. *ICTR* 11 January 1884, p. 48; *BITA* 1889, p. 28.

[177] RC Depression, evidence Smith, p. 65.

1885, when two French steelworks joined, and in July that year they acted in concordance with the cartel for the first time over a tendering case for the Italian government.[178] This meant that the cartel now controlled almost all of the European rails market. The IRMA's registered office in London did not act as a sales syndicate, but simply managed the quota distribution book and made sure that the established minimum price of £4 13s. per ton ex works, which applied both internationally as well as within Great Britain, was adhered to.[179]

The representatives of the national cartels took over the management of the IRMA within a rota system; decisions required a three-quarters majority vote by the members.[180] Tenders by each entitled works were in turn, as in Germany, protected by higher prices set by the other works.[181] The relevant national group was responsible for discipline within the cartel. Any export business that was brought about by circumventing the IRMA or was contracted by one of the works which had not joined the cartel, as, for instance, Osnabrücker Stahlwerk's deal with the Danish Railways cited above, was added to the quota for the relevant national group.[182]

Discipline within the cartel on the part of member works seems to have been excellent. Although internal disputes about individual quotas are occasionally mentioned, crass infringements of the terms are not.[183] On the other hand, the sales expectations which had been established when the cartel was formed had turned out to be excessively optimistic. The presumed total annual exports of one million tons for the three countries was not reached even approximately during the two and a half years of the IRMA's term[184] (see Table 20). In spite of stable prices, the rails trade proved one of the poorest performers in the whole of British heavy industry,[185] whereas in Germany the recession still remained within bounds.

It had by then become very difficult to secure orders for rails in Britain. Big

178 *ICTR* 22 May 1885, p. 693; 3 July 1885, p. 14.

179 RC Depression, evidence Smith, p. 60.

180 Mannesmann-Archiv, P1.25.35: (Phönix) Protokoll der Aufsichtsratssitzung am 5 September 1884. These minutes contain the information that Director Thielen was going to London for two months, to assume the management of the IRMA as the German Group's representative. For the voting procedure in the cartel, see Werkarchiv Krupp-Bochum, BV 12900 Nr. 14: (Bochumer Verein) Protokoll der Aufsichtsratssitzung am 28 Februar 1884.

181 RC Depression, evidence Smith, p. 60.

182 GHH-Archiv, 300 108/11: Berichte für den Aufsichtsrat, Lueg to Haniel on 31 January 1884.

183 One of the driving forces in this dispute appears to have been the Hörder Verein, which had been pushed relatively late into the international cartel and insisted on a review of the quotas. It eventually secured some success when the German quotas as a whole were raised. See above, note 171.

184 The estimate of 1 million tons of rails exports by the participating groups had provided the basis for the decision to found the IRMA. GHH-Archiv, 300 108/11: Berichte für den Aufsichtsrat, Lueg to Haniel on 31 January 1884. For Great Britain this would have meant annual exports of 660,000 tons, whereas these amounted to only 525,000 tons in 1884 and 460,000 tons in 1885.

185 *ICTR* 15 August 1884, p. 211.

works like Dowlais and Bolckow Vaughan could not utilise even 50 per cent of their capacity. Consequently Bolckow Vaughan shut the steelworks down for several months at a time in 1884 and 1885 and processed their existing contracts smartly in the remaining time, to economise on costs.[186] It was even claimed that the railmakers' customers had compacted to starve the cartel out.[187] People were also complaining that their production costs had risen due to inadequate levels of production.[188]

In January 1886 all this finally resulted in Bolckow Vaughan announcing the imminent demise of the cartel on 31 March 1886, in which, however, it was supported by only one other British works.[189] At the firm's subsequent General Meeting, the following reason was given for this step:

> We forced up the price, no doubt, under that arrangement, 10/- or 15/- per ton, but we had the power of producing 4000 tons per week, and we were limited to producing 1000 or 1500 and with a large establishment it involved considerable loss to us in the opinion of the majority of the board.[190]

The result of terminating the rails cartel, a measure which no one on the board objected to in the slightest,[191] was a catastrophic fall in the price of rails. In Great Britain and on the export markets, it fell to *c.* £3 10s. (70 marks).[192]

In Germany, the situation did not have as oppressive an effect during the IRMA's term of office. On the one hand, the cutback in the production of rails was less drastic and on the other, the Railmakers' Association in Germany was able to maintain a comfortable price, compared with the IRMA, of over 130 marks per ton which, thanks to the agreement, was not contested by any foreign competition.[193] Even after the IRMA had collapsed, internal prices could still be stabilised at 108–15 marks per ton, which allowed the price difference of 45 marks per ton compared with the export market to be defended.[194] The only expectations not to be fulfilled were those of a few

186 BSC-Northern RRC, loc. no. 04902: directors' report year ending 31 December 1884, p. 14; directors' report year ending 31 December 1885, p. 7. Glamorgan CRO, loc. no. D/DG E3: report on balance sheet for year ending 29 March 1884, p. 2.

187 *ICTR* 8 August 1884, p. 179.

188 Glamorgan CRO, loc. no. D/DG E3: report on balance sheet for year ending 29 March 1884, p. 7.

189 BSC-Northern RRC, loc. no. 04898–04900: minute book 7, board meeting 18 December 1885 and 29 January 1886.

190 BSC-Northern RRC, loc. no. 04902: directors' report year ending 31 December 1886, Henry Pochin, p. 23.

191 F. W. Bolckow, the founder of the firm, was one of those supporting the cartel agreement. Ibid., directors' report year ending 31 December 1886, F. W. Bolckow, p. 14.

192 *Berg- und hüttenmännische Zeitung* 54 (1886), *Montanproductenmarkt*, 17 December 1886, p. 528 and 46 (1887), *Montanproductenmarkt*, 18 March 1887, p. 113.

193 HA-Krupp, FAH III B 194: Von Fried Krupp erzielte Durchschnittspreise für Schienen.

194 *Berg- und hüttenmännische Zeitung* 46 (1887), *Montanproductenmarkt*, 18 March 1887, p. 113.

Table 20. *The production and export of acid and basic Bessemer steel rails: Great Britain and Germany, 1882–1886*

Year	GB (1,000 t.)[a]	Percentage IRMA-exports[b]	G (1,000 t.)[c]	Percentage IRMA-exports[d]
1882	1,236	(59.4%)	532	(32.9%)
1883	1,097	(68.7%)	474	(35.4%)
1884	785	67.2%	400	35.1%
1885	707	68.5%	422	36.8%
1886	730	(70.3%)	392	(40.2%)

The values in brackets relate to exports which would have been subject to IRMA regulations, had there been any. In Great Britain's case, this included exports to the colonies.

a *BITA* 1889, p. 28.

b Ibid., p. 33.

c Statistisches Jahrbuch des Deutschen Reiches (1907) II, pp. 90–3.

d Ibid., pp. 265–9. The figures for Germany include *c*. 4 per cent mounting parts, which were not taken into account when allotting rails quotas.

works, which hoped to be able, with the IRMA, to expand production and thus reduce production costs. GHH, for instance, had only joined the IRMA to achieve this effect.[195]

The rails market now no longer presented any big opportunities for expansion. The railway networks in the industrialised countries were already pretty complete and the demand for replacements was stagnating because the steel rails were more resistant in spite of the increasing traffic. The railways' specific consumption of rails fell and thus lost its function as a pace-setter for the growth of the heavy industry. Towards the end of the eighties the rails works experienced a final economic boom, with the construction of the Argentinian railways, which collapsed abruptly, wholly in the style of the railroad boom in the USA, in 1890 when the big banking crash came.[196] By then production of acid Bessemer steel in Great Britain had passed its peak. In Germany this had already been the case by 1882, because the Thomas process was more widely distributed there. Although the original Bessemer process had played the pioneering role in superseding the puddling process, this was, however, entirely thanks to the blossoming of railway construction and to the great suitability of acid Bessemer steel, which was not otherwise much prized, for making rails. As with the construction of railways, it was visibly losing its relative importance towards the end of the nineteenth century. From now on,

[195] GHH-Archiv, 300 108/11: Bericht für den Aufsichtsrat am 9 März 1884.

[196] A. G. Ford, Overseas Lending and Internal Fluctuations 1870–1914, in: A. R. Hall (ed.), *The Export of Capital from Britain 1870–1914*, London 1968, p. 89.

new markets were needed to maintain the expansion of the steel industry, markets which had hitherto been retained by the more versatile puddling works. However, these new markets were only to be won with new steel qualities comprising the advantages of mild and easily wrought puddled iron.

5 New processes and new markets

If the first phase of the new steel industry, which was distinguished by a uniform product, still bore the familiar features of unmistakable British superiority and of diligent imitation on the Continent, maintained by tariffs and cartels, the following phase of diversification, however, brought forth two clearly differentiated autonomous structures. While the terms of the institutional frameworks in both countries continued to be retained as far as possible – state protectionism and collective restrictions on competition in Germany and, after the brief episode with the IRMA, free competition and entrepreneurial autonomy in Great Britain – their production mechanisms developed along quite different directions. Whereas capacities in Germany were further extended for the newly opened markets mainly by using the Thomas process, in Great Britain mainly the open-hearth process was used instead. Much significance has been attributed to this difference for explaining British steel production's lower growth rate during the following decades, claiming that the British steel industry would have been able to maintain its lead on the world markets had it applied the obviously successful 'German' production technology. Before we deal with the question of entrepreneurial prescience in both countries, let us first investigate how these different structures and their underlying conditions and motives came into being.

1 Basic Bessemer steel

(a) The introduction of the Thomas process

When assessing the great economic importance of the Thomas process above all for the German steel industry, it is often overlooked that, in technical terms, it involved only a slight modification to the original Bessemer process. It was possible for existing Bessemer works to engage in a trial operation according to the Thomas process without going to great trouble and above all without incurring notable costs. All that needed to be done was to line an existing converter with dolomite and to provide equipment for heating and charging lime. In this way, the interested works could very quickly obtain a clear view of the new working methods and undertake a preliminary estimate of the conversion costs. The many plants which had been closed during the wave of rationalisation in the seventies were just what was needed for these trials. This was the case, for instance, with Brown, Bayley & Dixon in Sheffield,[1] with Hörder Verein,[2] Rheinische Stahlwerke,[3] Dortmunder Union,[4] GHH[5] and Patent, Shaft and Axletree in Staffordshire.[6] In plant with three converters, the third, which would otherwise have acted as a reserve, could serve for these trials, as, for instance, at the Aachener Verein[7] and the Bochumer Verein.[8] Gienanth in Kaiserslauten contracted to test pig iron from Saarland on Stumm's authority.[9]

Unlike when the original Bessemer process was introduced, the breadth of the trials meant that right from the start an open discussion developed within the industry about technical and economic information pertaining to the Thomas process. After initial uncertainties, which were primarily attributable to problems with the fireproof lining in the converters, P. Trasenster of the Ecole des Mines in Lüttich was able to present the first reliable computation of costs in the summer of 1880. Assuming that a regular operation could be maintained, he estimated the conversion costs for basic Bessemer steel at 21.68 marks per ton inclusive of the patent fees of 3.20 marks per ton. For the purposes of comparison, he had taken 13.20 marks per ton as the conversion costs for Bessemer steel, thereby referring to the trend in the more cheaply producing Ruhr works. Furthermore, given that the Thomas process involved rather more wastage than the acid Bessemer process, Trasenster considered that altering works for the new process could only be justified in places where phosphoric pig iron was at least 16 marks per ton cheaper than common Bessemer pig iron.[10]

1 *ICTR* 23 April 1880, p. 460.

2 Commissions-Bericht über den derzeitigen Stand der Entphosphorung des Eisens im Bessemer Converter nach Thomas's-Gilchrist's patentierten Verfahren. (P. Tunner, F. Kuppelwieser, A. Krautner), in: *Zeitschrift des berg- und hüttenmännischen Vereins für Steiermark und Kärnten* 12 (1880), p. 232.

3 Ibid., p. 240.

4 HA-Krupp, WA IV 835: Generalversammlung Dortmunder Union am 10 Dezember 1881.

5 GHH-Archiv, 300 108/10: Berichte für den Aufsichtsrat, September 1882.

6 *ICTR* 2 June 1882, p. 627.

7 Macar, p. 168.

8 Werkarchiv Krupp-Bochum, BV 12900 Nr. 12: Protokoll der Verwaltungsratssitzung am 19 Mai 1880.

9 Thyssen-Archiv, RSW, 640-00 C 1 c: Stumm to Rheinische Stahlwerke on 19 May 1880.

10 According to the *Zeitschrift des berg- und hüttenmännischen Vereins für Steiermark und Kärnten* 12 (1880), p. 506. It is extremely probable that Trasenster in Liège was going by the conversion costs of the Cockerill Bessemer works on the spot. The low conversion costs would accord with expectations from this 'rapid' works, built to an American design. Re Bessemer plant and organisation of production at Cockerill's, see pp. 73 and 76 above.

Unlike the previous cost computations and barely credible claims made by the inventor, Trasenster's expertise was very well regarded, once it had been put around in the trade journals that the basic Bessemer works at that time still had no idea of their own costs.[11] Half a year later the Rheinische Stahlwerke undertook its own internal computation of costs, which confirmed Trasenster's figures. In these, the conversion process cost 20.27 marks per ton at the turn of the year 1880/1.[12] On top of this came a patent fee of 1 mark per ton. However, only Rheinische Stahlwerke and Hörder Verein were able to benefit from this low fee, having together acquired from Sydney G. Thomas the patent rights for Germany and Luxemburg. As licensers, they demanded 2.50 marks per ton from the other works.[13]

Once Bolckow, Vaughan & Co. in Great Britain, Rheinische Stahlwerke and Hörder Verein in Germany and the Witkowitzer Stahlwerk in Austria had demonstrated the practical feasibility of the Thomas process in the course of 1880, its diffusion now depended solely on the price difference between acid Bessemer and basic Bessemer pig iron in any particular place. Trasenster's investigation thus came to the conclusion that it would pay to introduce the Thomas process above all in the east of France and in Alsace-Lorraine. Initial trials with the original Bessemer process had been broken off there on account of the prohibitive cost of transporting the raw materials. In his view, the Ruhr, Sheffield and Belgium were at break-even points. Astonishingly enough, he considered Cleveland, initially in line to derive the greatest advantage from the Thomas process,[14] to be below the break-even point, which always referred to the cost of the acid Bessemer process on the same site and was not to be regarded as absolute. In South Wales and Cumberland, the absence of suitable ores meant that the basis for the Thomas process was missing.

Other experienced observers reached similar estimates even without conducting detailed investigations. A correspondent for *Iron* concluded, after a critical appraisal of the Gilchrist-Thomas process, which dwelt especially on the difficulties of conducting it in Great Britain, that it might well pay in Eastern France and in Westphalia, but not in Great Britain, not even in Cleveland – since low-phosphorus iron ore was obtainable there in sufficient quantity and at favourable prices. There was no reason why the Bessemer industry in Cumberland and Wales should entertain any anxiety. In Belgium,

11 GHH-Archiv, 300 108.8: Protokoll der Sitzung des Aufsichtsrats am 15 Februar 1881. *Zeitschrift des berg- und hüttenmännischen Vereins für Steiermark und Kärnten* 12 (1880), p. 494.

12 Thyssen-Archiv, RSW, 640-00 C 2 k: Beantwortung der mir am 17ten Januar 1881 vom Herrn Commerzienrathe Meyer zu Celle gestellten Fragen.

13 Thyssen-Archiv, RSW, 640-00 C 1 i: Denkschrift betr. die bei der Vergebung der Licenzen für das Thomas-Verfahren beobachteten Grundsätze.

14 According to *Zeitschrift des berg- und hüttenmännischen Vereins für Steiermark und Kärnten* 12 (1880), p. 509.

the *Moniteur des intérêts matériels* urged paying careful attention to the new process, but pointed out that at the moment it was not worth introducing.[15]

As with his previous estimate of conversion costs, Trasenster's investigation was confirmed in this matter too by further developments. In the Minette regions of Eastern France and Alsace-Lorraine, the Thomas process was introduced without further ado and new works immediately started to be built for it.[16] Until now, these regions had been supplied with steel from Rhineland-Westphalia and from Le Creusot, but now they could perform this themselves. Since basic Bessemer steel was initially only viewed as a substitute for acid Bessemer steel, this meant that the old puddled rails works there would not only win back their market but, furthermore, could also surface as dangerous competitors in what had hitherto been steel districts.

Given these prospects for the effects of the Thomas process, people in the Ruhr were anything but enthusiastic about it. Carl Röchling, who at that time was still on the Board of the Rheinische Stahlwerke, and had contributed considerably to the firm's financial recovery after its collapse in 1877, did not expect to benefit at all from the success of the Thomas patent which they had just acquired.

> In my opinion A3 [pig iron with a high phosphorus content, U.W.] should nowadays be obtainable in Luxemburg, Ottange and at de Wendel at *c.* frs. 45 = M. 36 per ton, and when the converters have been set up according to these rates, as is certainly happening at de Wendel, then I would like to know how Westphalia with its high cost prices for pig iron, which are bound to fluctuate between M. 15–20 more, will remain prosperous. De Wendel previously made *c.* 40,000–50,000 tons of iron rails and, in my opinion, before 2 years are past, will be on the German market with an annual production of 80,000 tons of steel rails, and what is worse, will have to be in order to be able to survive.[17]

The Rheinische Stahlwerke even made a short-lived bid to impose a production limit on all licensees for the Thomas patent, so that they themselves could derive the greatest benefit from the invention.[18]

The Rheinische Stahlwerke's fears were shared by others in the steel sector. At the Bochumer Verein's general meeting in October 1879, Louis Baare was already predicting hard times ahead in the event of the Gilchrist-Thomas process becoming viable in the Saar and Lorraine regions.[19] His opinion was

[15] Both according to ibid., pp. 452f.

[16] Ibid., p. 509. R. M. Daelen, Neuanlagen von Eisen- und Stahlwerken in den Minette-Districten, in *Stahl und Eisen* 2 (1882), pp. 527–31.

[17] Thyssen-Archiv, RSW, 640-00 A 1: C. Röchling to F. Goecke on 18 June 1879.

[18] Ibid., Suermondt to F. Goecke on 15 May 1879.

[19] Werkarchiv Krupp-Bochum, BV 126 00 Nr. 1: Protokoll der Generalversammlung am 28 Oktober 1879. At the Administrative Board Meeting on 24 March 1880 the view was once again strongly upheld internally that the Thomas process would harm the Ruhr industry in

shared by the *Rhein- und Ruhr-Zeitung*; the general opinion among the existing Bessemer works in the Ruhr, on the other hand, was that the new process was economically uninteresting.[20] Their sceptical attitude was refuted by the high licence take-up in Germany during the initial years. Already in 1879, prior even to the successful first charge at Hörde and the Rheinische Stahlwerke, de Wendel, de Dietrich and Stumm all acquired rights.[21] The two Lorraine works of de Wendel and de Dietrich had some time ago already installed a few Bessemer plants, which however, were not viable on such an unfavourable site. On the other hand, it did mean that they had experience of their own in the technology and organisation of refining by converter, and were thus able to start the new process straightaway. Given that the trade journals were agreed that they were sitting on the most favourable site of all for the Thomas process, there seemed to be very little risk involved.[22] Though Stumm, in Saarland, did not have a Bessemer plant, it joined up with Gienanth instead, since the latter had a converter a short distance away in Kaiserslautern, which had only been worked sporadically and could be employed for trials using pig iron from Dillingen.[23]

The success of the first charge on the Hörde and Ruhr site led to the Lorraine ironworks Ars acquiring the licence immediately afterwards.[24] The next, in January 1880, was the Burbacher Hütte in Saarbrücken, which derived extraordinarily good-value phosphoric pig iron from its Luxemburg mines in Esch.[25] Apart from Röchling, which was then still working with the Rheinische

favour of works along the Saar and in Lorraine. Ibid., BV 129 00 Nr. 12: Protokoll der Verwaltungsratssitzung am 24 März 1880.

20 *Rhein- und Ruhr Zeitung* on 25 November 1879.

21 Thyssen-Archiv, RSW, 640-00 C 1 b: Thomaspatent, Vertrag mit de Wendel, 6 June 1879, fixed price 180,000 marks: 640-00 C 1 c: Thomaspatent, Vertrag mit Stumm und Konsorten (Aktiengesellschaft Dillinger Hüttenwerke und Gebr. Gienanth, Kaiserslautern), 15 June 1879, fixed price 90,000 marks; 640-00 C 1 d: Thomaspatent, Vertrag mit de Dietrich & Cie, Niederbronn, 30 July 1879, fixed price 90,000 marks.

22 In any case, nothing was ever known about a basic Bessemer works at de Dietrich in Niederbronn. A summary by Schrödter for the Verein Deutscher Eisenhüttenleute mentions for 1901 only one single small open-hearth furnace there. *Gemeinfassliche Darstellung des Eisenhüttenwesens*, pub. by Verein Deutscher Eisenhüttenleute, 4th edn, Düsseldorf 1901, pp. 136–41.

23 Thyssen-Archiv, RSW, 640-00 C 1 c: Stumm to Rheinische Stahlwerke on 12 October 1879. In it, Stumm described the Bessemer plant at Gienanth as an 'insignificant establishment' in which simple 'technical observations and trials' could be set up. His evaluation accorded thoroughly with the actual importance of the Gienanth works within the German steel industry.

24 Thyssen-Archiv, RSW 640-00 C 1 e: Thomaspatent, Vertrag mit Lothringer Eisenwerk, Ars/Mosel, 15, 21, 27 November 1879 und Nachtrag am 19, 21, 23 August 1880, fixed price 90,000 marks.

25 Thyssen-Archiv, RSW, 640 00 C 1 f: Thomaspatent, Vertrag mit Luxemburger Bergwerks und Saarbrücker Eisenhütten Act. Ges. zu Burbach bei Saarbrücken, 30 January 1880, fixed price 120,000 marks. Re the Burbacher Hütte's (in Esch, Luxemburg) very cheap sources of pig iron, see Müller, *Überzeugung*, pp. 21f.

Stahlwerke, and thus for the time being did not need a licence of its own, there were no further licence takers from the south-western part of the Reich.[26] The interested works in this district had immediately realised their opportunity to take on steel production, thereby underlining yet again the Ruhr's fears.

The Ilseder Hütte was the first of the works not in the south-west to acquire the Thomas process, having supplied Hörder Verein with some of the pig iron for its trial of the process.[27] Like Lorraine pig-iron, Ilseder pig-iron was highly phosphoric and could be produced very cheaply. That it was extremely suitable for the Thomas process had been demonstrated by the trials in Hörde, and the next step for the Ilseder Hütte was to take over the manufacture itself. In so doing, Ilseder Hütte was basically acting just as the Georgs-Marien-Hütte in Osnabrück had done ten years previously, when it constructed its own steel-works on site to process its own Bessemer pig iron. The Ilseder Hütte's competition was so feared in the Ruhr that, in their case, the two licensers imposed a higher patent fee of 3.50 marks per ton, compared to the usual 2.50 marks per ton, as well as totally forbidding the production of heavy railway rails. Furthermore, Ilseder Hütte had to pay 300,000 marks, the highest premium to be fixed, to conclude the contract, since their future production was estimated considerably higher than even de Wendels'.[28]

It was only after this that the first works in the Rhineland-Westphalian area, namely the Aachener Verein, joined in, on 30 January 1880. In any case, at Aachener Verein, the introduction of the Thomas process was based on the premise that there was a greater price difference between Bessemer and phosphoric pig-iron from Aachen than was actually the case, and they soon

[26] This only altered when the Rheinische Stahlwerke's project to build, together with Röchling, a blast-furnace works on the Aldringen ore fields had collapsed. Röchling then acquired the licence for the Thomas process, in order to be able to work there under his own management. Thyssen-Archiv, RSW, 640-00 C 2 s: Thomaspatent, Vertrag mit Karl Röchling, 25 January 1889, fixed price 75,000 marks.

[27] Thyssen-Archiv, RSW, 640-00 Csk: Thomaspatent, Vertrag mit Ilseder Hütte, 24 January 1889, fixed price 300,000 marks, increased patent fee 3.50 marks/t.

[28] The patent fee was shortly reduced to 2.50 marks per ton for production amounting to over 21,430 tons, on condition that no rails weighing over 12 kg per metre (i.e. no railway rails) were to be made. Thyssen-Archiv, RSW, 123 00 4 56: Protokoll der Aufsichtsratssitzung am 25 Februar 1884. It appears from a letter by Director Massenez of the Hörder Verein that a fixed price of 20,000 marks for every 10,000 tons of presumed annual production was demanded of the licence-taker. In the case of the Ilseder Hütte, this would have meant a presumed annual production of 150,000 tons. At this period such quantities of steel could be manufactured in Germany only by Krupp. The Ilseder Hütte was thus credited with becoming as important using the Thomas process as Krupp had been using the Bessemer process. The level of the fixed price was, however, always a matter for negotiation, and consequently did not reflect an 'objective' capacity estimate, but the negotiating room was determined rather by the anticipated level of production. Thyssen-Archiv, RSW, 640-00 C 1 g: Massenez an Goecke am 27 Dezember 1879.

complained bitterly about this to the licensers, who had apparently given rise to false expectations.[29]

The Bessemer works in the Ruhr and the Osnabrücker Stahlwerk initially conducted a purely obstructive policy. They submitted proposals for invalidating the Thomas patent, at the same time as trying to get their own slightly modified variants patented as their own invention.[30] When this failed, they agreed that they would only negotiate jointly for the Thomas patent.[31] Behind this lay, not a desire to introduce the process as soon as possible, but much more their concern to secure the rights as cheaply as possible at a time when the process still did not involve any economic benefit. They considered that it would be thoroughly possible to develop the process further to make it cheaper; the low-cost acquisition of the patent, which in the present economic boom would not have presented a great burden, would then be to their advantage. However, the licensers were obviously speculating along the same lines and consequently did not enter into this transaction.

Their pact collapsed when Bochumer Verein broke away from their united front in March 1880 and took out a contract for the Thomas patent on the existing terms.[32] This surprising move did not, however, signalise any reduction in the cost of the Thomas process for the Ruhr works, but was based, as with the Aachener Verein, on a straightforward miscalculation. Their

29 Thyssen-Archiv, RSW, 640-00 C 1 g: Thomaspatent, Vertrag mit Aachener Hütten-Aktien-Verein Rote Erde, 30 January 1880, fixed price 40,000 marks. The Rheinische Stahlwerke had set their sights at Aachen on a price difference for pig iron of 30 marks per ton in favour of basic Bessemer pig iron. In fact, it amounted to only 18–20 marks per ton. However, even with this lower price difference, the Thomas process would, in Kirchdorf's (the Aachen director) opinion, have been profitable, if the fireproof lining to the converter had not worn out so unusually quickly. Thyssen-Archiv, RSW, 640-00 C 1 g: Kirchdorf to Rheinische Stahlwerke on 4 January 1880.

30 Thyssen-Archiv, RSW, 640-00 A-1: Goecke to board on 30 August 1879. There was nothing new about this manner of proceeding against seemingly important patents. In January 1878 Bochumer Verein, together with the Dortmunder Union and at that time the Hörder Verein as well, had appealed against the grant of a patent for Krupp's process for removing phosphorus using combined Bessemer and reverberatory furnaces. At the same time they agreed to provide the means for a trial furnace in order to be able to register a patent for their own slightly altered form of the process. The minutes of the board meeting give the impression that the whole thing was a routine matter, in that challenging foreign patents was one of the tasks designated to the board. Werkarchiv Krupp-Bochum, BV 12900 Nr. 11: Protokoll der Verwaltungsratssitzung am 29 Januar 1878; and similarly when the Thomas patent was announced, BV 12900 Nr. 12: Protokoll der Verwaltungsratssitzung am 30 Mai 1879.

31 Werkarchiv Krupp-Bochum, BV 12900 Nr. 12: (Bochumer Verein) Protokoll der Verwaltungsratssitzung am 5 Januar 1880. Thyssen-Archiv, RSW, 123 00 2.21: Protokoll der Aufsichtsratssitzung am 14 Januar 1880; 640-00 C 1 i: Denkschrift betreffend die bei der Vergebung der Licenzen für das Thomas-Verfahren beobachteten Grundsätze, von F. Goecke, 19 Januar 1895; 640-00 A-2: Anlage II zum Vertrage vom 11 März 1880 (= Contract with Bochumer Verein, see note 32).

32 Thyssen-Archiv, RSW, 640-00 C 1 h: Thomaspatent, Vertrag mit dem Bochumer Verein, 11 March 1880, fixed price 75,000 marks.

decision was based, on the one hand, on the substantial increase of prices for Spanish low-phosphorus iron ore, and on the other, on the technical manager's assurance that he would be able to make forty charges a day using the current Bessemer plant according to the Thomas process, as he had previously done according to the Bessemer process.[33] The latter conclusion proved deceptive. The technical manager, who had otherwise demonstrated his great organisational talents and had set the fastest Bessemer works in Europe on its legs, had to retract his statement as early as May, and was able to guarantee only 2,000 tons a month instead of the 5,000 to 6,000 tons. In the meantime, however, considerable funds had already been raised for materials and a lime-kiln, and the purchase of an orefield in Lorraine had been concluded.[34] A tremendous row broke out within the firm, which immediately decided to 'unburden' the technical manager of some of his responsibilities.[35] Since the fat was now in the fire, it was decided after a pause for thought, to continue with preparations for the Thomas process in spite of the loss-making trials.[36] The second Bessemer plant, which had been shut down, was renovated for this purpose and fitted out with bigger converters.[37] In the certainty that working the Thomas process would not become cheaper than the original Bessemer process, Bochumer Verein then finally started regular operations in April 1881, as the third works in the Ruhr to do so.[38]

Apart from the Rheinische Stahlwerke and Hörder Verein, no works in the Rhineland-Westphalia area had until this time considered adopting the Thomas process to be economically sensible. The Aachener Verein too, which had owned the rights for some time, shied away from making the changeover. The two licensers had granted it the advantage of paying Thomas a patent fee of only 1 mark per ton. However, this apparently still did not represent any

33 For a retrospective view, see Werkarchiv Krupp-Bochum, BV 12900 Nr. 12: Protokoll der Verwaltungsratssitzung am 5 Januar und 12 Mai 1880.

34 Ibid.: Protokolle der Verwaltungsratssitzungen am 28 Februar/23 März und 12 Mai 1880. 162,000 marks had already been paid out for the limekilns and material alone.

35 Ibid.: Protokoll der Verwaltungsratssitzung am 19 Mai 1880.

36 To this end, technicians were convened from Hörder Verein and Rheinische Stahlwerke to start up the Thomas process in Bochum. Ibid.: Protokoll der Verwaltungsratssitzung am 10 August 1880.

37 Protokoll der Verwaltungsratssitzung am 29 September 1880. Compare Macar, pp. 166ff.

38 Our version of the circumstances which led to Bochumer Verein taking on the Thomas process, differs considerably from that given by W. Däbritz in *Bochumer Verein Festschrift*, p. 215. Däbritz presents the acquisition of the Thomas patent and the adoption of the Thomas process as a well-considered successful action without any reservations on the part of the firm management. This, in our opinion, is in blatant contradiction to the sources. The attitude taken by the board to the adoption of the regular Thomas operation, 'even if the basic operation does not turn out cheaper than the acid operation [Bessemer], it is still desirable, since the production as a whole will be raised by it' is, in our opinion, redolent more of fatalism than of breaking new ground. Werkarchiv Krupp-Bochum, BV 12900 Nr. 12: Protokoll der Verwaltungsratssitzung am 28 April 1881.

advantage over the previous Bessemer process, since both works continued to operate their Bessemer plant as extensively as possible, and milled only a small proportion of their rails out of basic Bessemer steel.[39] This must have been a sure sign to their neighbouring competitors that, for the time being, the acid Bessemer process still brought in the more profitable business – a sign which the two works were naturally not the slightest bit interested in giving, since they had promised themselves very considerable incomes from the sale of the patents, which they had already secured in the case of the works in south-west Germany. Simply in order to obtain these incomes, it was already considered advisable to maintain a show-works and to work on perfecting the process. The favourable results of the Thomas process at Rheinische Stahlwerke and Hörder Verein were quite categorically determined by their income from patents which, in the final account, totalled over 3.4 million marks for each of the two firms.[40] Proof that it was possible to remove phosphorus in the converter was enough to induce the Maxhütte, for instance, to secure all the patent rights for Bavaria. Since they disposed of their own low-phosphorus iron ore, they did not consider using the Thomas process themselves. They were, nevertheless, determined to remove any threat to their extremely lucrative monopoly with the Bavarian State Railway, and permitted themselves to pay more than 300,000 marks to this end.[41]

In Great Britain, according to Trasenster's deposition, there was no site where the Thomas process could have had a clear cost advantage over the Bessemer process. Sheffield was the only place he ranked level with Westphalia, attributing to the Thomas process the same chances of success there as the Bessemer. There, too, was started the second trial operation with the new process, after Bolckow Vaughan. Under Thomas's direction, Brown, Bayley and Dixon started up the Thomas process using redundant plant, almost at the same time as Hörde and the Rheinische Stahlwerke.[42] As it was, this unhappy firm's fate was not of a nature to encourage any imitators. It went into liquidation in January 1881. The attempt to found the firm again in the summer of 1881 collapsed. It was finally bought by private individuals and, commencing in spring 1882, was started up once again as a conventional Bessemer works. The source of the bankruptcy was not linked to the

39 More acid than basic Bessemer steel rails were still being produced by the Rheinische Stahlwerke until the business year 1884/5. Thyssen-Archiv, RSW, 640-00 D b) 1: Aufstellung der Schienenproduktion 1879/80–1893/4. The Commissions-Bericht, p. 233, claimed that Hörder Verein was operating two of the three converters in its biggest plant according to the original Bessemer process, in order to be able to meet its current orders for rails.

40 Feldenkirchen, *Kapitalbeschaffung*, p. 45.

41 Thyssen-Archiv, 640-00 C 1 i: Thomaspatent, Vertrag mit Maximilianshütte, München, 6 April 1880. See too note 13.

42 Thyssen-Archiv, RSW, 640-00 A 1: Gilchrist-Thomas to Pastor, 31 October 1879; 640-00 A 2: Gilchrist-Thomas to Pastor, 2 March 1880. *ICTR* 23 April 1880, p. 460.

introduction of the Thomas process, but was, as a few years previously with the Rheinische Stahlwerke, the result of excessive stockpiling of pig-iron, purchased at boom prices.[43] The collapse of the pilot-works in Sheffield, which had been actively supported by Thomas, and the impossibility of convincing creditors of the Thomas process's commercial success so as to deflect their demands, must nevertheless have constituted a heavy mortgage. Furthermore, the leading works in the district, Charles Cammell and John Brown, had acquired iron mines in Spain, which they intended to exploit profitably in the long term using acid steelmaking processes.[44] Apart from which, the preceding years had shown that Sheffield was very unfavourably sited for selling steel rails and, in the event of a slump, would be the first to come under pressure. For this reason, new investment in this region was duly avoided, and even the latest rails boom of the years 1880–2 was taken up only within their existing capacity structure.[45]

Since in Great Britain, as in Germany, it had first of all to be established whether the Thomas process could manufacture a milder steel than that made by the acid Bessemer process, its widespread restriction to rail-making and the limited market associated with this had to be lifted before the new process could make its actual breakthrough. As a mere substitute competitor to the original Bessemer process, it was of no interest, since in the meantime the iron-ore shortage had stopped being the dominant issue and the British coastal works could not be underbid in their production costs for steel rails. Bolckow Vaughan also went on working its Bessemer plant using pig iron made of Spanish iron ore, and was building additional plant for the Thomas process.[46] Unfortunately, comparative computations for production costs have not been preserved in this firm's archives. For this reason, Trasenster's claim that the acid Bessemer process had by all means been cheaper in Cleveland cannot be directly checked.

Since Bolckow Vaughan had acquired extensive iron mines both in Spain[47] and, when the puddling works were flourishing, in Cleveland, the firm found it

43 *ICTR* 7 January 1881, p. 21; 18 March 1881, p. 332; 29 July 1881, p. 141; 2 December 1881, p. 704 and 31 March 1882, p. 363.

44 Flinn, thesis, pp. 81f., 98ff., 121. Flinn considered this link with low-phosphoric iron-ore supplies as an effective obstacle to the introduction of the Thomas process. Ibid., pp. 32f. His assessment is not shared here. It was due far more to the fact that the Bessemer process, as with the Siemens-Martin process later on, produced better results in Great Britain. See Chapter 5, s. 2. Having had a cheap and secure access to their own low-phosphoric iron ore, they did not even have to think about changing their steelmaking process. Compare the discussion about retaining the Bessemer process at Krupp's, pp. 76.

45 *BITA* 1880, pp. 28ff. *BITA* 1882, pp. 34–7.

46 BSC-Northern RRC, loc. no. 04902: report of directors, year ending 31 December 1879, 31 December 1880, 31 December 1881 and 31 December 1882.

47 Ibid.: report of directors, year ending 31 December 1875. The Spanish mines had been acquired in association with John Brown, Sheffield. See too J. K. Harrison, The Development of a

necessary to turn both low-phosphorus and phosphoric iron ore to account. Within the space of a few years the firm had accumulated enormous supplies of raw materials, turning the works into the biggest coal, iron and steel producer in the world.[48] These supplies naturally restricted the firm's freedom of movement and obliged it to turn them to account in places where cheaper alternatives according to different processes were also available. With the decline of the puddling process in Cleveland, the need for local iron ore was also considerably reduced. Thus, Bolckow Vaughan had decided to purchase eight blast furnaces as early as April 1879, in order to find an adequate outlet for their own iron mines, since their previous clients had not wanted to renew their former delivery contracts and selling on the free market did not appear profitable. The management had reckoned, in the event of having to reduce operations, on losing at least £20,000 a year.[49] Given that the iron ore was now to be processed into pig iron, it was necessary to find an outlet or process for it, in order not to feel the effect of these losses elsewhere. The preconditions for this were not very favourable, since the Cleveland pig-iron producers, to whom Bolckow Vaughan also belonged, had for some time been trying to combat the falling price of pig iron by means of concerted restrictions on production.[50] Of the shut-down quotas for blast furnaces established by the Cleveland Ironmasters' Association (CIA) the only ones to be excluded were those whose pig iron was processed to steel. This had previously only been possible with phosphorus-free pig iron. Using the Thomas process, it was now possible for Bolckow Vaughan to remove a part of its production of phosphoric pig iron, which it had increased due to lack of an outlet, from the quota allocation and endprocess it to steel.[51]

The flight from the iron-ore market had led to new investment in the end-processing stages amounting to several hundred thousand pounds, which the firm had assigned to the Thomas process very early on, and which, at this level,

Distinctive 'Cleveland' Blast Furnace Practice 1866–1875, in C. A. Hempstead (ed.), *Cleveland Iron and Steel, Background and 19th Century History* (The British Steel Corporation) 1979, pp. 81–115, here p. 112.

48 This was proudly announced in its Business Report for 1882. BSC-Northern RRC, loc. no. 04902: report of directors, year ending 31 December 1882.

49 BSC-Northern RRC, loc. no. 04898–04900; minutes book 5, Colliery Manager's Report, board meeting 28 February 1879.

50 The organisational framework for this was provided by the Cleveland Iron Masters' Association (CIA), which had been founded in 1866 to enable them to deal together with wage disputes with the workers' representatives. During the boom in the early seventies, the CIA scarcely appeared, but in 1876, however, it was reactivated on account of new pay disputes. BSC-London, CIA MSS: minutes book 1, meeting 18 December 1866, Rules of the Association; minute book 2, Rules as adopted at a meeting held 27 June 1876.

51 BSC-London, CIA MSS: report for year ending 30 June 1878 and report for year ending 30 June 1884. Minutes of meeting 17 January 1884 and 12 January 1885.

were passed only after determined opposition by some of the shareholders.[52] This controversial strategy was strengthened in April 1880 by a report about the firm's coalmines. In this it was stated that as much coal as possible should be processed into coke for the blast furnaces, since it could not be sold as household coal in sufficient quantities to be profitable.[53] Bolckow Vaughan had thereby become the prisoner of its huge raw-material production of around 1.5 million tons of phosphoric iron ore and two to three million tons of coal.[54] Even processing these materials into pig iron did not help, due to the restrictions on sales by the CIA and the poor prices obtained for Cleveland pig iron since the decline of the puddling works. The unparalleled forced construction of its basic Bessemer works with six 15-ton converters in the years 1880–1[55] must thus be regarded primarily as a strategy for ensuring that the raw materials were turned to account, and not so much as evidence that the Thomas process was particularly profitable in Cleveland.

As it was, when E. Windsor Richards, then director general of Bolckow Vaughan, at the Cleveland Institute of Engineers' annual meeting in November 1880, extolled the great productive capacity of the first basic Bessemer plant to go into production, he failed to mention production costs at all. The critical observer for the *Kärnter Zeitschrift* was thus completely correct in pointing out that 'All deductions derived from this assumption fall down, as soon as one considers that the new plant demonstrates just as much progress, as indeed it quite undoubtedly does, if it were dedicated to the Bessemer process according to the old method.'[56]

By 1881, the annual report to the Board was already clearly less enthusiastic about the introduction of the Thomas process. It said, looking back, that 'the process cost us, no doubt, a great deal more labour, anxiety and money than we expected it would when we commenced operations'. And since directors generally do not want to worry their shareholders, it added the proviso that they would nevertheless not regret having done so.[57] Remarkably enough, people at

52 By the end of the business year 1880 alone, £306,000 had already been paid out for blast furnaces and steelworks. BSC-Northern RRC, loc. no. 04902: report of directors, year ending 31 December 1880, p. 17. There, too, are the first heated debates about the Thomas process' prospects for success.

53 BSC-Northern RRC, loc. no. 04898–04900: minute book 5, Mr Armstrong's Report on the Company's Collieries, board meeting 16 April 1880.

54 BSC-Northern RRC, loc. no. 94902: report for directors, year ending 31 December 1880, p. 22.

55 Ibid., report of directors, year ending 31 December 1882, p. 15. J. K. Almond, Making Steel from Phosphoric Iron in the Converter – the Basic-Bessemer Process in Cleveland in the years from 1879, in C. A. Hempstead (ed.), *Cleveland Iron and Steel, Background and 19th Century History* (The British Steel Corporation) 1979, pp. 205f.

56 *Zeitschrift des berg- und hüttenmännischen Vereins für Steiermark und Kärnten* (1880), pp. 491f. and p. 494.

57 BSC-Northern RRC, loc. no. 04902: report of directors, year ending 31 December 1881, p. 12.

these often very heated general meetings continued to refer exclusively to the big profits, which the process was supposed to yield in the future.[58] At the opening of the Cyfarthfa Bessemer works in South Wales too, Edward Williams, the head of the Bolckow Vaughan steelworks, emphasised above all the security of his works' supplies in the event of a continental war, which was not indicated at this time, but he was silent on the subject of costs.[59]

The professional world remained sceptical. *Iron* set its sights on the original Bessemer process as before, and, in February 1882, it could still comment that in spite of all the technical success with the Thomas process, there was still no 'positive evidence' that it was also a commercial success.[60] This view was apparently shared by the management of the firm Darlington Steel & Iron Co. and of the Erimus Steel Works, both in the immediate neighbourhood of Bolckow Vaughan, since they went against their original intentions and operated their new steelworks, built in 1880, according to the acid Bessemer process after all.[61] It was shared too by the most influential man in the Cleveland iron industry, I. L. Bell, who abandoned yet again his plans for a basic Bessemer works.[62] They were justified when Bolckow Vaughan published a tardy self-critical assessment at its general meeting, held to discuss the business year 1886. Only then did the internal conflict about the correct strategy for the production of steel surface. One of the directors, Pochin, accused his colleagues at the General Meeting of having previously failed to make further investments in Spanish low-phosphorus iron ore, which had lost the firm big profits.[63]

His accusation could not easily be dismissed. Henry Davis Pochin, originally in the chemical industry in Manchester, was one of the best-informed persons in the British steel industry, in which he had invested his fortune.[64] He was on the board of two other Bessemer works in addition to Bolckow Vaughan, the traditional John Brown & Co. in Sheffield, which together with Bolckow Vaughan owned iron-ore mines in Spain, and the Tredegar Iron Co. in South Wales, which also processed Spanish iron ore. Apart from which he was

58 Ibid.: report of directors, year ending 31 December 1881 and 31 December 1882.

59 *ICTR* 20 July 1883, p. 84.

60 *Iron* 3 February 1882.

61 Almond, *Steel Production*, pp. 173f.

62 BSC-Northern RRC, loc. no. 04889: (Bell) minutes of board meeting 31 December 1880. The construction of yet another basic Bessemer steelworks was considered there, but in any case it never came to anything. I. L. Bell was director of the North Eastern Railway and later also of the Barrow Hematite Steel Co. He thus had very precise and first-hand knowledge both as a steel producer and as a steel consumer. His decision against adopting the Thomas process was therefore all the weightier, in that he had considerable problems selling his pig iron and also belonged to the CIA.

63 BSC-Northern RRC, loc. no. 04902: 22nd general meeting for the year ending 31 December 1886, p. 17 (Pochin).

64 About Pochin's far-ranging involvements, see Carr and Taplin, p. 86.

involved in the Sheepbidge Coal & Iron Co., in Derbyshire, which also owned further Spanish iron mines jointly with John Brown & Co.[65] In Cleveland he was further involved with Palmer's Shipbuilding and Iron Co., which started production of open-hearth steel in 1885, based on Spanish iron ore.[66] There were indeed few men in Great Britain who disposed at first hand of such comprehensive insider knowledge about the comparative profitability of the different processes for producing steel as did Pochin.

Finally, Arthur Cooper, the works manager of the North-Eastern Steel Co., admitted, looking back in the nineties, that the big railway companies were always more ready, as they still are, to buy acid or hematite steel rails (= acid Bessemer rails, U.W.) and indeed at a lower price, than to obtain basic steel rails, due to the great stocks of hematite iron ore which are present on the West Coast, and on account of the low price at which Spanish hematite ore is delivered both to South Wales and to Middlesbrough. Consequently, the basic process (Thomas process, U.W.) was developed only to a slight extent in Great Britain.[67]

Only one other basic Bessemer works, the North-Eastern Steel Company run by Cooper, was built in Cleveland in the neighbourhood of Bolckow Vaughan. In any case, this firm, which was founded on 9 July 1881, was in essence the creation of two inventors, S. G. Thomas and P. C. Gilchrist, who had otherwise encountered very little enthusiasm for their process in Great Britain, and were forced to promote its general use very much at their own expense. Their basic Bessemer plant was equipped with four 10-ton converters. The works, in which rails and blocks were mainly to be produced, was built according to the most modern ideas. It came into operation in the summer of 1883 and went on producing until 1919, without eliciting a great deal of respect.[68] Nothing is known about their production costs.

When the Cleveland iron industry did change over to steel, in the end it chose the open-hearth process. Bolckow Vaughan and the North-Eastern remained the sole producers of basic Bessemer rails in Great Britain and soon tried to penetrate into other processing areas with their basic Bessemer plants. The fact that this did not bring the success they hoped for had its origins in the composition of Cleveland pig iron, which, though to some extent suitable as the primary material for railmaking, was not suitable for mild steel, which was in far greater demand.

[65] Flinn, thesis, p. 121. In the seventies, long before it built a Bessemer works, Tredegar had already smelted Spanish ore in its blast furnaces. Glamorgan CRO, D/DG C.4: (Tredegar Iron Works) Cost of Pig Iron. Sept. 1878–Oct. 1879.

[66] *BITA* 1885, p. 40.

[67] Here according to Thyssen-Archiv, RSW, 640-00 Db) 4: Der Thomasstahl in seiner Bedeutung als Schienenmaterial, Entwurf nach den Beschlüssen der Commissionssitzung am 29 Mai 1895, p. 7.

[68] Almond, *Making Steel*, pp. 207–11 and pp. 217ff.

(b) *Expanding into new markets: mild steel*

The production of this very desirable mild steel in the basic Bessemer works was countered initially by an insufficient understanding of the metallurgical process in the converter. The steel workers thought at first wholly along the lines of the original Bessemer process, in which silicon was a valuable fuel and phosphorus an impure element. They consequently used a pig iron containing, as well as phosphorus, as much silicon as possible. In any case, however, this was harmful. Silicon attacked the converter's basic lining and so had to be bound by adding large quantities of lime. In addition, there was always a trace of silicon left in the finished steel, which made it harder insofar as it acted like a higher carbon content. Although basic steel produced in this way was suitable for rails, and the innovation did at first aim at this obvious market, it was not suitable for wire, plates and tubes and the many other products which for the time being were still produced from puddled iron.

It was only in 1880 when the Hörder Verein came to realise that a pig iron with the lowest possible silicon content, but one rich in phosphorus and, if possible, also in manganese, was the easiest to refine in the basic Bessemer converter, and which furthermore turned into mild steel without any further efforts, that the way was pointed for the Thomas process's future development. In any case, only those sites where the production of this kind of pig iron at a competitive cost was possible, were able to participate in the further expansion of basic steel production. Cleveland, which had originally seemed to promise the greatest prospects, did not appear to be one of them.

Although the basic Bessemer works there had immediately grasped the advantages of low-silicon pig iron, they did not succeed in manufacturing it.[69] Reducing the silicon content increased the sulphur content of Cleveland pig iron, which made it useless for steelmaking.[70] Iron ore with a manganese content, which was introduced successfully at Hörde to counteract the sulphur, was not available in Cleveland. This apparently had an effect also on their ability to produce mild steel for shipbuilding plates. In view of the strongly contracted market in steel rails, both the basic steel producers there started producing 'soft steel' for use in shipbuilding in 1884.[71] Whereas the material delivered from the North-Eastern did at least meet some of the shipyards'

69 Windsor Richards in a discussion in the Iron and Steel Institute on 6 May 1880, in: *JISI* 1880 part 1, p. 100.

70 In Cleveland, 1 per cent silicon content involved 0.16 per cent sulphur, and 0.5 per cent silicon content as well as 0.24 per cent sulphur. The first variant had too much silicon for the Thomas process, and the second too much sulphur. Almond, *Making Steel*, p. 202.

71 BSC-Northern RRC, loc. no. 14800–14900: (Bolckow, Vaughan & Co.) minutes book 7, board meeting 23 May 1884. Almond, p. 210. B. Martell, The Present Position occupied by Basic Steel as a Material for Shipbuilding (lecture in the Institution of Naval Architects), in: *Iron* 19 August 1887, pp. 176f.

requirements, although its variable quality offered too little security, the catastrophically poor quality of the steel delivered by Bolckow Vaughan led, on 17 December 1885, to a general prohibition by Lloyd's against the use of basic steel for shipbuilding.[72] This meant that both works were still to a great extent limited to the rails market. Cleveland had been clearly devalued as a site for the Thomas process, in contrast to its initial high hopes.

In a lecture to the Society of Arts in April 1882, in which they referred in particular to the production of mild steel from pig iron with a high manganese content, Thomas and Gilchrist thus no longer pointed to Cleveland, but to Scotland, Lincolnshire and Staffordshire as the most favourable sites for the Thomas process in Great Britain.[73] The pig iron in these districts was rich in phosphorus and manganese but poor in sulphur, and for this reason was suitable for the production of mild steel according to Hörde's example. Cleveland pig iron was supposed, on the other hand, to be less suitable than Staffordshire pig iron 'because it contained three times as much sulphur, and in order to make it suitable they were obliged to add much manganiferous ore, which was expensive and added to cost'.[74]

Shortly afterwards, basic Bessemer works were built in Scotland and Staffordshire, which were intended right from the start for the production not of rails, but of mild steel only, and they followed the Hörder production methods faithfully. In Staffordshire Thomas and Gilchrist, together with Alfred Hickman, founded the Staffordshire Steel & Ingot Co. Ltd in Bilston with a capital of £70,000.[75] Hickman had several blast furnaces for the production of pig iron in Bilston, whose existence was threatened by reversal in the fortunes of the puddling process. The production of pig iron in Staffordshire had fallen steadily during the previous decade. In 1881 only a third of all the blast furnaces were still working.[76] In view of this situation, which threatened his works, Hickman allowed trials using the Thomas process to be conducted in the closed Bessemer plant in the nearby Patent, Shaft & Axletree Co.[77]

In the autumn of 1884, the new works started up with three 6-ton converters.[78] After initial difficulties over the use of siliceous pig, they changed over, as in the Ruhr, to very cheap 'cinder iron', in which up to 80 per cent of

72 Whereas Martell does not name the firms, though their identities emerge clearly from his lecture, *Stahl und Eisen*, the organ of the Verein Deutscher Eisenhüttenleute, was not deterred from naming the luckless British competition. Basischer Stahl im englischen Schiffbau, in: *Stahl und Eisen* 7 (1887), pp. 614ff.

73 *Thomas und Gilchrist*, p. 375.

74 P. C. Gilchrist in a discussion at the South Staffordshire Mill and Forge Managers' Association, in: *ICTR* 1 December 1882, p. 600.

75 *ICTR* 18 August 1882, p. 158; 1 September 1882, p. 212.

76 *Mineral Statistics*, 1881, p. 77. *ICTR* 2 November 1883, p. 549.

77 F. W. Harbord, *The Thomas-Gilchrist Basic Process, 1879–1937*, in: *JISI* 1937 part 1, p. 84. *ICTR* 2 November 1882, p. 600.

78 *ICTR* 1 December 1882, p. 600.

the iron input fed to the blast furnace consisted of old puddling slag. Vast quantities of this old puddling slag or 'tap cinder' lay close at hand, since the area had been a centre for the wrought-iron manufacture, and cost only *c.* 25s. for a whole bargeload if it was not simply given away in return for the transport costs.[79] Most of it was smelted into pig iron, from which the desired 'soft steel' was made in the basic Bessemer converter. This steel was sold above all to the tinplate manufacturers of the region, who since the early eighties had been going over to steel instead of puddled iron for their starting material.[80] After a truly catastrophic start – the steam boiler plant exploded shortly after the works' opening[81] – the firm proved thoroughly capable of survival and expansion. It was only in the 1920s, when the works were taken over by Stewarts & Lloyd's, that the Thomas process there came to an end.[82] The demise of the puddling process meant that its essential raw material stopped being supplied.[83] A further blast-furnace works in the Midlands to be threatened by the dwindling demand for wrought iron was the Lilleshall Iron Co. in Shropshire, which started producing mild basic steel with two converters for the needs of the district in 1885.[84]

In Scotland, too, wholly similar circumstances led to the construction of two basic Bessemer steel works. Before it was known that the old puddling slag was excellently suited for the production of basic steel, plans to found one basic Bessemer steel works there had already fallen through. The planned 'Caledonian Steel Company' had not succeeded in raising the required capital because too few shares had been assigned.[85] Here too, however, the conditions were altered by the Scottish blast furnaces' growing sales problems and the possibility of processing their previously valueless slag-heaps. The Glasgow Iron Company in Wishaw, which had already had to shut down one of its blast furnaces, decided in 1883 to build a basic Bessemer steel works with 7-ton converters, similar to the Staffordshire Steel & Ingot Co., which started up in 1887. This decision had been preceded by trials conducted in the works using a small 3-ton converter, in order to test the utility of pig iron made out of puddling slag and a mixture of local iron ore.[86] The Glasgow Iron Co. owned slag-heaps accumulated from thirty years' work, whose iron content was estimated at between 150,000 and 200,000 tons.[87] This waste, which had

79 Harbord, *Thomas-Gilchrist*, p. 86.
80 *ICTR* 18 January 1884; 20 June 1884, p. 766.
81 *ICTR* 5 June 1885, p. 764.
82 Carr and Taplin, p. 225.
83 Harbord, *Thomas-Gilchrist*, p. 86.
84 *Iron* 16 October 1885, p. 353.
85 *ICTR* 6 October 1882, p. 363; 14 March 1883, p. 321.
86 Ibid. and *BITA* 1887, p. 115.
87 *JISI* 1884 part 1, p. 201.

previously only been an inconvenience, was to be processed in the unemployed blast furnaces into pig iron suitable for the basic Bessemer converter.

This pig iron had what was for cinder iron a very low phosphorus content of little more than 2 per cent, but otherwise it was similar to Cleveland pig iron, apart from its low sulphur content, which was much valued following the unfortunate experiences in Cleveland. However, it became apparent in the course of operations that, as in Cleveland, its silicon content was too high.[88] For this reason the works had to be closed down again in 1894.[89] It may be supposed that the supplies of slag also vanished at this point, so that they were faced with a drastic rise in the price of pig, making it impossible to operate at a profit any longer. On the other hand, in terms of quality, the steel fulfilled all expectations. In 1887 it was the first basic steel to pass the stringent tests set by Lloyd's Register and was given exceptional permission to be used for specific purposes in shipbuilding.[90]

Merry and Cunninghame in Glengarnock was the second works in Scotland to decide to build a basic Bessemer steelworks with four 10-ton converters, in order to secure an outlet for their blast furnaces which had been built in the forties.[91] Of the existing nine blast furnaces, three or four were to be employed for producing the necessary phosphorous pig iron. Thus the annual production envisaged for the steelworks was not very high, going by what four converters could consume. R. M. Daelen, one of the leading German steelworks specialists made the following comment about the whole plant: 'in general the constructions are kept as simple and economical as possible, which could be done because fewer demands were made on its productive capacity than in the case of new plant in the Middlesbrough district' (in Cleveland, U.W.).[92] The steelworks came into operation in 1885 and supplied the tinplate manufacturers in South Wales in particular. Once the prohibitive McKinley tariffs had almost completely wiped out tinplate exports to the USA, the firm immediately went over to producing shaped steel girders for the construction trade.[93]

The annual capacity of both Scottish basic Bessemer steelworks, with seven converters with capacities of 7 to 10 tons between them, was now set at the modest level of 100,000 tons, thereby still not achieving even the same capacity as the North-Eastern in Cleveland, which, with four 10-ton converters, had an annual capacity of 125,000 tons.[94]

88 Payne, *Colvilles*, p. 71. *ICTR* 14 March 1883, p. 321.
89 The plant was shortly afterwards equipped for the Bessemer process. Payne, *Colvilles*, p. 71.
90 Martell, p. 177.
91 Payne, *Colvilles*, p. 69.
92 R. M. Daelen, Über neuere Bessemerwerke, in: *Zeitschrift des VDI* 29 (1885), p. 1016.
93 Payne, *Colvilles*, p. 69.
94 Daelen, *Bessemerwerke*, pp. 555 and 1018.

The four basic steelworks described above had assumed a special role insofar as they were not set up specifically for using phosphorous iron ore, as was generally the case with the Thomas process. By processing puddling slag they were employing a waste product which had almost no value in the traditional iron districts, but which occurred only in limited quantities. Since they were actually driving the puddling works out of the market with their mild steel, in the long term they were depriving themselves of their basic raw material.[95] Thus, they could from the start only be provisional in nature. The more they produced, the briefer their life expectancy was. Given their limited future prospects, the works were from the start not organised for the sort of mass production which would have been possible at the time, in terms of available technology. They were consciously looking for a niche in the market which could guarantee the continued existence of their threatened blast-furnace works. The requisite capital was limited to what was absolutely necessary. This is particularly clear in the case of the Staffordshire Steel & Ingot Co. and the Glasgow Iron Co., which contented themselves with producing steel ingots and billets.[96] In this way, the whole of the Glasgow Iron Co. works, with three converters including the steam boiler plant and the mill for ingots and billets, and thc building too, were calculated to cost £30,000.[97] Compared with that, the Phönix works in the Ruhr had to pay *c*. 800,000 marks (£40,000) for its basic Bessemer plant alone, which also had three converters, but this did not include the mill etc.

The Caledonian Steel and Iron Co., which collapsed during its inception phase, was to have introduced the Thomas process to Scotland in great style and at the peak of its technical potential it required a base capital of £250,000,[98] which was almost four times the total sum raised in Staffordshire. It correspondcd to the amount raised for the North-Eastern, which was founded with £200,000, with construction costs including land purchase in the order of £140,000.[99] However, the necessary security for such an undertaking was apparently lacking in both Staffordshire and Scotland. The construction of these four basic Bessemer works based on slag thus provides no clue as to whether there had been any change during the eighties in people's basic opinions about the profitability of the Thomas process as compared with the alternative steelmaking processes. In steel production, people kept where possible to processing low-phosphorus iron ore from Cumberland and Spain.

95 Harbord cites this to explain the disappearance of the British basic Bessemer steel works at the beginning of the twentieth century. Harbord, *Thomas-Gilchrist*, p. 86.

96 See note 80. The Glasgow Iron Co. further processed their billets in their Motherwell mill. This rolling mill had originally been connected to a puddling works. However, it could be employed without further ado for mild basic steel. *ICTR* 14 March 1883, p. 321.

97 *ICTR* 14 March 1883, p. 321.

98 *ICTR* 6 October 1882, p. 363.

99 Almond, *Making Steel*, pp. 207ff.

In spite of initial fears, the Thomas process did not make any headway against the open-hearth process in the manufacturing of mild steels either; this process will be discussed at length below.

In Germany, the Rhineland-Westphalian front, which had hitherto been almost wholly united, collapsed once Hörder Verein had demonstrated that the usual Westphalian pig iron was best suited to the production of mild steel in a basic converter, if Siegerland iron ore with a manganese content and slag were loaded into the blast furnace. Early in 1881 and in brief succession, the three steel producers with their own blast furnaces and puddling works contracted for the licence with Rheinische Stahlwerke and Hörder Verein, and acquired the Thomas patents.[100] Phönix, GHH and Dortmunder Union were now, just like Hörder Verein, in a position to process their puddling slag, which had previously merely been a nuisance, by means of a slightly altered blast furnace process, as well as produce their own Thomas pig iron, which was best suited for making mild steel.

Since GHH and Dortmunder Union both owned two Bessemer plants each, they could introduce the Thomas process without having to give up the acid Bessemer rails business, which was at the time very lucrative. The Dortmunder Union set up, alongside its running Bessemer plant, a plant from the Heinrichshütte, which had been closed long ago, and started working it on a trial basis as early as January 1881 – two months before acquiring the patent rights.[101] GHH, however, delayed starting the Thomas process. It was so extraordinarily well provided with contracts for acid Bessemer rails that it was only able to meet them with the help of its second plant, which was closed until 1881.[102] At GHH, they had until then still not understood how to operate their Bessemer plant as intensively as at Dortmunder Union. Although both firms' Bessemer works were almost identical in construction, the Dortmunder Union's production capacity was always assessed higher.[103] It was only when the rails boom had passed its peak, thus freeing the capacity of its second

100 Thyssen-Archiv, RSW 640-00 C s 1: Thomaspatent, Vertrag mit Dortmunder Union, 26 March 1881, fixed price 75,000. Ibid., 640-00 C 2 m: Thomaspatent, Vertrag mit GHH, 5 April 1881, fixed price 60,000. Ibid. 640-00 C s n: Thomaspatent, Vertrag mit Phönix, 8 April 1881, fixed price 50,000 (marks).

101 HA-Krupp, WA IV 835: Bilanzen, Jahresversammlungsberichte verschiedener Werke, Aktiengesellschaften, Banken, etc., Dortmunder Union, Generalversammlung am 10 Dezember, 1881, pp. 12f.

102 GHH-Archiv, 300 108/8: Protokoll der Aufsichtsratssitzung am 30 März 1881. According to reports by the AR, a total of 48 charges had to be made with two plants in 24 hours, in order to meet the demand at the lowest possible costs. During the following months, however, scarcely more than 42 charges per day were achieved. Rapid driving was not one of the GHH's strengths. GHH-Archiv, 300108/9: Berichte für den Aufsichtsrat, Juli 1881 bis Juni 1882.

103 Macar, pp. 150–3. This was expressed in the Dortmunder Union's higher cartel quota in the Schienengemeinschaft (Rails Association) and in the higher fixed price for the Thomas patent. See note 100.

Bessemer plant, that GHH began making basic steel in the autumn of 1882.[104] The additional costs of changing a Bessemer plant over to the Thomas process totalled 40,000 to 50,000 marks, and were described as a 'comparatively slight expenditure'.[105]

In Germany, then, unlike Great Britain, the discovery of mild basic steel did not at first give rise to any new foundations. Instead, the existing steelworks took the first available opportunity to abolish the restriction to railway rails as almost the only mass product, and to get out of a market which had been almost completely frozen into cartels. Here, too, the opportunity of straightaway processing large supplies of puddling slag in cheaply fitted-out plant gave the works the time and experience they needed to procure new markets by diversifying production. Unlike Great Britain, they were not squeezed out by sharper competition from the open-hearth works. Unlike there, too, they had in the Minette and Siegerland iron ores a local base ore which was suitable for making mild steel. Thus, knowing that their supplies of slag would run out did not mean the same as losing their basic raw material. When the phosphorous ores finally came out of Sweden in the nineties, and the less valuable Minette ores thus lost their significance for the Ruhr, this was just a bit of luck and not the result of any strategy formulated in the early eighties. At that time, unimpeded access to the Minette area lay right at the heart of the common policy conducted by the Rhineland-Westphalian works.[106] Once high-value Swedish iron ore came in, however, they immediately stopped this pressure, and previous adherents of the Mosel canal project were transformed into its equally determined opponents.[107] The decisive factor here, as opposed to Great Britain, was that the production of basic steel was considered from the start as a permanent arrangement on an ore base, and not as a temporary and lucrative way of making money out of waste.

(c) Optimising the material input[108]

In the meantime the Thomas process's starting problems had been overcome. The Directors' Report of 26 September 1882 for Rheinische Stahlwerke commented that 'The Thomas process has, in the course of one year, given rise

104 GHH-Archiv, 300 108/10: Protokoll der Aufsichtsratssitzung am 9 November 1882.

105 GHH-Archiv, 300 108/9: Berichte für den Aufsichtsrat, pp. 73–6.

106 Feldenkirchen, *Eisen- und Stahlindustrie*, pp. 73–6.

107 Gertrud Milkereit, Das Projekt der Moselkanalisierung, ein Problem der westdeutschen Eisen- und Stahlindustrie, in: *Beiträge zur Geschichte der Moselkanalisierung*, Cologne 1967, *passim*.

108 The following section must unfortunately be limited to a discussion about the rationalisation process in Germany, since the British sources are silent on this theme. It may, however, be concluded from the development of the basic Bessemer plant's production capacity at Bolckow Vaughan that there were at least similar processes operating.

to no further considerable technical advances, but its operation has become wholly regular.'[109] An important part of this regular operation had been provided by further developments to the converter's fireproof lining, which had initially presented the greatest problems in the course of operations. At the Aachener Verein, this had constituted grounds enough for once again interrupting trials with the Thomas process. Although pig iron proved to be more expensive there than originally supposed, the Thomas process would nevertheless have been profitable had the fireproof lining to the converter lasted better.[110] At the Rheinische Stahlwerke this problem had finally been solved by applying a mixture of dolomite and tar instead of the usual pure dolomite bricks. This proved not only very much more durable and reliable, but was also much cheaper to manufacture.[111] The durability of this fireproof lining soon gave them an advantage over the English basic Bessemer works, as was confirmed by Gilchrist when he visited the Rheinische Stahlwerke in April 1883.[112] Unfortunately, he did not discuss the extent to which the greater wear and tear to linings in England was due to the higher silicon content of pig iron there.

Although it was not possible to achieve the same production capacity with the Thomas process as with the Bessemer process, considerable advances were made up to 1882. For this reason alone, it was not possible to realise the same production quantity, because in the Thomas process 18 to 20 per cent of burnt lime had to be added to the converter, which reduced its capacity for pig iron accordingly.[113] At Bolckow Vaughan their large quantities of slag were the reason for setting up 15-ton converters instead of the 10-ton converter usual for the Bessemer process, which delivered almost the same amount of steel. As the silicon content of the pig iron fell, so too did the amount of slag, so that the German works could count on a more favourable input ratio. The converters in the Thomas process could not, however, take on more than 80 per cent of the quantity of pig iron possible in the acid Bessemer process.[114] When one takes this into account, the 5,000 tons production, which the Dortmunder Union could already aim at in March 1882 using one basic Bessemer plant with two

109 Thyssen-Archiv, RSW, 126 00 10: Bericht des Vorstandes der Rheinischen Stahlwerke für die ordentliche Generalversammlung am 26 September 1882, p. 5.

110 Thyssen-Archiv, RSW, 640 00 Clg: Kirchdorf (Aachener Verein) to Rheinische Stahlwerke on 4 January 1881.

111 Thyssen-Archiv, RSW, 126 00 9: Bericht des Vorstandes der Rheinischen Stahlwerke für die ordentliche Generalversammlung am 7 Oktober 1881, p. 7.

112 Thyssen-Archiv, RSW, 126 01 11: P. C. Gilchrist to Rheinische Stahlwerke on 21 April 1883.

113 Hermann Wedding, *Die Darstellung des schmiedbaren Eisens in praktischer und theoretischer Beziehung*. Erster Ergänzungsband. *Der Basische Bessemer- oder Thomas-Process*, Braunschweig 1884, p. 195. GHH-Archiv, 300 108/9: Berichte für den Aufsichtsrat, 18 April 1882.

114 Wedding, *Darstellung*, p. 195.

9.5-ton converters, was very considerable.[115] It involved more than twenty-six charges per day and was thus faster than the acid Bessemer plants at Krupp's or GHH in the same period.[116] In that month, Dortmunder Union made *c.* thirty-four charges a day in the same works using acid Bessemer plant of identical construction.[117] Given the same organisational and technical conditions, the acid Bessemer process thus still retained the lead. The wearisome work-rate of 1880 only two years beforehand when nine to ten charges at most were made in a day, and the night was spent doing repairs,[118] had nevertheless been superseded. Continuous day and night production had very quickly prevailed.

When the works introduced the Thomas process they were able to fall back on the technological and organisational potential of their existing Bessemer operation. Long-drawn-out learning processes, as when the organisation of production for the Bessemer process was developed, were no longer necessary. People now had precise ideas from the start about the standard of production organisation which could be attained, and the way to achieve it. All that was disputed, as the quarrels within Bochumer Verein demonstrated, was the speed with which this could be achieved. In this, the standard of the organisation of production achieved in Bessemer operations played a considerable role. A 'rapid' works in the original acid Bessemer process such as Dortmunder Union was able to report success earlier than other works. GHH, always rather ponderous, took two years to achieve the same production tempo in August 1884, which Dortmund had already demonstrated in March 1882.[119] In any case, Bochumer Verein won back its pioneering role in the Ruhr after a very short interlude. The 'unburdening' of its technical manager bore fruit. He lived up to his reputation as one of the best organisers in the case of the basic Bessemer operation as well, and by early 1882 he could already manage up to forty charges using the old Bessemer plant, which had been altered for the Thomas process.[120] His own doubts notwithstanding, he then succeeded in keeping to his original pronouncement, which had made the administrative

115 GHH-Archiv, 300 108/9: Berichte für den Aufsichtsrat, 18 April 1882.

116 According to figures produced by the head of his Bessemer works, Krupp could make with all five plants a maximum of 17,750 tons a month. The converters could take 5.6 tons, resulting in just 26 charges a day. HA-Krupp, WA VII f 484: Bessemerwerke, Stahlproduktion, p. 7. The highest average monthly performance at GHH was in March 1882, when 21.73 charges per day were achieved, 300 108/9: Berichte für den Aufsichtsrat, März 1882.

117 See note 115. A further point to note is that one of the two plants in Dortmund came from the Heinrichshütte, where it had lain cold for years. In Dortmund, only the old 5-ton converters had been replaced by bigger 8-ton converters, as in Bochum, which, however, were filled with 9.5 tons input. Wedding, *Darstellung*, p. 193.

118 *Zeitschrift des berg- und hüttenmännischen Vereins für Steiermark und Kärnten* 12 (1880), pp. 233 and 241.

119 GHH-Archiv, 300 108/9: Berichte für den Aufsichtsrat, August 1884. GHH had adopted the process in September 1882.

120 Macar, p. 167.

board decide in favour of the Thomas process. The conditions of entry for the Thomas process had thereby been re-established.

On account of the longer and more frequent repair work to the basic Bessemer converters, Bochumer Verein had set up a third converter in addition to the two which were already established. This was to achieve a situation whereby, as with the Bessemer process, one converter was always working. Given a length of thirty minutes per charge, this organisational principle gave rise to a possible maximum forty-eight charges per day.[121] In the case of Bochumer Verein, this had very nearly been achieved. The third reserve converter had contributed considerably to this good result, since it helped prevent the big repairs problems during the first stage of the Thomas process from slowing down the converter operation. The working tempo of the Bochumer 3-converter plant can thus not be directly compared with the Dortmunder 2-converter plant, as was the case with the Bessemer process at the end of the seventies, when the repair problems there were almost solved. Nevertheless, the work tempo in Bochum must have been regarded as unusually fast. Aachener Verein, which, just like Bochumer Verein, had enlarged its old Bessemer plant with a third converter and re-equipped it for the Thomas process, could achieve only twenty-eight to thirty charges per day.[122] Nor, indeed, were they as experienced in the Bessemer process there as in Bochum.

All the basic Bessemer plants mentioned here had come into being mostly as a result of very slight modifications to existing Bessemer plants. This meant that the capital requirement had been slight and, in any case, was far less than that needed for new plant. Furthermore, the production organisation in these plants had within a short space been raised to a standard only slightly inferior to that in acid Bessemer plant. The decisive factor in the Thomas process's rapid breakthrough in the Ruhr was the existence there of so many closed Bessemer plants from the seventies. The memory of those experiences meant that managers were far more hesitant about erecting new works, except in the south-west of the Reich. Consequently, the energetic expansion of basic steel production at Bochumer Verein and Dortmunder Union appears to have been especially impressive in that, though both these works were recognised as having an efficient acid Bessemer operation, they decided against extending it in favour of the Thomas process.[123]

We cannot agree with Feldenkirchen's opinion, that 'the devaluation of the Bessemer plants, most of which were built at the beginning of the 1870s,

121 Ibid. Werkarchiv Krupp-Bochum, BV 12900 Nr. 12: Protokoll der Verwaltungsratssitzung am 28 September 1880.

122 Macar, p. 168.

123 C. Lueg from GHH stressed this in his comprehensive report about the Thomas process. GHH-Archiv, 300 108/9: Berichte für den Aufsichtsrat, 18 April 1882.

following the introduction of the Thomas process', had brought the Ruhr steelworks into economic difficulties.[124] On the one hand, the devaluation of the plant did not result from the introduction of the Thomas process, but from the preceding rationalisation of the Bessemer operation, which effected an increase in capacity. On the other hand, the value of these plants rose again following the discovery of the Thomas process, since they were available, subject to minimal alterations, for the new branch of production. The majority of these 'old' plants was at least as modern as, if not better than, many new constructions from the early eighties. Their greatest advantage was, however, that they were well known to the technical management and the operating personnel, and thus could be brought back into blast at full tilt, almost in their present state. In addition, they offered the inestimable advantage in the initial phase, that they could be worked according to both the Thomas and the acid Bessemer processes. Innovation costs and risks were thereby correspondingly slight and were of immediate assistance in averting the economic difficulties which could ensue from the changeover.

If we look at how the concept of the basic Bessemer works altered in the course of the eighties, it is clear that the existence of previous Bessemer plant very certainly prevented the Ruhr works from investing at the start of the decade in works which, a mere three or four years later, would be deemed superannuated. Looking at the early new constructions in Great Britain, Belgium and the south-west of the Reich, we find once again all the characteristics of a Bessemer works specialised for a single product. There is nothing extraordinary about this, since these works had initially also been planned for the production of steel for rails. With the opening of the new market for mild steel, however, the demands on the organisation of production in the steelworks had also changed; this altered production organisation was soon reflected in a new concept for basic Bessemer steelworks.

At the beginning of the eighties, with the advent of the new Bessemer works in South Wales and the first basic Bessemer works, the development potential of Bessemer's basic concept of two or three converters and a central circular casting pit, was exhausted. Alexander Holley developed this concept further, resulting in a considerably speeded-up refining operation, which was, however, limited by the time needed by the pit crane to teem a charge of liquid steel into ingot moulds. When the Thomas process was introduced, this problem was rendered more acute on two fronts. On the one hand, the changeover was used to bring bigger converters into play. Since the time taken for refining did not depend on the quantity of steel, the new 10-ton converters delivered twice as much steel as the old 5-ton converters. In order to achieve the same number of charges a day, the pit crane had to process twice as much

124 Feldenkirchen, *Eisen- und Stahlindustrie*, p. 95.

steel. To do so, it needed, unlike the converters, twice as much time. The teeming process could not be speeded up, even when teeming into larger ingot moulds, above 300 kg per minute, since the ingots acquired defects when the steel was squirted in at speed.[125] A 5-ton charge could accordingly be teemed in seventeen minutes, which meant that teeming as well as changing over the ladle took just about as long as blasting a charge in the converter and the further addition of *Spiegeleisen*. Thus, to teem a 10-ton charge, the pit crane needed at least thirty-four minutes which, including changing over the ladle, came to a good three-quarters of an hour. This meant that the maximum number of charges using a big converter was limited to thirty-four a day, whereas with a 5-ton converter under otherwise identical conditions, about fifty charges a day were possible.

Unlike the transforming processes in the blast furnaces and converters, enlarging the receptacles for shaping the steel, the ingot moulds, did not result in a corresponding increase in capacity. Furthermore, casting mild steel, which could be made without any problem in the basic converter, took even longer than casting hard Bessemer steel. This could only be improved by providing more casting equipment. Bochumer Verein, followed by the Bochumer Gesellschaft für Stahlindustrie and by Hoesch, had already shown the way of the future with the Bessemer process, when they enlarged the circular casting pit by means of extended casting trenches.[126] Unlike the circular pit with its central crane, in the casting trenches the steel could be teemed simultaneously from a number of waggons. The casting capacity could be increased quite simply by adding more waggons. Constructional bottle-necks, as presented by the closed, circular casting pit, now no longer existed. The casting trenches could, as the need arose, simply be lengthened.[127]

This new concept involved a strict spatial division between the individual work stages in the steelworks. Each stage was given its own separate part of the building, and conditions were established to enable smelting the pig iron and Spiegeleisen, conversion, casting and repairs to take place as independently of one another as possible. This was to allow each of these activities to be organised along optimal lines, as well as enabling the entire works to react less sensitively to disturbances in the separate compartments. A communications network, which acted simultaneously as a buffer between the individual compartments, was provided by a system of rail transport, which was to a great extent autonomous. The basic Bessemer works was sectioned off into several buildings.

125 Daelen, *Fortschritte*, p. 388.

126 Macar, pp. 164–7. See too above p. 94.

127 Daelen, *Fortschritte*, p. 318: he emphasised these advantages in his fundamental article about bottle-necks in steelworks. The bottle-neck in the teeming process had in the meantime become more disruptive than the repairs to the fireproof materials.

Given such an arrangement, it was now no longer possible for the works manager to get an overview, in the actual sense of the term, unlike in the original compact plant, which combined all aspects of the process in one hall.[128] It presupposed a works organisation which allowed each individual compartment to operate on its own account. The way to this system had led via the 'rapid process' which had been developed in Bessemer plants since the late sixties. It had thereby become increasingly impossible for the works manager to direct and control every individual stage in the working process, as Bessemer still hoped when he envisaged his process. Rather, he had to rely increasingly on the individual smelters, teemers etc. themselves knowing what to do at the decisive moment. Following the drastic de-qualification of the personnel involved in steel production when puddling ceased, what was now needed was a new competence on the part of the foremen at least, in order to establish the conditions for the rapid process. On the basis of these newly taught skills, it then became possible to consider sectioning the steelworks into different, mainly self-sufficient departments. These notions were consequently translated into buildings for the first time in Hörde and above all Peine, and put into operation in 1882.[129]

In this refining shop, the basic converters were set up in a row. There were no cranes at all. Instead, a steam-driven casting car carrying a ladle ran on rails in front of all the converters. It took over the finished charges from the converters and transported them in its ladle out of the converter shop into the casting shop, where the steel was teemed into moulds in the extended casting trenches alongside the railway. This removed an impediment common to the converter and casting operations; furthermore, the enormous heat generated in previous Bessemer works, which had made it almost impossible for workers to remain in the casting pit for a great many charges, was considerably reduced.[130] Basically, the Hörder Verein's basic Bessemer works differed from the Peine rolling mills shown in Figure 14 only in that it was designed for four instead of six converters, and its casting trenches were still included in the converter shop.[131]

Extending this kind of works was much easier than before. By separating the converter and casting equipment, both could be increased or enlarged in relation to existing conditions, independently of one another. The transport system, which, unlike the fixed cranes, was available for all the converters, could be adjusted along predetermined steps in accordance with the works'

128 In the case of Phönix's old Bessemer works, they had furthermore set up an extra podium in the middle of the shop, since the many cranes and bogies would otherwise have prevented the works manager from getting a general overview. Macar, p. 115 and planche 13.

129 Beck, *Geschichte*, p. 662. Macar, pp. 158–64.

130 Daelen, *Fortschritte*, p. 339.

131 Macar, p. 159.

THOMASHÜTTE.
der Actien-Gesellschaft „Peiner Walzwerk" in Peine.

Giess Hütte

Schmelz-Hütte

Giess Hütte

Giess Grube

Eisenbahn für die Giesspfanne

Frisch-Hütte

14 The Peiner Walzwerk's Thomas works in Peine

rising total production. In the same way individual converters could be added or even removed without making any alterations to the casting equipment.

The Hörder Verein started with two converters in its new works and extended it in stages to four converters.[132] In Peine they started with three converters, adding a fourth in the course of the eighties.[133] Subsequently, both works were further extended by enlarging their converters from 10 tons to 15 tons or 18 tons.[134] Unlike the previous Bessemer works, individual converters and casting cars could also be put out of operation without breaking the sequence of steel production. The works had acquired greater adaptability (see Figure 14).[135]

The advantages of this new steelworks concept lay rather in its greater adaptability where extensions and alterations to steelmaking were concerned, than in any reduction in running costs. The compact old-style plant had the advantage of very short transport routes and consequently low energy requirements. If one put up with their limited output, then there was no reason why they should not continue to exist economically for a long time yet alongside the new bigger plants. The last basic Bessemer plant to be built in Germany with a central crane and a circular casting pit was at the Phönix works, erected in 1883, and it went on working for almost 80 years in the Ruhr. Apart from electrifying the block cranes, it had not been necessary to make any essential alterations to keep it competitive.[136] This shows how mature converter plant built according to Bessemer's and Holley's concept already were by the beginning of the 1880s. They presented units, which had been optimised to a high level, and whose production costs were hard to underbid. The extreme specialisation and integration of the individual parts of the plant, which were splendidly married up, provided the conditions for its most cost-effective production. As long as this plant was used for making one product only, ingots for rolling steel rails, they were scarcely ever surpassed. This was especially the case in the USA, where this type of plant was brought to perfection. It was only with the advent of the Thomas process with its greater product variability that its immobilism was felt.

The steel works' increasing product differentiation, which liberated them from their almost exclusive dependence on the rails market, allowed the firms to insist on plant which, instead of being optimised for one special product, could produce a wider spectrum at, on average, lower costs. The age of simple

132 Hermann Wedding, Die Fortschritte des deutschen Eisenhüttenwesens seit 1876, mit besonderer Berücksichtigung des basischen Verfahrens, in: *Stahl und Eisen* 10 (1890), pp. 939 and 941.

133 Wedding, *Darstellung*, p. 193.

134 Beck, *Geschichte*, pp. 1030f.

135 Illustration taken from Wedding, *Darstellung*, Tafel III.

136 Hermann Brandi, Entwicklung der Thomasstahlerzeugung in Europa und die bauliche Ausgestaltung der Thomaswerke, in: *Stahl und Eisen* 74 (1954), pp. 1263f.

cost comparisons was thus over. As earlier with the wrought-iron works, it was now necessary to distinguish between steel ingots of different sizes and compositions.[137] Furthermore, the optimal lot size, which in the Bessemer works practically coincided with their total production, now played a major role. It is therefore hardly surprising that the annual production of the new plant at Hörde and Peine was not ahead of all the other German basic Bessemer works, ranking instead below conventional works. In 1887 the Aachener Verein, Phönix and Dortmunder Union at least, produced more or just as much with their three converters each.[138] In 1890 Phönix could make up to 33 charges a shift in its plant, thus reaching the high 'American' tempo of the most rapid Bessemer works, whereas Peine could achieve only 28 charges per shift at most with one converter more.[139] The special attribute of this new generation of steelworks did not, in the first place, reside in any increase in their production capacity.

(d) The importance of the Thomas process in Germany at the end of the eighties

By 1890, the structure of the German steel industry had to a great extent already been determined by the Thomas process. Basic steel, being made predominantly in mild qualities, was well suited to most steel users in Germany, who did not make the same demands on the quality of the steel as the British shipyards did. Indeed, German 'Thomas steel' was as little capable of meeting their quality requirements as was the British.[140]

It was only in the case of commercial iron products, wire and intermediate products (excluding the hard sorts of steel) that the Thomas process achieved its best results in technical as well as price terms. This market was by far the biggest for German heavy industry, whereas in Great Britain it came after rails and shipbuilding materials. During the eighties, wire even pushed rails out of their first place in exports by the German iron-processing industry. Since 1886, wire made of basic steel was cheaper than that made of wrought iron, in virtue of which it soon took over the market.[141] The same thing happened with *Handelseisen* (commercial iron) and intermediate products although the Reich

[137] As, for instance, in the comparative inquiry by C. D. Wright, United States labor commissioner, here according to *Iron* 22 August 1890, p. 164. Carnegie and Gladwyn went a consequential step further in their handbook on the manufacture of steel. Due to the multitude of different ingots they only calculated the cost of the liquid steel in the ladle, that is, before it was actually teemed. David Carnegie and Sidney Gladwyn, *Liquid Steel, its Manufacture and Cost*, New York 1913, chapt. 17.

[138] According to Beck, *Geschichte*, p. 1008.

[139] Wedding, *Fortschritte*, p. 941.

[140] Martell, p. 176.

[141] *Statistisches Jahrbuch für das Deutsche Reich* (1907) 2, pp. 90–3 and pp. 265ff.

statistics do not enable a direct comparison to be made in this instance, since the entries for commercial iron products made of *Schweisseisen* (wrought iron) and *Flusseisen* (Thomas-, Bessemer- and open-hearth steel) do not match up. Bar iron, one of the most common intermediate products from wrought iron, was accounted to commercial iron products, whereas the associated steel billets or plates, which were often put to the same use, were listed separately with bars and ingots. Since 1886 the latter group had formed the largest share, and was also the cheapest form in which steel was sold.[142] On the other hand, what was listed in the statistics as commercial iron products made of *Flusseisen* only became cheaper than the same made of wrought iron in 1890. If one adds commercial iron products, wire and ingots, billets etc. together, by 1890 this group already included more than half of the steel produced in Germany, whereas rails furnished only a quarter, and steel plates only just 10 per cent.[143] The structure of the demand for steel in Germany was thus considerably different to that in Great Britain, where the demand for these three groups of products was about equally strong and for this reason alone had led to a differently structured steel industry.

The fact that the Thomas process, by producing mild steel, enabled the steelworks to diversify their products over and beyond the railways' requirements also endangered the stable cartel conditions of the seventies. These had allowed the Schienengemeinschaft (railmakers' association) and its adjoining small-tyre and wheelmakers' association to regulate practically the entire internal steel sales.[144] Even the up and coming production of railway sleepers, which for the time being were still made of wrought iron, but were soon to be produced from open-hearth steel following the Bochumer Verein's example, had by 1887 already been divided up into fixed quotas by the German Schwellengemeinschaft (sleepermakers' association) by 1877.[145] The wheelmakers' and sleepermakers' associations were, in this situation, no more than appendices to the mighty railmakers' association. In the early eighties, after the Prussian railways had been nationalised, they had *de facto* only one tenderer

[142] Ibid.

[143] Ibid.

[144] The Räder- or Radsatzgemeinschaft was always mentioned only in connection with the Schienengemeinschaft. Thus, before the Schienengemeinschaft was re-formed in 1876, the railway companies' tenders for wheels were only discussed among the larger works along general lines. HA-Krupp, WA IV 1045: Briefwechsel zwischen Otto Adams, Bochum, und C. Hagemann und O. Schnabel, Essen btr., Preisvereinbarung Krupp–Bochumer Verein 1874/75. It emerges from this that agreements were always made for individual transactions only and not on the basis of firm quotas. However, distributions according to firm quotas were made here too, in the eighties at the latest, which even reveal different rates for steel tyres and wheels. HA-Krupp, WA I 209: Notizbuch C. Uhlenhaut, pp. 4f., Beteiligungen. Werkarchiv Krupp-Bochum, BV, 12900 Nr. 14: (Bochumer Verein) Protokoll der Verwaltungsratssitzung am 26 Juni 1884.

[145] *Die Burbacher Hütte 1856–1906*, Saarbrücken 1906, p. 49.

and few clients left to deal with. In comparison, the slight quantities of steel which were produced outside the cartel and not for the railways were wholly insignificant. A degree of regulation had thereby been imposed on the German market, which would hardly ever be achieved again in peacetime. However, the foundations for this were removed when the Thomas process diversified steelmaking.

The new mass products made of basic steel, especially wire and semi-finished products, found a widely dispersed outlet among an unmistakably large number of takers. It was now no longer possible to consider regulating the market for producers and consumers along generally acceptable lines. What was more, wire and unfinished wares were also on offer from the wrought-iron works in large quantities and, unlike rails, differences in quality apparently did not play an important role here. In any case, there are no points of reference for a discussion about the relative advantages of iron or steel nails, girders, horse-shoes, shafts, ploughshares, etc., which could have compared even approximately with the debates about iron and Bessemer rails. It is much more likely that habit and direct price comparisons determined buying policy here.

The conditions for forming a successful cartel of the new basic steel products were thus far less favourable than they had been ten years earlier when the Schienengemeinschaft had been formed by the Bessemer works. A way had to be found to integrate the numerous wrought-iron works into a common alliance, without, from the basic steel works' point of view, freezing the steel-works' share, which tended to rise, over a long period. The attempt in 1885 to form a Verband deutscher Walzwerke (Alliance of German Rolling Mills) with its seat in Berlin, which was to regulate the entire production of steel and puddled iron, insofar as it was not already included in existing alliances, lasted only until the statutes were drafted.[146] They never reached agreement because, significantly enough, four big wire makers whose industry was just changing over from iron to steel, withheld their consent.[147] The year after, however, when the market had finally decided in favour of steel wire, this did not stop them forming a wire convention, which protected the slight price advantage of steel wire over iron wire, and was sufficient to drive the latter out conclusively in the early nineties.[148] A determinedly stable cartel came into being, as in the case of the rails market ten years previously, only when the few steelworks dominated the market. In the case of intermediate products, it took a few years longer to drive the wrought-iron works out, which meant that no further

146 This draft is published in Heymann, pp. 325–37.

147 *Berg- und hüttenmännische Zeitung* 44 (1885), *Montanproductenmarkt*, 9 October 1885, p. 433.

148 *Berg- und hüttenmännische Zeitung* 44 (1886), *Montanproductenmarkt*, 29 January 1886, p. 46. In 1893, the production of iron wire still amounted to just 58,000 tons, whereas in the same year 395,000 tons of steel wire were pulled. *Statistisches Jahrbuch 1907*, pp. 264–7.

binding agreements came into being here in the nineteenth century.[149] In this case too, one of the most essential preconditions for a successful cartel had first of all to be established; an equality of interests, which arose out of equal manufacturing conditions.

Thus, once they controlled the internal market with one product, the steelworks then attempted to replace free competition with set prices and quantities. This was supposed to prevent 'ruinous competition', which had been excluded from the previous markets by means of contractual agreements but which had used the Thomas process to spread into the new markets, from destroying the precarious equilibrium within the industry. This equilibrium did not rest on the economic efficiency of individual firms, as expressed in low production costs, but on the works' technical production capacity, which reappeared in the cartel quotas, irrespective of any cost comparison. In consequence, it cannot be positively stated whether the strong increase in the steelworks' capacities began immediately before the steelworks' alliance was founded in 1904, creating once again the need for a cartel, which led to disciplinary action against outsiders, or whether it was already intended to set up positions for the battle over the allotment of quotas. In any event, the formation of this second comprehensive cartel for the German steel market was not unexpected.[150]

(e) The exceptions: Krupp and Osnabrücker Stahlwerk

As the only Bessemer works with their own blast furnaces in the Rhineland-Westphalian district, Krupp and the Osnabrücker Stahlwerk had refrained from acquiring patent rights. Both of them disposed of their own low-phosphorus iron ore for the Bessemer process – Osnabrück on site and Krupp in Northern Spain. Both Krupp and Osnabrück had belonged to the group of Rhineland-Westphalian steelworks, which wanted to deal privately only with patent owners,[151] and Osnabrück had, for its part, furthermore challenged in law the grant of the Thomas patent, though without success.[152] For this reason it is very dubious whether it ever at that time planned to introduce the Thomas process. It is much more likely to have belonged among the protagonists of a hard obstructionist policy similar, perhaps, to conditions in the USA, where all the Bessemer works had together purchased all the patents for the de-phosphorising process in order to block them.[153]

149 Heymann, pp. 254f.

150 Compare Feldenkirchen, *Eisen- und Stahlindustrie*, pp. 143 and 264.

151 Mannesmann-Archiv, P 1.25.35: Protokoll der Aufsichtsratssitzung am 18 November 1879. To this group belonged: Krupp, Dortmunder Union, Bochumer Verein, Hoesch, GHH, Osnabrücker Stahlwerk, Königs- und Laura-Hütte, Phönix.

152 Müller, *Georgs-Marien*, pp. 81f.

153 Temin, *Iron*, pp. 183f.

Due to its location, the Osnabrücker Stahlwerk was tied to processing low-phosphorus pig iron from its parent firm, the Georgs-Marien-Hütte.[154] It would have had to procure the right sort of material for the Thomas process from the Ruhr area, as it already had to do for its fuel.[155] Given that all the raw materials needed for steelmaking would have had to be transported to Osnabrück in this way, the high freight costs would have scuppered any sort of competitive production there. Consequently, Osnabrücker Stahlwerk remained limited to its acid Bessemer production according to its previous schedule of 40,000 to 50,000 tons per annum, which was almost without exception rolled into rails.[156]

Krupp too evinced no interest in changing his five Bessemer plants over to the Thomas process. Together with Dowlais, Consett and the Spanish firm of Ybarra, Krupp owned the Orconera Iron Ore Co., which exploited the most lucrative deposits of low-phosphorus iron ore in Spain close to Bilbao.[157] Krupp and Dowlais both had a claim on a third of production, and Consett and Ybarra on a sixth each. The iron ore was delivered at cost plus 1s.7d. free on board to each shareholder up to the quantities stipulated in the contract: 200,000 tons each for Krupp and Dowlais and 100,000 tons each for Consett and Ybarra. Any extra deliveries were calculated according to market prices.[158] The Carlist risings in Spain during the seventies had prevented Orconera from delivering as much as had been envisaged, so that it was only in the eighties that Krupp was able to profit fully from the cost advantage of having his own ores.[159] From 1882 at the latest, Orconera was in a position to deliver more than the quantities contracted and, from this source alone, Krupp could reckon firmly on enough cheap iron ore for *c.* 100,000 tons of Bessemer pig iron a year.[160] On top of this came this exceptionally profitable company's dividends.

154 While the official merger of the two firms, which had cooperated from the start, only took place in 1885, the Georgs-Marien-Hütte had already held the majority shares since 1882. Furthermore, the Osnabrücker Stahlwerk was used for training the directors of the Georgs-Marien-Hütte and had thus always been a sort of branch concern. Müller, *Georgs-Marien*, p. 82.

155 Ibid., pp. 82ff. The small puddling works with eight furnaces for making sleepers had been closed down due to superior competition on the part of sleepers made of basic steel. Sleepers were not produced from acid Bessemer steel, since it was not mild enough for this purpose, nor indeed for many others.

156 Ibid., pp. 77f. Contrary to original expectations, the local coal supplies were not suitable. The freight costs which this entailed for the firm were considered a very great burden.

157 Manuel Gonzalez Portilla, *La formacion de la sociedad capitalista en el pais vasco (1876–1913) 1*, San Sebastian 1981, p. 69 *et passim*. Flinn, thesis, p. 112.

158 Glamorgan CRO, D/DG C.3: Orconera Iron Ore Co. Ltd., Krupp contract. Flinn, thesis, pp. 123f.

159 In 1880 Orconera was still unable to deliver the total quantity of 600,000 tons stipulated in the partners' contract. Flinn, thesis, p. 125.

160 Glamorgan CRO, D/DG E.3: report for the year ending 31 March 1884. Flinn, thesis, p. 289, claims that Krupp had received between 600,000 and 800,000 tons a year from Orconera. However, this is totally exaggerated. Given that 2.1 tons of iron ore were needed for 1 ton of

The £50,000 share capital per shareholder was offset by an average annual dividend of £13,229 in the eighties.[161]

This cheap supply of raw materials for the acid Bessemer process, which Krupp alone enjoyed in Germany, meant that he operated under terms different to those for the rest of the works in the Ruhr. In boom times especially he alone could reckon on a secure supply of raw materials at low prices, calculated in advance. His entire blast furnace and steelworks operation could be optimised for a long-term tied-up combination of raw materials. Thus Krupp was able, during the rails boom in the second half of 1881, when the competing works had just decided to change over to the Thomas process, to produce his Bessemer pig from Spanish ore at an average price of 59.00 marks per ton, whereas GHH, for instance, although its Oberhausen site was equally good for transport, had to expend on average 63.59 marks per ton for its.[162] The market price lay at the time as high as 65 to 69 marks per ton.[163] During the business year 1884/5 the corresponding costs at Krupp's amounted to 47.56 marks per ton, and at GHH to 52.22 marks per ton. During the same period, GHH's own Thomas pig iron cost, in comparison, 42.02 marks per ton.[164] If one then adds patent costs of 2.50 marks per ton and the difference in conversion costs between the Thomas and Bessemer processes, it becomes obvious that, at the time, the Thomas process was of benefit to GHH, but not to Krupp. Table 21[165] provides a hypothetical comparison for July 1885.

pig iron, this would have meant a production of 300,000 to 400,000 tons of Bessemer pig iron, while the other ores included are scarcely taken into account. Krupp never achieved such a production in the period under question. Even in 1891, when Orconera produced one million tons of iron ore, and there was thus no shortage, Krupp only received 220,000 tons. HA-Krupp, WA I 209: Notizbuch Uhlenhaut, p. 68.

161 Glamorgan CRO, D/DG E 2: profits from 1 January 1864–31 March 1889. Dowlais had the same number of proprietary shares as Krupp.

162 HA-Krupp, WA I 859: Umfang und Selbstkosten der Schienenproduktion 1875–91. GHH-Archiv, 300 108/9: Berichte für den Aufsichtsrat, 1881. The July figures are missing from our computation of the GHH's average costs for pig iron, which, however, can scarcely have altered the picture.

163 Bundesarchiv, R 13 I, Nr. 170: Bericht des Geschäftsführers des VDESI an die Generalversammlung am 14 Dezember 1882, p. 7.

164 HR-Krupp, WA I 859: Umfang und Selbstkosten der Schienenproduktion 1875–91. GHH-Archiv, 300 108/12: Monatsberichte für den Aufsichtsrat.

165 Data for GHH and Rheinische Stahlwerke according to GHH-Archiv, 300 108/13: Carl Lueg to Vahlkampf on 21 August 1885. Wedding, *Darstellung*, p. 193. Krupp-Archiv, WA I 859: Umfang und Selbstkosten der Schienenproduktion 1875–91. The price difference for Krupp included in this table is naturally only hypothetical, since Krupp did not make any Thomas pig iron. It is based on the difference between the production costs for Krupp's Bessemer pig iron and the GHH's Thomas pig iron. The GHH's blast furnaces in Oberhausen and Krupp's Mülheimer Hütte and Johannishütte were sited not far from each other, so that the cost of their raw materials at the mine must have been about equal. Wedding assumed incorrectly that a patent fee of 3.50 marks per ton was paid. This level was, however, levied exclusively from the Ilseder Hütte and has been corrected for our comparison. As licensees, the Rheinische Stahlwerke only had to pay Thomas a fee of 1 mark per ton, whereas all the other works apart

GHH would subsequently have derived an advantage of 3.12 marks per ton by using the Thomas process, without taking the patent fee into account, and 0.62 marks per ton when taking it into account. With Krupp, on the other hand, the Thomas process would only have turned out more favourable than the Bessemer process if the excess costs of conversion had not exceeded 3.71 marks per ton; this, however, was unthinkable as was demonstrated by the comparison with Rheinische Stahlwerke, whose production using the Thomas process was recognised as opportune. Since in Krupp's case the patent fee has yet to be taken into account, he cannot under any circumstances have felt any compulsion to change his steelworks over to the Thomas process.

Given that Krupp also possessed the most efficient rails mill in Germany, the general result was that his production costs for Bessemer steel rails lay constantly below the export price, until the International Rail Makers' Association collapsed in 1886.[166] This was not at all the rule in Germany. Before the international cartel had been formed, other works had already reverted to their practice in the seventies and had sold abroad at prices below their average production costs.[167] For these works, the Thomas process presented the opportunity for un-coupling their steel production from the rails market and for replacing their 'losing price' exports, which were always considered precarious, with making wire or semi-finished goods.[168] Krupp, on the other hand, was able to follow the prices dictated by the British works on the foreign markets.[169] What was more, these foreign markets almost always preferred acid Bessemer rails to basic Bessemer rails.

The railways' terms of delivery had been formulated with regard to the original Bessemer process and their experiences with Bessemer rails. There was no need to adjust these delivery terms for the new basic steel, as long as enough acid Bessemer steel could be secured at a price not above that for basic

from Ilsede had to pay the 2.50 marks demanded by Rheinische Stahlwerke and the Hörder Verein. When calculating the lowest costs of Thomas pig iron as opposed to Bessemer pig iron, an input ratio of 1,100 kg (Bessemer) to 1,165 kg (Thomas) was established for the pig iron, reflecting the higher wastage incidental to the Thomas process. Thomas und Martinwerke, in: *Stahl und Eisen* 6 (1886), p. 657.

166 HA-Krupp, WA VII f 941: Schienenwalzwerk; FA H III B 194: Verwaltung Allg. 75-01, Kosten und Rentabilität.

167 GHH-Archiv, 300 108/9: Bericht für den Aufsichtsrat, Dez. 1881. Werkarchiv Krupp-Bochum, BV 12900 Nr. 14: (Bochumer Verein) Protokoll der Verwaltungsratssitzung am 4 Oktober 1883 und am 29 November 1883. Müller, *Georgs-Marien*, p. 82.

168 This was, for instance, the declared strategy of the GHH. GHH-Archiv, 300 108/9: Bericht für den Aufsichtsrat, Mai 1882.

169 The production costs for Bessemer rails at Krupps in the business year 1884/85 amounted to 86.56 marks/t. whereas the IRMA at the same time guaranteed a price of 93 marks/t., ex works. HA-Krupp, WA I 859: Umfang und Selbstkosten der Schienenproduktion 1875–91. The GHH, on the other hand and in spite of its Thomas process had costs of 109.64 marks/t., and thus lay considerably above the cartel price for exports. GHH-Archiv, 300 108/12: Berichte für den Aufsichtsrat 1884/5.

Table 21. *Varying cost inducements for introducing the Thomas process to the Ruhr in 1885*

	Excess costs of conversion in the Thomas converter	Reduced costs of Thomas pig compared with Bessemer pig
GHH	5.31 (+ 2.50 patent)	8.43
Rhein. Stahlwerke	4.45 (+ 1.00 patent)	
acc. to Wedding	4.60 (+ 2.50 patent)	
Krupp		3.71
(all prices in marks per ton)		

Bessemer rails. This meant that basic Bessemer rails only made a breakthrough with railway companies which derived an income from providing transport for basic steelworks and thus had an interest, over and above the actual purchase price, in the production of basic rails. In Great Britain, the North-Eastern railway was such a company, since Cleveland's two basic steelworks lay within its district.[170] In Germany, they were the whole Prussian State railways and the Reichsbahn in Alsace-Lorraine.[171] Export contracts, on the other hand, often insisted expressly on the use of acid Bessemer rails, and it cost the German basic Bessemer works a great deal of trouble over the years to convince the railway companies that their steel was compatible.

Since Krupp's intention was to employ his rail rolling mills to capacity by turning out *c.* 100,000 tons per annum for the export market in particular, and he was able to produce the acid Bessemer rails it wanted as profitably as before, he did not at first have any motive for reducing his acid Bessemer steel production and replacing it with the Thomas process. Owning Spanish ore meant that his manufacturing conditions resembled those of British works in one important point more than they did those of his immediate neighbours. He thus continued, as in the 1860s to orientate his investment plans according to British examples as well. In 1882 his plan to combine blast furnaces and steelworks was resumed. As in 1874–5, the idea was to bring the converter operation up to the technical level of the newest works in South Wales. At the same time, the site in Essen was to be abandoned on account of its unfavourable transport conditions, and both its steel and rolling works were to be moved to the blast furnaces on the Rhine. There, they would on the one hand save on

170 A. Cooper of the North-Eastern Steel Co., according to Thyssen-Archiv, RSW, 640-00 D b) 4: Der Thomasstahl in seiner Bedeutung als Schienenmaterial, Entwurf nach den Beschlüssen der Commissionssitzung von 28 Mai 1895, p. 7.

171 Ibid., pp. 17f.

freight expenses, and on the other be able to work directly from the blast furnace.

A proposal by Longsdon envisaged the erection of a Bessemer plant with three 10-ton converters and the removal of the existing rails mill. The total costs for this were estimated at 1,420,700 marks, although the old works' existing machinery and equipment were to be re-employed as far as possible.[172] A further estimate for a plant with two 10-ton converters and a cogging mill, without moving the rails mill, started at 920,000 marks.[173] Thus, the plans at Krupp's were, as before, in terms of a lavish, modern plant. Although these plans were never realised, they do show clearly how Krupp continued to be orientated by the Bessemer practice in South Wales. The relationship of just one Bessemer plant, with two or three 10-ton converters and the existing rolling mills, with an annual capacity of 100,000 tons allows us to conclude that he had in mind a forced production similar to that in the new works there.

Although Krupp did not envisage carrying out this project in the near future and continued to operate his Essen works, his alternative investment strategy was once again very 'British' – extending his open-hearth steelworks. In this he, together with his arch-rival the Bochumer Verein, was ahead of the competition in the Ruhr, and the two former crucible steel producers made sure of their lead in the quality steel range, whereas the rest of the works in the Ruhr were at first extending their mass steel base with new Thomas works.

When Krupp finally, in the mid-nineties, did start a new steelworks along the Rhine in order to avoid the expense of freighting pig iron to Essen and of re-smelting it there, the preference was given to the Thomas process. This was not due to a tardy perception on his part that retaining the Bessemer process had been a short-sighted mistake. The plant in Essen was now almost thirty years old and, although it had been modernised in the sixties, it was basically still the early type with small 5-ton converters. In the meantime, however, the newer basic and acid Bessemer works were operating with converters up to three times its size. Thus rebuilding was in any case due, as the plans at the start of the eighties had already shown. However, it was only when Krupp could no longer keep up with the rail prices set by the competition, that the long-delayed decision was made.[174]

172 HA-Krupp, WA VII b 125: Estimate of costs for the erection of three 10-ton crucibles utilising the machinery etc. of the then existing Bessemer works. Transferring the preparing rolls and rail rolling mill together with the workshops belonging thereto. 'Crucibles' should be understood as meaning converters. At Krupps too, *Tiegel* were frequently referred to in German texts, when converters were actually meant. It is possible that this unusual form of speech was a result of the secrecy surrounding the early Bessemer operations, when it was still called 'wheel work' to disguise it, at a time when Krupp was still making wheels out of crucible steel.

173 HA-Krupp, WA IV 980: Kostenanschlag über ein Bessemerwerk, direkt vom Hochofen arbeitend, mit Vorwalzwerk, 16 February 1882, Schmitz.

174 Feldenkirchen, *Eisen- und Stahlindustrie*, p. 181.

The security of future access to sufficient quantities of raw materials proved decisive in choosing the process. Security, however, was something the Spanish iron-ore mines, which were still highly profitable, could no longer offer. It was already clear that supplies in Northern Spain would soon run out. In order to regulate the supply of this ore over a longer period, since it was, above all, indispensable for high-quality steel, the Orconera partners had decided in 1887, together with a neighbouring company, to restrict mining operations.[175] Since Krupp was anyway the Orconera's biggest client, there was no question of raising his quota. His fear was rather that the other partners would insist on resuming parity shares in the future. Any expansion in steel production at Krupp's could not be borne by Spanish ores. On the other hand, they were indispensable for making artillery and ship-building steels. Thus, the production of Bessemer steel had to be throttled, and not extended by modern plant. Krupp was nevertheless looking for an ore base for his new works not in his own country, and certainly not in the Minette district; instead, he adapted his works for processing Swedish iron ore, which could only be carried out according to the Thomas process. As it was, his attempt to buy into the ore mines in North Sweden, in order to achieve the same security of supplies as he had hitherto enjoyed with Spanish ores, proved unsuccessful.[176]

2 Open-hearth steel

(a) The early development of the Siemens-Martin process

The Siemens-Martin process's early development, in which its specific advantages over the converter process had not yet come to bear, lasted until the late seventies. During this time it competed, mostly unsuccessfully, with the acid Bessemer process in the production of rails. Although Siemens-Martin rails were equal in value to Bessemer rails, they were always a bit more expensive to produce. Their decisive disadvantage at this time was that the open-hearth works were not able to match the wave of rationalisation in the Bessemer works by taking equally cost-effective measures. This meant that the unexpectedly successful rationalisation of the Bessemer works was primarily responsible for the fact that initial sales expectations for Siemens-Martin steel were not fulfilled, thus restricting it to market niches. The assumptions on which these expectations had been based and the specific limits of the process which prevented them from being realised will be briefly examined in the following pages.

175 Flinn, thesis, p. 128.

176 Ulrich Wengenroth, Raw material ventures: multinational activities in acquiring and processing of iron ore before the First World War, *European University Institute, Colloquium Papers 158/84*, Florence 1984, pp. 18f.

Where the organisation of production in the steelworks was concerned, it differed most significantly from the Bessemer process in the duration of its charges. This depended on the size and composition of the ingredients and took several hours.[177] Given the extra small repairs to the furnace bottom which were necessary after each charge, it was not possible to make more than two to three charges per day. In 1880 Jeans qualified ten charges as 'a fair week's work'.[178] This meant that the open-hearth furnace was considerably slower than even the puddling furnaces, which it however surpassed in capacity and in the last resort in productivity too. Since the refining process, in this case, occurred of its own accord, the quantities of steel which could be produced in an open-hearth furnace were not limited by a workman's physical abilities. Unlike the Bessemer process, however, it was practically impossible to increase production without enlarging the whole furnace. Repairing the bottoms and re-heating the furnaces between charges did not take half as long as the charge itself. Macar mentions three to four hours for repairs and re-heating, compared with seven to eight hours for the charge, as being the average in the seventies.[179] Even if all interruptions had been eliminated, its capacity would have expanded by just 50 per cent.

The enormous potential for rationalising the Bessemer process at the end of the 1860s set a 20-minute charge time against an interruption of two and a half hours; this could never be the case with the Siemens-Martin process, and there was never any question of 'rapid driving' in an open-hearth works. Consequently the Siemens-Martin process developed along quite different lines to the Bessemer process. There is scarcely any comparison between the forms taken by their separate production organisations and the way they developed. This, however, applied only to the actual refining operation. Where teeming the ingots and subsequent rolling were concerned, of course, the same conditions applied as in the Bessemer and Thomas works, since in this case it did not matter which refining vessel the liquid steel came from. For this reason big open-hearth works corresponded entirely to converter steelworks in their manner of endprocessing the steel.

A few difficulties arose when determining the costs of production. On account of the multitude of usable raw materials, and especially the constantly changing proportion of scrap metal, it was not possible to draw up any cost accounts comparable to those in Bessemer works, which would have been characteristic of the whole industry. In the case of the Bessemer and Thomas processes, this was only possible because very narrow limits had been set on the raw materials employed in these processes. In particular, the silicon and phosphorus content of the pig iron was specified within narrow limits, as determined by the heat requirements of the refining process. Since the heat in

[177] Osann, p. 425. [178] Jeans, *Steel*, p. 101. [179] Macar, p. 173.

the Siemens-Martin process was not necessarily derived from burning off the impurities in the pig iron, they had far greater freedom in this matter. The possibility of being able to combine three different ingredients – pig iron, scrap metal and iron ore – within generous limits, opened up wholly new perspectives of cost adjustment, as compared with the refining process in the converter. At the same time, the conversion costs lost some of their impact, precisely because it was now no longer necessary everywhere to start from the same raw materials.

The cost of the raw materials and the length of the refining process varied according to the ratio of the different ingredients. Refining a charge of pig iron and iron ore took longer than a pig iron and scrap charge, but the raw materials were a bit cheaper. The situation became completely obscured when the ratio of all three possible raw materials was altered, as was the rule. The optimum costs for steel production turned out to be quite different, depending on the prevailing prices for pig iron, scrap, iron ore and coal (duration of refining proccss!). Although there are no indications that the composition of the charges was altered at short intervals in a works, in order to follow the price movements as closely as possible, it is clear that there was a considerable difference between regions.

In the case of Dowlais, which processed iron and steel scrap from the works in open-hearth furnaces, its manufacturing costs for ingots compared with acid Bessemer ingots are given in Table 22.[180] It is not clear from this schedule of costs what the proportion of scrap-metal costs to conversion costs was. There is also no reference to the costs calculated for their own scrap. Table 23[181] shows a breakdown of accounts for the Martin (i.e. scrap-based) process at the beginning of the eighties. It relies, however, on the German market price for calculating the raw materials and as a result is probably set too high. On the other hand, the conversion costs ought to be approximately accurate, since they have been calculated from observations made in several works.

After subtracting the pig iron/scrap iron input, this gave conversion costs of 17.74 marks/ton inclusive of the renewed credit entry for scrap. The three items 'Wages and overhead costs, expendable items, fireproof materials etc. and Redemption payments' together add up to 15.29 marks per ton and thus agree very nicely with the corresponding figures in a French account of 1876, which

180 Glamorgan CRO, D/DG E3: report on balance sheet for year ending 29 March 1884, appendix, pp. 28–9. It is remarkable that shortly before the start of the wave of rationalisation in the seventies, open-hearth steel ingots were for a year on average cheaper than Bessemer ingots. The same effect has been observed at Krupp's, which operated according to the Dowlais model. Krupp, p. 174.

181 Macar, p. 172. Small, insignificant adding-up mistakes in the original have been corrected.

earmarked 16 marks per ton for them.[182] A British source from South Wales had given £1 1s.2d. per ton for the same expenses in April 1872.[183] According to Daelen, conversion costs in Great Britain seldom lay above 20 to 25 marks per ton. He was referring to a figure provided by Holley, who reckoned on 23 marks per ton, of which 4.50 marks per ton was for coal, for the end of the sixties in Landore.[184]

One should not attribute too much significance to absolute figures and their differences, since the methods for obtaining them are not always completely above suspicion, and the accounts are based on different compositions of charges. They are here merely to serve as the basis for a rough estimate: assuming that scrap metal was about the same price as pig iron, which is attested by the available cost accounts, making steel in the open-hearth furnace was not cheaper than in the converter; nor, however, was it much more expensive. The price lists at Dowlais come to nearly the same conclusion. Consequently, open-hearth steel which was manufactured from pig iron and scrap, could only seldom compete in terms of price with Bessemer steel. On the other hand, the Siemens process using pig iron and iron ore – and, above all, the pure iron-ore process which he developed to produce steel directly from iron ore – was deemed competitive at the beginning of the 1870s under specific conditions. For this process only were pure large-format open-hearth works built in Great Britain, which were supposed to compete with the Bessemer works in the production of steel rails. However, this proved a premature hope, and the works soon reverted to using scrap as well. We will take a closer look at this case using the Steel Company of Scotland as our example.

The construction costs for an open-hearth furnace were scaled according to its capacity. Table 24[185] gives the overall costs for different big furnaces, including the gas generators.

This agrees more or less with costs at Krupp's, where 21,000 marks were accounted in 1874 for a furnace with a capacity of 5.5 tons.[186] These costs were realistic, if the furnace was to be used as single units or only for smelting iron and steel. They were minimal costs, which did not include any mechanical aids and, above all, no casting equipment. They cannot, therefore, be compared with the construction costs for Bessemer works. A complete plant, inclusive of

[182] *Zeitschrift des VDI* 20 (1876), col. 484f. Since the cost accounting reproduced there deals not with the scrap but the ore input, it is as a whole not comparable. This is why we have in our comparison left out expenditure on coal, since this was necessarily higher with the pig iron and iron-ore process.

[183] *JISI* 1872 part 2, p. 180. Accordingly the same restrictions apply as in note 182.

[184] R. M. Daelen, Zusammenstellung verschiedener Äusserungen über den Herdofen und das Herdofenschmelzen, in: *Stahl und Eisen* 12 (1892), p. 15.

[185] Macar, p. 178.

[186] Krupp, WA IV 56: Beantwortung der von der Commission für Bessemerwerksangelegenheiten pp. gestellten 27 Fragen, Eichhoff, p. 137.

Table 22. *Production costs for Bessemer and open-hearth ingots at Dowlais, 1871–1880*

Year (until 31 March)	Acid Bessemer ingots (£ s.d.)[1]	Open-hearth ingots (£ s.d.)
1871/72	6 07. 0.3	7 03. 3.4
1872/73	7 08. 1.3	9 01. 5.4
1873/74	8 00. 3.7	7 17. 5.2
1874/75	6 16. 2.1	7 03. 8.3
1875/76	5 15. 1.7	5 14. 8.7
1876/77	4 15. 7.5	4 19. 7.7
1877/78	4 07. 8.1	4 10. 1.1
1878/79	3 14.10	3 18.11.4
1879/80	3 07. 7.1	3 12. 6.7

[1]Pounds, shillings, pence and decimals of pence.

Table 23. *Analysis of production costs for 1 ton of open-hearth steel*

1,055 kg pig iron, scrap etc.	at 70.00 marks/ton	73.85 marks
550 kg coal	at 7.25 marks/ton	3.99 marks
Wages and overhead costs		6.17 marks
Expendable items, fireproof materials etc.		8.00 marks
Redemption payments		1.12 marks
		93.13 marks
minus 22 kg scrap	at 70.00 marks/ton	1.54 marks
		91.59 marks

Table 24. *Construction costs for various large open-hearth furnaces*

Capacity	Daily production	Construction costs
500 kg	1,000 kg	8,000 marks
1,500 kg	3,000 kg	11,000 marks
5,000 kg	10,000 kg	18,000 marks
10,000 kg	20,000 kg	28,000 marks

casting equipment, which was intended exclusively for the production of steel, came considerably more expensive. It would generally consist of several furnaces as well. In the year 1879, Consett set a price of £10,000 for two 10-ton furnaces, which corresponded to £5,000 per furnace. When these works were extended in 1886, ten further furnaces of the same type were calculated at only £40,000, corresponding to £4,000 per furnace.[187] Consett was first fitted out with two 10-ton furnaces, which amounted to a production capacity of 150 tons of steel manufacture per week, and thus lay just below the production capacity of an almost equally expensive Bessemer plant from the late seventies.

This meant that building an open-hearth works was no cheaper than an equally efficient Bessemer works. Since the mills for both processes were the same, there was no significant difference in this respect either. The Steel Company of Scotland's open-hearth works, which was fitted out mainly for the production of rails with an annual capacity of 40,000 tons, had cost a total of £250,000 by the time it was finished in 1876.[188] In comparison, the Osnabrücker Stahlwerk's Bessemer works, which was also erected mainly for rail-making, was equipped with two plants from Galloways and estimated to have an annual capacity of 50,000 tons at its completion in 1873, had cost about 5.5 million marks.[189] The lion's share of the costs naturally fell to the plant for endprocessing the steel ingots, and these determined the entire investment outlay to a much more decisive degree than the refining method employed in each case. At the start of the eighties too, open-hearth works for the production of ship plates were built to the same scale. David Colville's works with five 10-ton furnaces and an annual capacity of about 25,000 tons had cost £215,000 by its completion in 1883.[190] Thus, the type of finished product, as long as it was a milled product, did not make a great difference to the costs of construction.

In any case, the Siemens-Martin process, as opposed to the Bessemer or Thomas processes, offered the possibility of taking up steelmaking at very low investment costs with a single furnace. A 500 kg furnace costing 8,000 marks could already produce high-quality steel, which would likely have good sales opportunities in a limited local market. Consequently, the Siemens-Martin process spread to many small works, most of whose annual production comprised only a proportion of the capacity of a single converter plant.

Once it had been established, as it soon was, that open-hearth steel was no cheaper to make than Bessemer steel, big works like the Steel Company of

187 BSC-Northern RRC, loc. no. 4687: minutes, board meeting 6 September 1879 and minutes of board meeting 16 February 1886.

188 BSC-Scottish RRC, loc. no. UA 77 box 9: directors' report for the year ending 31 August 1876.

189 Müller, *Georgs-Marien*, pp. 75f.

190 Payne, *Colvilles*, p. 59.

Scotland or Landore, Wilhelm Siemens's works in South Wales, did not find any imitators in the seventies. Direct competition with Bessemer steel was avoided and instead the plentiful scrap was processed into high-value steel castings, which competed with expensive crucible steel. The low construction costs entered above for individual open-hearth furnaces without lavish transport and casting equipment, were after all applicable in this field of manufacture. It meant that small and middling works, especially foundries, were in a position to procure the means of casting steel and manufacturing special steel qualities with a low investment outlay.

Since the early open-hearth furnaces did not permit the use of phosphoric ores, it was not possible to smelt every sort of scrap without drawing distinctions. The most suitable sort, it soon emerged, was Bessemer steel. Scrap of this type occurred in large quantities in the Bessemer rail works – especially in the form of rail ends. Thus, the annual report of the British Iron Trade Association for 1879 said:

> Besides the extensive independent applications of the open-hearth process for the production of high-class steel, it has, indeed, almost come to be regarded as a necessary, or at all events a highly convenient adjunct to the Bessemer process.[191]

It is therefore not surprising that in the course of the seventies many Bessemer works set up open-hearth works to process their waste. In Great Britain these were the Barrow Hematite Steel Co. and the Bolton Iron & Steel Co.,[192] and in Germany, they were Krupp, Bochumer Verein, Dortmunder Union and Phönix.[193]

However, it was not long before warning voices were heard, pointing out that the simplicity with which scrap could now be processed had led to greater indifference in the Bessemer works about the quality of the steel produced there. The generosity with which products, which had not turned out absolutely 100 per cent, were being in some works thrown out as scrap for their open-hearth furnace, could in business terms hardly be justified any longer.[194]

All the same, it was not only the possibility of exploiting scrap which played a role in determining whether to take on the process. The quality gap between Bessemer and crucible steel, which was still there as before, could now be closed by open-hearth steel. High-quality steel castings could thus be produced from a mixture of steel scrap and Bessemer pig iron at prices only slightly above those for Bessemer steel. The slowness of the process, which caused it to fall very far behind the Bessemer process where the mass production of steel

191 *BITA* 1879, p. 44.
192 Ibid., p. 46.
193 *Krupp*, p. 173. Däbritz, p. 175. Macar, pp. 174–7. The figures supplied there for GHH and Gesellschaft für Stahlindustrie in Bochum were of recent date.
194 Macar, p. 177.

Table 25. *The production of open-hearth steel in Great Britain and Germany, 1873–1879*

Year	Great Britain	Germany
1873	77,500 tons	15,255 tons
1874	90,500 tons	15,802 tons
1875	88,000 tons	15,909 tons
1876	128,000 tons	21,046 tons
1877	137,000 tons	42,798 tons
1878	175,500 tons	52,026 tons
1879	175,000 tons	57,057 tons

was concerned, allowed it to concentrate on quality steels, which could not be produced with sufficient certainty in the Bessemer converter on account of the great rapidity of its conversion process. During a discussion in the Iron and Steel Institute in 1877, in which they mentioned the use of ferro-manganese, new at the time, for producing mild steel, especially for plates, Snelus, the director of the West Cumberland Steel Co., stressed that using it in an open-hearth furnace was considerably simpler and the results surer than in a Bessemer converter, and that mild steel could thus be produced much more reliably according to the Siemens-Martin process.[195] The Railway Companies were insisting increasingly, on grounds of quality, on using open-hearth steel, which was only slightly more expensive, for their axles and steel tyres, which had previously mostly been made out of Bessemer steel.[196]

It was only when these possibilities were consciously exploited and the attempt to compete with Bessemer steel in the mass production of railway rails was abandoned, that secure and expanding markets opened up to the Siemens-Martin process. These markets remained limited during the 1870s in any case, since a mass product comparable to the Bessemer works' rails had not as yet been found. Top-quality steel castings were in too little demand to be able to achieve high production figures similar to those for rails. The total production grew correspondingly slowly, as shown in Table 25.[197]

The German production of open-hearth steel towards the end of the seventies was to a great extent covered by Krupp and the Bochumer Verein

195 *JISI* 1877 part 1, p. 91.

196 The Rheinische Stahlwerke based the construction of their open-hearth works expressly on the railway companies' new demands. Thyssen-Archiv, RSW, 126 01 13: Bericht des Vorstandes für die ordentliche Generalversammlung vom 8 Oktober 1885, p. 3. The crucible steel shops were broken up at the same time, having been rendered redundant by the open-hearth furnace.

197 *BITA* 1880, p. 34.

together.[198] In 1879 alone, the Bochumer Verein produced just 35,000 tons[199] and Krupp's average for the business years 1878/9 and 1879/80 was a good 16,000 tons.[200] These two works just about monopolised the manufacture of first-rate crucible steel in Germany. They were now able to build on this and prepare part of the crucible steel product, namely high-value axles and steel tyres, out of open-hearth steel. In the process, the production of crucible steel at Bochumer Verein was almost entirely superseded by the Siemens-Martin process.[201]

This was due to the fact that the Bochumer Verein was the only German open-hearth works to launch an extensive new product on the market. In 1879 it set up a rolling mill for steel railway sleepers which had a monthly capacity of 2,000 tons. Already in its first year it had amassed orders for over 20,000 tons from the railways.[202] However, sleeper sales were never to become a mass market like the rails business. After a few years of hesitant use, primarily in trials conducted by the trade, the railways even went back in part to using wooden sleepers,[203] so that this market, while very important to the Bochumer Verein, was not to the open-hearth works as a whole.

At Krupp's too, they thought that open-hearth steel would soon supersede crucible steel, and that it would not be possible to make full use of their new crucible steel smelters, which also had a Siemens regenerative gas furnace. In this event, they were to look into the possibility of rebuilding, in order to be able to set the crucible furnaces up for the Siemens-Martin process, if needed. In that case, the gas producers at least were to be taken over unaltered.[204] Thanks to plenty of military contracts, Krupp was nevertheless able to maintain the level of his crucible steel production, and even to raise it in the eighties.[205] Thus, unlike the Bochumer Verein, his works did not undergo an internal shift. In any case, open-hearth steel gradually came to replace puddled steel as input for crucible furnaces. It was, however, primarily employed for cast and wrought pieces for civilian purposes and increasingly for steel plates

198 *Reichsstatistik 1873–1879*. Beck, *Geschichte*, p. 1057, produces figures which tend to be considerably lower, and anyway mentions no sources for them. As it is, the quantities mentioned there for the entire Reich towards the end of the seventies were exceeded by Bochumer Verein on its own, see note 23, and cannot for this reason be correct.

199 Däbritz, p. 175.

200 HA-Krupp, WA VII f 767: Produktionszahlen 1876–1887.

201 Däbritz, p. 175. The production figures for the open-hearth works are missing for the years 1873 to 1876, so that it is no longer possible to measure precisely the extent to which crucible steel was superseded, and which sank in this period from 8,500 tons to 720 tons per annum.

202 Werkarchiv Krupp-Bochum, BV 12600 Nr. 1: General-Versammlung, 28 October 1879, p. 7.

203 Werkarchiv Krupp-Bochum, BV 8300 Nr. 9a: Bericht des Bochumer Vereins an die General-Versammlung des VDESI-Nordwestliche Gruppe am 24 Mai 1870, pp. 9–10.

204 HA-Krupp, WA VII c 96. A. Krupp to Procura on 18 December 1873.

205 See note 24.

too.[206] During the seventies, Krupp had a firm outlet for top-quality plates made of wrought iron, whereas in the case of normal plates he was underbid by the competition.[207] He was thus keen to latch on to the good reputation of the former, and to add steel plates made of open-hearth steel to a market which was already accessible to his firm. Parallel to British developments at Landore and the Steel Company of Scotland, Alfred Krupp made arrangements in 1876 for setting up rolling lines for ship plates.[208]

In Great Britain too, a few firms dominated the market for open-hearth steel, although not in as extreme a form as Krupp and the Bochumer Verein in Germany. The Steel Company of Scotland, the Landore Siemens Steel Company and the Panteg Steel Works and Engineering Co. competed for more than half of the entire production in 1879.[209] As opposed to the market leaders in Germany, these three works had been conceived from the start as open-hearth works, and were not associated with existing works.[210] They relied neither on the production of Bessemer nor of crucible steel, and had to buy up the scrap they needed in the neighbourhood. The biggest self-processor of scrap was Krupp's old model Dowlais, which manufactured open-hearth steel from its own scrap and Bessemer pig, on a similar scale to Krupp.[211] The similarities in the production of steel in both steelworks, starting with their common iron-ore mines in Spain, are occasionally striking.

The only British crucible steel producers, like Krupp and Bochumer Verein in the seventies, to get involved in the large-scale production of open-hearth steel, were Vickers, Sons and Co. and Charles Cammell in Sheffield, with ten and six furnaces each.[212]

Among the works mentioned above, the Steel Company of Scotland is particularly significant in that, in spite of initial scepticism, it was erected in the seventies almost exclusively for the manufacture of rails with the intention thereby of competing with the Bessemer works. Since the SCS was in a key position for opening up new markets for the Siemens-Martin process, we will now investigate it in a brief case study.

206 Krupp, p. 161.

207 HA-Krupp, WA IX a 1,8: Verhandlung in den Conferenzen mit den Herren Vertretern des Etablissements am 11 und 12 September 1876.

208 Ibid.

209 *BITA* 1878, p. 42 and *BITA* 1879, pp. 44f.

210 Jeans, *Steel*, p. 103.

211 Dowlais was already producing 19,000 tons in 1877/8. Glamorgan CRO, D/DG E 3: report on balance sheet for year ending 29 March 1884, appendix, p. 29.

212 *BITA* 1877, p. 70.

(b) *The pace-setter: the Steel Company of Scotland*

In the 1870s and 1880s, the Steel Company of Scotland was the biggest and most efficient open-hearth works in Great Britain and had secured the most receptive markets for open-hearth steel. The firm had been founded in 1872 by the owners of the Tharsis Sulphur and Copper Company, first and foremost the brothers John and Charles Tennant from Glasgow, to exploit the waste products from the extraction of sulphur and copper. The founders had no personal experience of steelmaking and for a long time were outsiders among the heavy industrialists of Scotland.[213]

The Tharsis Sulphur and Copper Company derived its name from the extensive minefields in Southern Spain near Huelva, which it owned and which were the precursors of the later far more famous Rio Tinto mines there.[214] The iron ore from the Tharsis mines was an iron pyrite with a large sulphur and copper content. It was brought from Huelva to Scotland, where it was processed in the St Rollox chemical works, which were also controlled by the Tennants.[215] The sulphur was next removed from the ore to make sulphuric acid. Then the remains were subjected to a washing procedure in the works of the Tharsis Co. to extract the copper. A heavy reddish-purple dust remained as waste, consisting almost exclusively of iron oxide. Its appearance caused it to be known as 'purple ore' or 'blue billy'. Within a very short time Tharsis had accumulated many thousands of tons of this 'purple ore', which was increasingly difficult and, in all likelihood, expensive to dump.

Consequently, ever since the foundation of Tharsis Co. in 1866, the Tennants had been looking for a way to process this waste product, in order to solve at a stroke their storage problem and simultaneously to exploit the large quantities of iron contained in their 'purple ore'. Their search brought them together with Wilhelm, later Sir William, Siemens, who had just started trials for directly converting ore into steel, with the intention of thereby by-passing the blast furnace. This process, were he to succeed with it, seemed ideal for the Tennants' needs. Since their 'purple ore' could only be used as an extra ingredient in blast furnaces, they would have had to build enormous blast-furnace capacities in order to process it completely. Siemens on the other hand assured them that with his process, it would be possible to produce steel exclusively from 'purple ore'. The necessary capital investment for further processing thus appeared considerably smaller, and the anticipated end product would not have to compete with well-entrenched kinds of pig iron in

213 Payne, *Colvilles*, pp. 24ff.

214 S. G. Checkland, *The Mines of Tharsis, Roman, French, and British Enterprises in Spain*, London 1867, p. 16.

215 Payne, *Colvilles*, pp. 24f. The following description of the endprocessing follows Checkland, chapt. 12, Tharsis and Technology.

a firmly established market. Furthermore, they would have been able to limit themselves to buying in coal, instead of having to worry about additional supplies of iron ore. Indeed, their original intention was simply to rid themselves of large quantities of waste, and not to move up into heavy industry.

Although Siemens's first trials using the Rowan Bessemer works in Glasgow produced no satisfactory results,[216] the Tennants were sure that he would still manage to solve their persistent problems and, in February 1872, they founded the Steel Company of Scotland, which was to get an open-hearth works to process their 'purple ore', following the example of Siemens's own works in Landore and strictly according to his instructions. A rail rolling mill was planned to process the steel. Confidence in Siemens's abilities was so great that, already in July 1872, when he was on the spot to review yet again the overall plans for the works, a contract to deliver 20 to 25 thousand tons of 'purple ore' was drawn up with the Tharsis Co. for 1873.[217] For the time being, until Siemens's direct reducing process functioned, the plan was to work as at Landore with charges consisting of pig iron, scrap metal and iron ore. In which case, Siemens assured them, one-third of the pig iron could consist of cheap Scottish pig and the rest of Bessemer pig from high-grade haematite.[218] This would have made the pig iron input considerably cheaper than that for the Bessemer converter. However, they were then supposed to change over to the direct conversion of ore into steel as soon as possible, so that they could stop buying in scrap and pig iron. As it was, all these predictions turned out to be false.

The direct production of steel from 'purple ore' was achieved after long and expensive trials, but the steel ingots were, at £10 a ton at the turn of 1874/5, still considerably more expensive than the then market price for finished steel rails. The process was straightaway shelved,[219] thereby removing the SCS's actual business basis. It remained a works which depended on purchasing almost any kind of raw materials. A French ironworker, who visited the works in 1875, gave the composition of the charges as follows:[220]

Pig iron	3,000 kg
Steel waste	1,000–1,200 kg
Iron ore	1,500–1,800 kg

The large supplies of 'purple ore' could not be processed in this way, and

216 Checkland, p. 120. Payne, *Colvilles*, p. 25.

217 BSC-Scottish R.R.C., loc. no. UA77 box 13: minutes, board of directors, first meeting 14 February 1872 and minutes, board meeting 31 July 1872. Payne, *Colvilles*, p. 25.

218 See below, p. 230.

219 Rails prices at the SCS in the second half of 1874 were £9 11s.1d., and production costs for steel according to the direct open-hearth method were £9 19s.4d. BSC-Scottish RRC, loc. no. UA 77 box 13: minutes, board meeting 18 August 1874 and 12 January 1875 and minutes, board meeting 12 January 1875.

220 Pelatan, p. 33.

over the years they had to be gradually written off.[221] The contracts with Tharsis Co. were cancelled.[222] The Tennants and their fellow founders thus owned a large steel rails works operating according to a process which was not otherwise considered competitive in the industry. What had been planned as a temporary solution at its start, remained the rule, and it was now up to them to make the best of the unfortunate situation.

Profitable operations were not possible under these conditions – nor had they been anticipated in this form, since considerations about profitability had been based on the accuracy of Siemens's claims. Although their income from the rails trade covered the cost of materials in 1875, it did not cover overhead costs and interest payments. 'It has been with considerable difficulty, and only by accepting very bare prices, that orders could be obtained sufficient to keep the works fully employed.'[223] Even when Siemens reduced his patent fee by 20 per cent, as well as cutting a further 30 per cent off for rails, this situation did not change. Given that the SCS price for rails at the time was just £10 per ton,[224] the remaining 5s.7d. in patent fees was not a considerable sum of money, but it could, however, make the difference between financial success or failure in agreements which came close to the costs margin and apparently even below it, so that they were especially important from the perspective of the SCS, in their threatened situation. What was more, the payments were going to the man whom they held morally responsible for their disappointed expectations.

Consequently, Siemens was approached with reference to the false expectations which he had aroused, with a view to further reductions or cancellations of his patent fees. To render these demands more emphatic, a statement was to be drawn up

> showing the losses to the Company resulting from – the contract for purple ore, the construction and erection, and the working of the Rotator Furnaces [for producing steel directly from iron ore, U.W.], the increased cost of using hematite pig iron only, in place of ⅓ Scotch, and ⅔ hematite, for steelmaking, also showing the loss sustained by the Company since the commenced operations, and the amount paid to Dr. Siemens for Royalty, Drawings, & c, with the view of submitting this statement to Dr. Siemens, and asking him – in consideration of the serious losses from the causes referred to, and that the expectations held out at the formation of the Company, that the ores and the iron of the district could be used for steel making have not been realised – to make no further charge against the Company for Royalty until their business is in a better position.[225]

221 BSC-Scottish RRC, loc. no. UA 77 box 9: directors' report for the year ending 31 August 1876, p. 5 and 31 August 1877, p. 6.

222 BSC-Scottish RRC, loc. no. UA 77 box 13: minutes, board meeting 10 March 1875.

223 BSC-Scottish RRC, loc. no. UA 77 box 9: directors' report for the year ending 31 August 1875, pp. 5f.

224 BSC-Scottish RRC, loc. no. UA 77 box 13: minutes, board meeting 12 January 1875, 11 January 1875 and 18 August 1875.

225 Ibid., minutes, board meeting 11 April 1876.

The elegant solution to this conflict came into being in the summer of 1876 at Siemens's urging. His patent claims were settled by means of shares in the impaired firm and his income was thereby directly tied to the continued success of the enterprise.[226]

In this critical situation, the SCS took the bit between its teeth. Additional open-hearth furnaces were set up, in order to work the rails mills to full capacity in day and night shifts, instead of only by day as before. The four new furnaces had a capacity of 10 tons as opposed to the previous 6 tons. Through them the production capacity of the works was almost doubled to 40,000 tons a year, with a financial outlay of £9000.[227]

Thus this rationalisation strategy corresponded to that applied to the Bessemer works during the same period – condensing the time taken by the production process by more intensive use of the plant. In this case anyway, it was limited to the endprocessing work in the mills. On the other hand, this strategy was scarcely feasible with open-hearth furnaces. Here, enlarging the furnaces was all that could be considered, which, in the case of the Bessemer works, was of subordinate importance compared with 'rapid driving'. The SCS rolling lines were overworked and in order to diversify production, a steel foundry with two further furnaces and a plate mill works were erected. As opposed to rails, with which they merely hoped not to incur any further losses, these two branches were considered to have future prospects. 'There is considerable local demand for steel piston rods, crank-shafts, axles and other descriptions of forgings; and when trade generally, and especially the Clyde shipbuilding trade, revives, the Directors anticipate that a large and profitable business of this kind may be done.'[228]

People even believed that the production of steel plates would become 'very profitable'. The basis for such optimism was provided by the trials conducted by the SCS at the instigation of Nathaniel Barnaby, Chief Naval Architect to the Royal Navy. In 1874 Barnaby had seen ship's plates made of mild steel in France, which had impressed him greatly.[229] They had been rolled from the *métal fondu* which was manufactured in Terre Noire and Creusot.

The precondition for making this mild steel was, as had already been established, the availability of ferromanganese, of which there was at the time none in Great Britain. The reason for this lay, ironically enough, with Tharsis Co. Henderson, a former director of Tharsis Co., had also worked at dressing the purple ore and had produced the first ferromanganese from it in combination with waste materials containing manganese from the St Rollox works. The

226 Ibid., minutes, board meeting 21 June 1876 and 30 June 1876.
227 BSC-Scottish RRC, loc. no. UA 77 box 9: directors' report for the year ending 31 August 1875, p. 3.
228 Ibid., pp. 4ff.
229 Payne, *Colvilles*, p. 34.

Tharsis Co. secured the patents for this process and put it on ice, upon which Henderson left the firm and set himself up independently.[230] Since his hands were tied in England by the patent rights, he joined up with M. Julien of the Terre Noire Cie. and sold him his ferromanganese process. In Terre Noire the manufacture of ferromanganese was further developed until manganese contents of 75 per cent were achieved and, above all, it was made cheaper by half. By this detour, ferromanganese came to Scotland, where it became a foundation stone of the SCS's recovery.

Since at this time steel plates were scarcely being used in the British shipbuilding industry,[231] a few efforts still had to be made to introduce the new product. The SCS was indeed sited in the best position imaginable for the purpose. In 1876 more iron ships were being built on the Clyde, near Glasgow, than in the whole of the rest of the world.[232] Its biggest potential market lay directly at its door. The development of ship's plates and their sale was conducted from the start in close cooperation with Siemens's Landore Steel Co. in South Wales. Even before the plate-manufacturing plant was ready, a provisional pact had been concluded with Landore 'to avoid competition in Siemens steel plates'.[233]

The first contract came from the Navy, at whose instigation the experiments for manufacturing plate had been undertaken. In connection with this contract, extensive tests were carried out on the steel plates delivered to the shipyard by John Elder Company in Glasgow, which built the ships.[234] From these tests, which led to open-hearth steel being recognised by the Board of Trade and Lloyd's, the future conditions of acceptance for shipbuilding plates were developed, which were published in the British Iron Trade Association's annual report for 1878.[235] Open-hearth steel was thereby accepted by the highest authorities as a material for shipbuilding, and generally binding criteria for assessing it were established, by which both consumers and potential producers could orientate themselves. The SCS was always valued highly for the pioneering work it had carried out in close collaboration with John Elder and Co.,[236] although it was no longer able to reap the fruits of it on its own.

The SCS hoped that the recognition granted to its own steel and its naval contract would provide a considerable impetus for selling steel plates for

230 Checkland, pp. 117ff.

231 Jeans, *Steel*, pp. 711–21.

232 Sidney Pollard and Paul Robertson, *The British Shipbuilding Industry – 1870–1914*, Cambridge, Mass. 1979. p. 62.

233 BSC-Scottish RRC, loc. no. UA 77 box 13: minutes, board meeting 2 August 1876.

234 BSC-Scottish RRC, loc. no. UA 77 box 9: directors' report for the year ending 31 August 1876, p. 4.

235 *BITA* 1878, pp. 44f.

236 *ICTR* 20 May 1881, p. 605.

civilian purposes as well.[237] However, the conditions for this were extraordinarily bad, both in 1877, when the plate mill finally started up, and in 1878 as well. The shipyards on the Clyde were all suffering to the same extent from a sales crisis and from a long series of bitter strikes by the skilled workforce. Furthermore, the Admiralty had decided in future to cover its requirements for steel plates from Landore.[238] Although the agreement with Landore was not suspended, and SCS secured a compensation payment from Landore for its lost contract,[239] its plate works remained underemployed. One important reason for this was that in the meantime, prices for ship-building plates made of wrought iron had fallen sharply.[240]

After losing the rails market, some wrought-iron works had gone over to manufacturing plate in order to keep their works busy, thereby considerably depressing the price of plate. For Landore, which had secured the naval contract, this was less important than it was for SCS, which had to dispose of their plate on the civilian market in competition with iron-plate prices.[241] They thus set about reducing their agreed price levels, without prejudicing their pact with Landore.[242] For plate delivered in 1878 for the first naval contract, £18 10s. per ton was still demanded, in accordance with the agreement. Shortly afterwards £14 per ton was still being demanded for a civilian transaction with the same shipyard, John Elder and Co.[243] Plate made from wrought iron could, on the other hand, be got for £6.[244] Since only 20 per cent weight could be saved by using steel, due to its greater strength, it was not, with this price differential, attractive enough for civilian purposes.[245]

The SCS only reduced its prices in 1879 and was now delivering steel plate at £9 per ton to the works on the Clyde.[246] The success of this measure proved decisive. It could already be announced in the directors' report for the business

237 BSC-Scottish RRC, loc. no. UA 77 box 9: directors' report for the year ending 31 August 1876, p. 4.

238 Jeans, *Steel*, p. 482.

239 BSC-Scottish RRC, loc. no. UA 77 box 13: minutes, board meeting 12 June 1878 and 19 June 1878. The settlement came to 2.5 per cent of the sales total.

240 BSC-Scottish RRC, loc. no. UA 77 box 9: directors' report for the year ending 31 August 1878, pp. 4f.

241 At Consett in Cleveland when making further calculations about plate for shipbuilding, it was expressly stipulated that they had to take as their basis, not Landore's higher prices, but the SCS's lower prices. Landore did not present any direct competition, but SCS did. BSC-Northern RRC, loc. no. 04687: minutes, board meeting 6 September 1879.

242 BSC-Scottish RRC, loc. no. UA 77 box 13: minutes, board meeting 27 November 1878.

243 Jeans, *Steel*, p. 747.

244 Donald N. McCloskey, *Economic Maturity and Entrepreneurial Decline: British Iron and Steel, 1879–1913*, Cambridge, Mass. 1973, p. 135, col. 6.

245 Jeans, *Steel*, pp. 744ff.

246 BSC-Scottish RRC, loc. no. UA 77 box 13: minutes, board meeting 8 October 1879. Re price developments compare *BITA* 1885, p. 44 and Appendix. D. Pollock, *Modern Shipbuilding and the Men engaged in it*, London, 1884, pp. 7–12 and pp. 21–4.

year 1878/9 that the works was now fully employed and that steel plates were in keen demand.[247] During the year, the demand grew to such an extent that the directors of SCS decided in December, in order to double their production capacity, to buy up the Blochairn works (a neighbouring wrought-iron works which had been shut down) and to refit it for the production of open-hearth steel.[248] The Blochairn Iron Works belonged to Hannay and Sons and, with fifty-four puddling furnaces, had been Scotland's biggest wrought-iron works. In 1872 they were producing around one-third of all Scottish wrought-iron, but in the following year it fell victim to the economic crisis and had to be closed.[249]

In November 1879 the SCS was able to pay out its first dividends.[250] It had never been in a position to do so as a rails producer. Although it managed, by overworking its rail mills and extending its production capacity in 1875/6, not to suffer any further losses in the rails trade, it relied in this respect on the support of the Scottish Railways. The rails purchases shown in the minutes of the board meetings were always concluded at prices higher than those listed by the BITA.[251] Where the railway board was concerned, the same considerations apparently played a role as those exhibited by A. Cooper for the eighties and nineties, namely that the railways always gave local steelworks the edge when handing out contracts, in order to safeguard their own income from freight.[252] Since the SCS was the only rails producer in Scotland during the seventies, it did not need to fear any local competition. In any case, the railways only allowed it to charge the minimum prices necessary for survival. Profits came only with ship's plates. In that same year too, possibilities arose for a second business with future prospects. The SCS entered into relations with the construction firm Arrol & Co. to produce the steel for the rail bridge over the Firth of Forth.[253]

In his works on the Tharsis Co., Checkland raised the following point: 'The Tharsis directors were without inhibiting preconceptions. It was they, and not

247 BSC-Scottish RRC, loc. no. UA 77 box 13: minutes, board meeting 9 December 1879, 17 December 1879 and 19 December 1879. See too Payne, *Colvilles*, p. 39.

248 BSC-Scottish RRC, loc. no. UA 77 box 13: minutes, board meeting 9 December 1879, 17 December 1879 and 19 December 1879. See too Payne, *Colvilles*, p. 39.

249 *Mineral Statistics 1872*, p. 123. S. Griffith, (Griffith's) *Guide to the Iron Trade of Great Britain*, London 1873 (reprint New York 1968), p. 163. T. J. Byres, *The Scottish Economy during the 'Great Depression', 1873–1896, with special reference to the Heavy Industries of the South-West*, B.litt. thesis, Glasgow 1962, p. 382.

250 6 per cent was paid out for 1879. BSC-Scottish RRC, loc. no. UA 77 box 9: directors' report for the year ending 31 August 1879, p. 7.

251 This is revealed by comparing the price lists in *BITA* 1879, p. 41 with the prices aimed at by SCS. BSC-Scottish RRC, loc. no. UA 77 box 13: minutes, board meetings 30 June 1876, 29 November 1876, 21 February 1877, 23 May 1877, 17 April 1878, 1 May 1878.

252 See above, p. 193, note 170.

253 BSC-Scottish RRC, loc. no. UA 77 box 13: minutes, board meeting 30 July 1879 and 4 August 1879. This also contains the written offer to Arrol & Co. dated 31 July 1879.

the great Scottish ironmasters of the day, who brought the age of steel to Scotland.'[254] However, when one sees how trustingly the Tennants entered into their adventure with the SCS, and invested several hundred thousand pounds in a branch of industry unknown to them, to say that they 'were without inhibiting preconceptions' is putting it rather mildly.

Krupp, who was certainly not hostile to innovation, was much more cautious in this respect. Siemens had visited Krupp's works in October 1872, when he also introduced to him his new process for winning steel directly from ore, which at that time was already being tested in Landore.[255] Once he learnt that Vickers in Sheffield was also busy building this kind of plant, Krupp immediately wanted to acquire the German rights, so as to become sole user for the immediate future.[256] Longsdon was commissioned to give an expert opinion on the new process, and construction work on Bessemerwerk III was broken off, thereby giving the latter its *coup de grâce*.[257] After it had been established that Martin steel, according to Martin's pig iron and scrap process, which had already been conducted at the works since 1869, was too expensive for manufacturing rails, the construction of Bessemerwerk III was originally intended to employ the rails mill to capacity – but now plans for an open-hearth plant according to Siemens's iron-ore process were adopted instead.[258]

Longsdon's expert opinion proved extremely favourable and he ended by recommending that 'We should lose no time . . . for the metal thus made will be very much cheaper than the Bessemer owing to saving the use of the Blast Furnace.'[259] Although people knew that Siemens was still having problems with his new process, like the Tennants, Krupp still believed that Siemens was the man to overcome such difficulties, even if it took him years to do so.[260] He therefore insisted on his own experiments as soon as possible, in order 'not to lose any time, since we are close to deciding the question: will Bessemer steel still be made in the future? Will we need Spanish iron as urgently in the future as we thought even half a year earlier? Will we still be using blast furnaces for our manufacture in the future, and will we need the new ones?' Until these questions were clarified, however, Bessemer steel was to go on being made and, where possible, made cheaper.[261] This led to the creation of the Bessemer Commission and to the rebuilding of the Bessemer works, as described in

[254] Checkland, p. 121.
[255] HA-Krupp, FAH II B 313: C. Uhlenhaut to A. Krupp on 4 October 1872.
[256] Ibid.: Ulhlenhaut to A. Krupp on 4 October 1872 and A. Krupp's reply on 14 October 1872.
[257] Ibid.: A. Krupp to Procura on 8 October 1873.
[258] HA-Krupp, FAH II P 112: Eichhoff to A. Krupp on 22 October 1873. In it, Eichhoff is still speaking out in favour of the Bessemer plant in Werk III, since it would have paid for itself, until the Siemens-Martin process at Krupp's was running properly.
[259] HA-Krupp, FAH II B 313: Longsdon's expert opinion to A. Krupp on 21 October 1873.
[260] A. Krupp's commentary on Longsdon's expert opinion (ibid.).
[261] HA-Krupp, WA IX a 187: A. Krupp to Procura on 25 October 1873.

Chapter 3. Apart from his experiments, the only concrete measure urged by Krupp was the construction of a foundry, which would be used anyway and, if needed, could later on use their own pig iron which the Siemens iron-ore process would render superfluous.[262] This, however, was all. Since the experiments did not confirm the low costs anticipated by Longsdon, further investment was excluded and, for the time being, they went on working with the pig iron and scrap process, using the waste produced by their own Bessemer crucible and puddled steel manufactures. To this end, six conventional open-hearth furnaces, each with a capacity of four tons, were set up between 1869 and 1873.[263]

Unlike the founders of the SCS, Krupp used the unfavourable outcome of his experiments as a reason for abandoning his plans to expand in this direction for the time being, and he directed all his energy to rebuilding and reorganising his Bessemer works. The Tennants, on the other hand, passed over the unfavourable results of their own experiments, which were set up on Rowan Bessemer works. It must be asked why these experiments were carried out at all. The SCS's financial results were revealing. The enterprise could only carry on with difficulty through the seventies, so that the Scottish Ironmasters had surely been justified in their reluctance towards steelmaking. The SCS's renown for having brought the Age of Steel to Scotland had been dearly bought. From the moment it became really profitable, the established iron industrialists entered onto the scene as well. Between 1879 and 1880, William Beardmore, David Colville, John Williams, the Mossend Iron Co. and the Summerlee Iron Co. built open-hearth works in rapid succession for making ship's plates.[264] By 1885, ten Scottish firms had introduced the Siemens-Martin process.[265]

The two other big open-hearth works in the 1870s in Great Britain, the Landore-Siemens Steel Co. and the Panteg Steel Works in South Wales, did not fare any better than the Steel Company of Scotland. Although they had not made the mistake of staking their hopes on the Siemens iron-ore process alone, they were equally dashed by the falling price of steel rails. They had no comparable cost reductions to set against the wave of rationalisation in the Bessemer works, as well as which on their location, unlike the SCS, they were faced with massive competition on the part of the Bessemer works for local rails contracts.[266] Landore was almost bankrupt by the mid-seventies. Looking

262 Ibid.
263 Krupp, p. 173.
264 Payne, *Colvilles*, p. 45.
265 Ibid., p. 46. *Iron* 20 February 1880, p. 134; 25 February 1881, p. 131.
266 Nearly ten years later, the then works manager at Landore explained that, after the Bessemer works had been rationalised, they had not been able to keep up with their rails contracts. James Riley, On recent improvements in the method of the manufacture of open hearth steel, in: *JISI* 1884, part 2, p. 443.

back on this period, the then general manager, James Riley, referred to their 'lingering, miserable existence', which the works was able to escape only by starting up production of mild ship's plates.[267] In 1878, James Riley went over to the Steel Company of Scotland, at Siemens's mediation, where he pushed through and directed the changeover to plates for shipbuilding.[268] Landore was saved for the moment by a similar changeover, which was associated with lucrative government contracts. Nevertheless, when the firm did go into liquidation in 1879, it was due, not to its steel production, but to rash coalmine purchases. As it was, the reconstruction of the firm proceeded during the same year without any great difficulty.[269]

The Panteg Steel Works were not much luckier. They were unable to compete successfully on the rails market, and failed to find new markets in time. They were consequently closed in 1879.[270] The works were only bought in 1882 by Wright and Butler, who broke up its rail-mill works and replaced it with an ingot mill producing small ingots for the manufacture of tin plates.[271] Wright and Butler had previously bought the Elba Steelworks near Swansea in 1878 and had fitted it up to produce ingots for tin-plate. The two works finally merged in 1889.[272] Wright had previously collaborated with Siemens in his experimental works in Birmingham, and had subsequently also been at Landore. He had supervised the construction of the open-hearth works at the Steel Company of Scotland and, together with Butler, the Panteg Steel Works as well.[273] By the end of the eighties, both men owned the biggest open-hearth works in South Wales, whereas Landore went to Mannesmann after its second bankruptcy in 1888.[274]

The production of rails from open-hearth steel was quite obviously unprofitable. The Steel Company of Scotland was alone in not closing its rails mill even in the eighties, and it ended by joining the IRMA as well.[275] This was only in order to secure its local market, where the SCS wanted to protect its monopolistic position. Contrary to the Bessemer works, the SCS no longer depended urgently on the rails trade, and sold shares in the quota allotted to it

267 Ibid., p. 443, compare too Payne, *Colvilles*, p. 38.

268 BSC-Scottish RRC, loc. no. UA 77 box 9: directors' report for the year ending 31 August 1879, p. 7. J. Riley was one of the few workmen who managed to rise to such a top position. Charlotte Erickson, *British Industrialists: Steel and Hosiery, 1850–1950*, Cambridge 1959, p. 13 and pp. 168f.

269 *Iron* 11 January 1879, p. 46.

270 Carr and Taplin, p. 114.

271 Ibid.

272 *BITA* 1889, p. 111.

273 Carr and Taplin, p. 114.

274 *Iron* 30 November 1888, p. 484, speaks about a 'disastrous failure'. In *Iron* 15 February 1889, p. 143 it is said that the works had been auctioned off for £101,000 way below its actual value.

275 BSC-Scottish RRC., loc. no. UA 77 box 14: minutes, board meeting 16 January 1884, 23 January 1884 and 6 February 1884. See above pp. 150–1 as well.

for 3s. a ton to Bolckow Vaughan, which was in great trouble due to its Bessemer and Thomas works working below capacity.[276]

(c) Rationalisation strategies and ideas for new works

The open-hearth works' efforts at rationalisation were intended to enable them to meet the downward pressure on steel prices exerted by the Bessemer works, but they looked rather modest compared with the enormous increase in productivity with the Bessemer process. As we have seen, the margin for speeding up the refining process by curtailing the interruptions between individual charges was very slight. Instead of optimising the duration of production, as in the Bessemer and Thomas works, the more obvious solution was to optimise the material inputs. The traditional rationalisation strategies in the ironworks, such as enlarging the vessels and reducing fuel consumption, were once again very important for the Siemens-Martin process. These were basically two measures which were closely involved with one another. Enlarging the furnaces generally entailed a reduction in the relative consumption of fuel.

The entire heat output from steel smelting can be divided into three groups:

(a) during the smelting
(b) chimney gas losses
(c) radiant heat losses

It was only the heat given out in (a) that served the actual purpose of the whole process, making steel. The heat given out in (b) and (c), on the other hand, was lost in the form of radiant or waste gas heat. The gas producers' heat losses need not be taken into account here, since they were not directly connected to the furnace's heat losses.

Comparing the three groups shows that the unproductive heat given out in (b) and (c) was greater in the furnaces of the time than the heat given out in (a) for smelting purposes. Were one then to enlarge the stove and thereby the weight of the charge by a given factor, the heat required for smelting would increase in the same proportion, but not the heat losses under (b) and (c). Osann had used as his example a calculation based on empirical data to show that if the capacity were multiplied by three, the radiant and chimney gas losses would only be doubled.[277]

Direct cost comparisons, as in Krupp's case for the Bessemer process, which could set a figure on the exact savings made by enlarging the furnaces, are unfortunately not available. Since the fuel consumption depended not only on the size of the furnaces but also on the composition of the charge, such data as

[276] Ibid.: minutes, board meeting, 24 June 1885 and 1 July 1885.
[277] Osann, p. 517.

do exist for comparing the amount of coal used in different works with different furnaces cannot be used to illustrate these savings. The composition of the charges varied too greatly and the time taken by the smelting is generally not mentioned. For this reason, we must be content with the initial theoretical considerations and statements presented by James Riley, the works manager at Landore, and those made later at the SCS, to wit that these fuel savings constituted one of the essential improvements to the Siemens-Martin process between the seventies and eighties.[278]

Towards the end of the seventies, both at Landore and at the SCS the original 6-ton furnaces were replaced by bigger 10-ton to 12-ton furnaces. The relative fuel consumption thereby fell and the production capacity of a furnace rose relative to its enlarged charge. The weekly production of these new, double-capacity furnaces had, according to Riley, been increased by almost a third. Whereas a weekly production of 50 tons using 6-ton furnaces was considered satisfactory until 1875, by 1884 a 12-ton furnace had to make 140 to 150 tons a week, that is about twelve charges instead of eight, to be so considered.[279] Given a capacity extension of just 50 per cent, using the furnaces more intensively thus played a minor role in doubling their capacity. However, this effect was probably considerably less than it appeared to be in Riley's account. After all, he was comparing the efficiency of the smaller furnaces at Landore with that of the SCS's 12-ton furnaces, given that in 1875 he was still employed at Landore, and only went over to the SCS in 1878. Although both firms were in constant close contact, freely exchanging their technical discoveries, the composition of their charges was nevertheless different. In the seventies Landore included about 70 per cent pig iron, which it got from its own blast furnaces, whereas SCS included only about 50 per cent pig.[280] In both cases, the remaining input was made up of more or less equal parts of scrap and iron ore. However, a higher proportion of pig iron lengthened the refining time, so that in 1875 the smaller furnaces in Landore could on that account alone make less charges per week than the bigger furnaces in Scotland could make later on. Shifting the emphasis from the production of rails and ingots in the seventies to the production of steel plates in the eighties also had the same effect. Making ingots for tin plate and rails produced 11 to 15 per cent scrap, which was smelted down again. The manufacture of plates, on the other hand, involved about 35 per cent scrap, so that far more self-produced scrap occurred in the production of plates than before. This meant that the proportion

278 Riley, p. 444.
279 Ibid.
280 For SCS see Pelatan, pp. 32–3. For Landore, see Jeans, *Steel*, p. 470.

of pig iron in the charge could be replaced by faster-smelting scrap – which succeeded in shortening the charge time.[281]

Part of the success of this rationalisation, for which Riley took the credit, possibly even the greatest part, must be attributed to the varying composition of the charges. This impression is also conveyed in descriptions by third parties. According to Pelatan's travel journal, a charge at the SCS in 1875 took, including casting the ingots, six to seven hours.[282] According to Jeans, on the other hand, Landore, towards the end of the seventies, required seven to eight hours just for the refining process, after the load had already been smelted.[283] All in all, they must surely have reckoned on eight to nine hours for the whole charge, including smelting the load and drawing off the steel. Starting from the relative length of charge determined by the raw materials in the order of 6:8 or 7:9 hours per charge in favour of the SCS, then the successful rationalisation process postulated by Riley, which was not derived exclusively from enlarging the furnaces, with its relationship of 6:9 in favour of the Scottish practice in 1884 is reduced to an insignificant remnant.[284]

The fact that the effect which the rationalisation efforts of the seventies and eighties had on capacity was based almost exclusively on enlarging the furnaces, was also of great advantage to the firms when planning ahead. The open-hearth works were spared the destabilising effect, which the rationalisation of the Bessemer works had on the whole industry in the seventies. Years later, the plant's production capacity still corresponded to the values which had been established when the furnaces were built. There was no way that a 'technological crisis' could develop from the effect which individual rationalisation measures had in increasing capacity. The feasible expansion of production went according to the firms' plans and did not get out of hand as it did with the Bessemer works. Consequently, the expansion phase of the Siemens-Martin process in the late seventies and eighties was not accompanied by widespread complaints about plant being worked below capacity. In spite of the crisis-ridden climate of the whole economy, the works grew in a very controlled manner along with the gradually increasing demand. This would not have been possible for the Bessemer works, due to the unexpected effect on capacity of their rationalisation efforts.

Apart from its effects on fuel consumption and production capacity, enlarging the furnaces would naturally have affected the labour input, and

281 This scrap occurred during processing the steel ingots to finished products. Discussion between John Head, representative of W. Siemens and J. Riley in Iron and Steel Institute, *JISI* 1885 part 2, pp. 406–9.

282 Pelatan, pp. 32f.

283 Jeans, *Steel*, p. 470.

284 The ratio of 6:9 corresponds to the ratio for the number of charges per week of 8 (Landore 1875) to 12 (SCS 1884), produced by Riley as his only standard measure.

would thus have been linked to the wages bill. Since the smelters only managed the refining process and added the extra ingredients, their numbers were to a great extent independent of the weight of the charge. In their case, problems arose only towards the end of the nineteenth century, when the open-hearth furnaces had grown very large and charging took far longer, leading to a greater loss of heat from the furnaces. From that time on, the need to speed up the charging process was met by fitting charging machines. It would have made less sense to increase the labour input in this connection, since the men would only have got in each other's way in front of the narrow furnace doors. As might be expected, no references are to be found to the effect that enlarging the furnaces entailed a proportional increase in the workforce. However, we can imagine that this could have been the case of the men operating the gas producers and above all, the casting personnel, who now had to cope with the furnaces' considerably larger production.

The labour-saving innovations thus concentrated on endprocessing the liquid steel, transporting it and teeming it into ingots. Basically, they involved taking over mechanical aids from the Bessemer works, whose casting and transport equipment for liquid steel were already highly developed and specialised by the end of the seventies. Since liquid open-hearth steel did not differ from Bessemer steel as far as its endprocessing requirements were concerned, all the equipment in the Bessemer works could be taken over unaltered. This would in any case only have made sense in places where they had achieved a production quantity comparable to that of the Bessemer works and capable of working to capacity the transport and casting equipment they had taken over. This was only the case with big open-hearth works. An impressive example of such an adaptation, alongside the earlier plant at Dowlais and the West Cumberland Co., for instance,[285] was the erection of a second SCS works in Blochairn in 1880, which was fitted out like a Bessemer works for the manufacture of a single product – ingots for rolling into steel plates (see Figure 15).

The liquid steel was collected in a conveyable ladle and taken to a central casting crane in the middle of the plant. From there on the rest of the works was built exactly like a Bessemer works of the same period, with two converters.[286] In Blochairn, the ingots were teemed from the casting crane in a circular casting pit, and were carried from there to the cogging mills. Since the works had been conceived from the start for an annual production of at least

[285] Re Dowlais see Pelatan, p. 56. Re West Cumberland Co., see Lancaster and Wattleworth, pp. 50–1. The plants corresponded as far as possible to the early Bessemer plant with two converters, as illustrated in Fig. 3. Siemens-Martin furnaces were simply installed in place of the converters.

[286] Compare the illustration of a British Bessemer works of 1882, which, apart from the converter, was built in exactly the same way down to the last detail, in: *Stahl und Eisen* 2 (1882), p. 552.

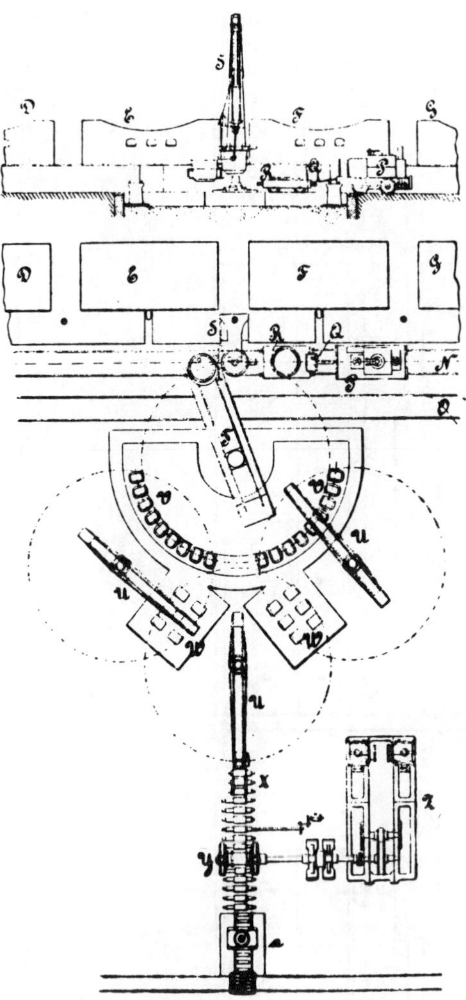

D, E, F, G = Siemens-Martin furnaces
S = central crane
T = casting
U = ingot crane

V = casting pit with ingot moulds
W = hot pits for ingots
X, Y = cogging mills
Q, P, R = locomotive for transporting liquid steel

15 The Steel Company of Scotland's open-hearth works at Blochairn

Société de Bochum.

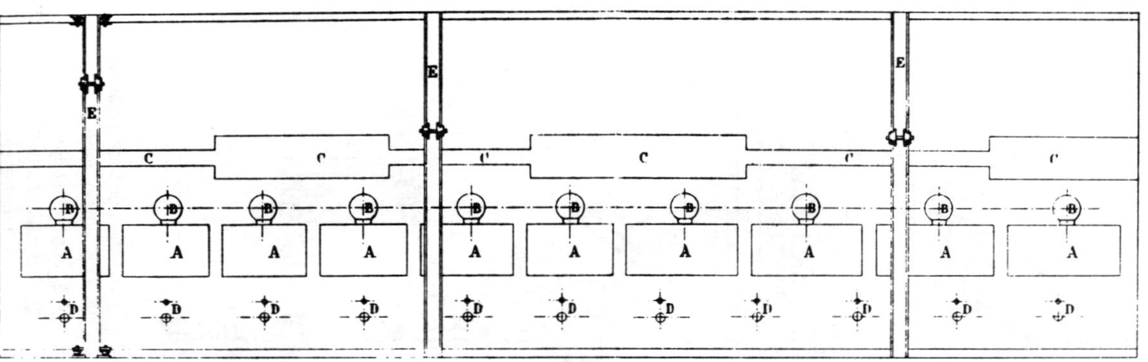

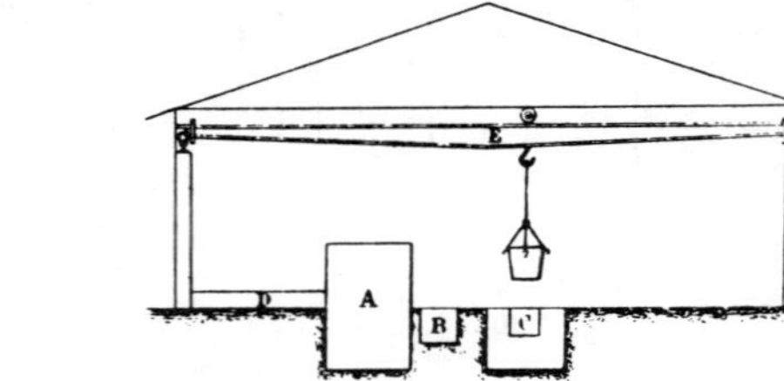

A = open-hearth furnace
B = space for casting ladle
C = casting trench
D = gas and air alternators
E = overhead crane

16 The open-hearth works of the Bochumer Verein

40,000 tons, they could count on using their lavish casting equipment to capacity.[287] A further precondition for its smooth operation was the way the casting crane and pit were optimised for making ingots of consistently uniform size. This arose from the sole production of plates. The uniformity of the product in the new works in Blochairn was the precondition which enabled the technology developed in the Bessemer works, also for a uniform product, to be taken over to good purpose (see Figure 15).[288]

As the plant in the Blochairn works was being publicly presented to the trade, the end of the circular casting pit with its lavish crane system was, however, already being envisaged in the converter steelworks. Starting with the basic Bessemer works, which came closest to the product variability possible in the open-hearth works, extended casting trenches with their steam-driven ladle carriages began to make headway. This solution was tailor-made for the open-hearth works with their long row of furnaces. In Blochairn too, the problem of how to convey the liquid steel along the row of furnaces to the casting crane was solved in this way. As we have already seen, it was the more adaptable variants, whereby the ladle carriages could immediately be set up for casting as well, which were then also established in the open-hearth works. At the same time, the ladle carriage could also be replaced by an overhead crane. The Bochumer Verein, having already shown how to increase production capacity in its Bessemer works by separating the conversion and casting operations through the introduction of an extended casting trench, acted exemplarily in this respect with its open-hearth works too. As judged by Macar, the Belgian engineer whom we have already quoted occasionally, the Bochumer works, which had by far the biggest production of open-hearth steel in Germany, was an *installation des plus remarquables*.[289] Its workshop was *c*. 80 metres long and 25 metres wide. Its ten open-hearth furnaces stood, as was usual, in a row.

As Figure 16 shows, the casting ladles were transported by three overhead cranes which teemed the steel into ingot or other moulds in casting trenches set parallel to the row of furnaces. The casting trench was greatly widened in three places to make space for casting larger steel ingots, a speciality of the Bochumer Verein's. Although the casting and refining processes had not yet been put in separate places in the Bochumer works, the problem of their functional coherence was resolved. Enlarging this works was thus made possible, as with the more recent basic Bessemer works, simply by extending the casting trenches and the crane path and by adding further furnaces, and did not require additional casting and transporting plant, which would have

287 Payne, *Colvilles*, p. 41.

288 Illustration taken from R. M. Daelen, Fortschritte in der Darstellung von Flußeisen und Flußstahl durch das Herdschmelzverfahren, in: *Zeitschrift des VDI* 28 (1884), p. 926.

289 Macar, p. 176. Illustration ibid., planche 19.

co-ordinated only with those furnaces which were to be added.[290] The principles of the organisation of production, with reference to moving the material and casting the steel, were the same as in the Peine basic Bessemer works described above, which pointed the way for future developments.

To the extent that the open-hearth works pushed forward their mass production of steel to rank alongside that of Bessemer and Thomas works, the equipment and the way the works were conceived were similar for all three processes. The Bessemer works acted as a model insofar as the basic experiences with the mass production and mastery of the technical and organisational levels had been encountered in them. The original Bessemer process influenced the other two decisively, which is why it has to be attributed far greater significance in the production of steel since the last third of the nineteenth century than its dwindling share of the total production would initially lead one to suppose.

(d) The rise to dominance of the open-hearth process in Great Britain

Once it had opened up the market for ship's plate and, to a lesser extent, for tin plate, the Siemens-Martin process began to develop along separate lines in Great Britain and in Germany (see Table 26).[291, 292] This applied not only to the industries' growth rates, but also to their preferred smelting procedures, which were closely associated with the former.

Whereas in Germany, as in the 1870s, the pig iron and scrap process with a preponderant proportion of scrap continued to dominate, in Great Britain the pig-iron, scrap and iron-ore process with a preponderant proportion of pig iron and only slight quantities of scrap came increasingly to the fore. With this process, iron ores were only included to speed up the decarburizing process for large quantities of pig. In the foreground was their need to exploit pig iron. Given the open-hearth works' enormous production, which had nearly reached the magnitude of converter steel production by the end of the decade and finally overtook it in 1894, producing on a scrap basis would no longer have been even thinkable.

The strongly expanding demand was borne above all by the need for ship's

290 Werkarchiv Krupp-Bochum, BV 12 900 Nr. 12: Protokoll der Verwaltungsratssitzung am 22 September 1881. The decision is taken here to build two additional open-hearth furnaces and to lengthen the plant accordingly.

291 *BITA* 1890, p. 40.

292 Figures according to Beck, *Geschichte*, p. 1059, who gives no source references for them, after having regretted on p. 1009 that there are no reliable statistics for open-hearth steel in Germany. He ignores the data provided by *Reichsstatistik*, which was unfortunately only entered up to 1882, before falling victim to a comprehensive review of its presentation. These figures are given here in brackets after Beck's figures. The fact that the figures for 1882, the last year, match, allows one to hope that Beck's number series are roughly accurate.

Table 26. *The production of open-hearth steel in Great Britain and Germany, 1881–1890*

Year	Great Britain	Germany
1881	338,000 tons	188,100 tons (124,906 tons)
1882	436,000 tons	152,500 tons (154,775 tons)
1883	455,000 tons	326,300 tons
1884	475,300 tons	313,100 tons
1885	583,900 tons	276,000 tons
1886	694,200 tons	210,600 tons
1887	981,100 tons	224,600 tons
1888	1,292,700 tons	408,700 tons
1889	1,429,200 tons	465,200 tons
1890	1,564,200 tons	388,000 tons

plate and tin plate, with the bulk of regional distribution correspondingly split between ship's plate in Scotland and the north-eastern coastline, and tin plate in South Wales.[293] As it grew increasingly export-orientated, the tin-plate industry moved from the Midlands to the Welsh coast, whence it supplied the American market.[294] The centre of shipbuilding industry was on the Clyde, and its demands radiated out as far as Cleveland.[295] Both industries had been managed so much more weakly in Germany that the tendency of the German production of open-hearth steel to lag behind can be explained on this account alone.[296]

A comprehensive detailed breakdown of the finished products made from open-hearth steel is lacking in Great Britain. The British Iron Trade Association had at its disposal only incomplete information from firms about the uses to which the raw steel was put, and even that for only a few years. Since at least 60 per cent of finished products were registered by the Association, the published figures do nevertheless allow one to say something about the trend.[297] The distribution of the production of open-hearth steel in Great Britain is shown in Table 27.

Thus by far the greatest role was played by the production of plates, which

293 *BITA* 1887, pp. 34f. *BITA* 1888, p. 36.

294 Kenneth Warren, *The British Iron and Steel Sheet Industry since 1840*, London 1870, chapts. 3 and 4, pp. 45–85.

295 Pollard and Robertson, p. 26 and p. 38.

296 In 1880 the shipyards in Germany consumed 17,000 tons of plate for shipbuilding. Pollard and Robertson, p. 39. The production of tin plate in the German Reich was only given a separate entry in the *Reichsstatistik* from 1887, also amounting in that year only to 17,000 tons. *Statistisches Jahrbuch 1907*, p. 265 and p. 267.

297 Due to the small quantity of German production, this break-down is wholly absent.

were mainly destined for shipbuilding, either as plates for ships or for the steamships' boilers. Before concentrating on this branch, let us briefly investigate the changeover of tin-plate production in South Wales.[298]

This industry's input was ingots ('bars') for milling blackplate. These were thin plates, which were turned by tinning into tin plate. The tin plate industry existed mainly on exports to the USA, which took around three-quarters of its entire production in the eighties. There were basically two quality grades for blackplate and its starting material, the bars. Charcoal bars were the best quality. Until the seventies, they were still being refined on primitive charcoal hearths from the days before the puddling furnaces. The poorer quality bars were called 'coke bars' and were made out of wrought iron.

In its search for sales opportunities, Landore had started selling, initially with the greatest difficulty, Siemens-Martin steel to the tin-plate manufacturers. They soon overcame the prejudice which they initially still felt against the new material, and Siemens-Martin steel was even equated with charcoal bars.

The Bessemer works were also trying to find additional outlets among the tin-plate manufacturers. Their steel ingots were also accepted, but were only equated with the poorer 'coke bars', and correspondingly fetched a lower price. Just as when the steel plates were assessed in the shipyards, so too did the tests conducted in the course of processing decide in favour of open-hearth steel. Wright and Butler were the first to recognise the new opportunity. In 1878 they altered the Elba Steelworks in Gowerton, which had been closed, to produce 'tin bars' by the Siemens-Martin process, and in 1882 they bought the Panteg Steel Works, which had gone bankrupt three years earlier, for the same purpose.[299] The sudden growth in demand for open-hearth steel led to many tin-plate works closing their own refineries and going over entirely to buying steel ingots. They themselves could scarcely consider becoming steel producers, since tin-plate manufacturing traditionally took place in very small concerns, capable of processing the produce of a few refining hearths, but not that of an open-hearth works.

Such small concerns were naturally vulnerable to economic fluctuations, and in spite of various attempts by the industry to form a cartel, there were regular bankruptcies during the price fluctuations of the early eighties.[300] In 1881 Dowlais, which was also a supplier to the tin-plate works, decided for this reason to take on the production of tin plate themselves, in order to make sure

298 The following presentation of the tin-plate industry is according to Carr and Taplin, pp. 114f.

299 See above p. 213.

300 *ICTR* 14 January 1881, p. 46; 16 December 1881, p. 766. Re collapse, *ICTR* 26 May 1882, p. 602; 12 January 1883, p. 45.

Table 27. *The uses of open-hearth steel in Great Britain, 1879–1886*

Year	Plates and angles	Rails	Various (incl. tin-plate bars)
1879[a]	below 15%	no data	no data
1880[b]	1/3	1/3	1/3
1882[c]	43%	11%	46%
1883[d]	53%	5%	42%
1885[e]	72%	3%	25% (15% bars)
1886[f]	50%	4%	46% (25% bars)

a *BITA* 1879, p. 45. W. Siemens's figures.
b *BITA* 1880, p. 35. Based on total production, estimate.
c *BITA* 1881, p. 46. Based on total production, estimate.
d *BITA* 1882, p. 40. Based on 62 per cent of total production.
e *BITA* 1883, p. 42. Based on 60 per cent of total production.
f *BITA* 1884, p. 44. Based on 70 per cent of total production.

of an outlet and in future to avoid making a loss from this branch's failures.[301] Hence part of the old shut-down puddling furnaces was demolished and three mill lines were set up there for blackplate.[302] The entry of a big steelworks into the tin-plate trade, which had been dominated by small concerns, turned competitive conditions upside down. The *ICTR* commented retrospectively that: 'The start taken by Dowlais in tin-plate had been one great cause of depression elsewhere.'[303] However, even for Dowlais this did not prove at all a profitable move. The sceptics in the tin-plate branch had been right. Dowlais did not get a successful grip on the production of large quantities of tin plate and in 1883 they abandoned the loss-making enterprise once again.[304] They now restricted themselves, as before, to supplying steel ingots to the established small enterprises.

For the open-hearth works in South Wales, their sales of 'tin-bars' in the course of the eighties were as overwhelmingly important to them as sales of shipbuilding plate were for Scotland and Cleveland, although quantitatively

301 *ICTR* 19 May 1882, p. 571. *ICTR* calculated that Ebbw Vale would for the same reason soon go over to the manufacture of tin plate.
302 Glamorgan CRO, D/DG E 3: report for the year ending 2 April 1881.
303 *ICTR* 30 March 1883.
304 The losses incurred by tin-plate manufacture were already being mentioned in Glamorgan CRO, D/DG E 3: report for the year ending 1 April 1882. On 28 April 1883 the tin-plating operation was definitely stopped. Ibid.: report for the year ending 29 March 1884. See too *ICTR* 26 October 1883, p. 518. There the reason for the failure of the 'dangerous experiment' at Dowlais was seen in the attempt 'to blend two distinct classes of workmen with varying rates and customs'.

there was no comparison.[305] The remaining production of the open-hearth works was spread among a large number of different manufactured steel products, to the extent that the British Iron Trade Association did not feel capable of breaking it down reliably.

> Open-hearth steel is now manufactured into such a variety of products, that it is all but impossible to show exactly the quantities produced of each separate description. This is becoming increasingly difficult, owing to the greater range of products that are embraced from year to year.[306]

Such a complaint had never arisen when covering the production of Bessemer steel – a further indication of how the two sorts differed in quality and applicability.

Open-hearth steel came into its own in the shipyards. After Lloyd's had issued its first stipulations for the use of steel in shipbuilding in 1877, they soon turned to costing steel ships. B. Martell, the Chief Surveyor at Lloyd's, calculated the different running costs and purchase prices for three current ship types – a screw steamer registered at 2,300 tons net for the India trade, a sailing ship registered at 1,700 tons for general trading purposes, and a screw steamer for transporting iron ore with a payload of 1,300 tons.[307] On the grounds that the steel ships weighed less with the same payload, or took a larger payload while weighing the same, he came out in all three comparative calculations with more favourable or at least equal running costs for the steel ship, costs which were not markedly affected by the higher insurance premium for the more expensive ship. The steel ship came out cheapest where transporting iron ore was concerned. Since it took less draught with the same payload than a comparable iron ship, it could set out again even before or after high tide, and could thus perform more trips in the course of a year. In the case of sailing ships, there was, according to Martell, also a clear advantage in running costs in favour of the steel variants, whereas in the case of the screw-steamer for the India trade, the steel ship proved cheaper only under specific conditions.

All this meant that the position of wrought iron as a shipbuilding material now hung merely on the price differential between iron and steel. The more the price differential between steel and iron plates was reduced, the more ship types became profitable in their steel variant. McCloskey has demonstrated this correlation in the case of Scotland, the centre of the shipbuilding industry, in a most impressive manner by setting the relative price of iron to steel plates against the proportion of steel ship tonnage produced (see Table 28).

305 *BITA* 1887, pp. 34f. See Table 27 on p. 224.
306 *BITA* 1885, p. 41.
307 B. Martell, *On Steel for Shipbuilding*, paper read at the Nineteenth Session of the Institution of Naval Architects, extract quoted in Jeans, *Steel*, pp. 744ff.

Table 28. *The penetration of steel in Scottish shipbuilding, 1880–1890*

	Price of steel ship plates / Price of iron ship plates	Proportion of steel ship plates in new ships build in Scotland
1880	1.43	0.15
1881	1.48	0.14
1882	1.39	0.25
1883	1.36	0.31
1884	1.34	0.44
1885	1.21	0.39
1886	1.24	0.62
1887	1.18	0.75
1888	1.06	0.91
1889	1.06	0.96
1890	1.06	0.96

The result suggests a high elasticity of substitution of steel for iron, so that the price differential suffices to explain why iron plates were pushed out of shipbuilding. McCloskey further shows that the swing was particularly obvious when the price relationship was 1:2. This relationship corresponds to the greater amount of iron used compared to steel for ships of equal sizes. He concludes that the shipyards were indifferent as to whether they used iron or steel, and that this confirms the decisive significance of the price difference between the two materials[308] (see Table 28).[309]

His view is one that was also voiced at the time. The *ICTR* published a leading article in September 1885 dealing with developmental trends in the shipbuilding industry, which stated that:

> It is shown that on the Clyde a relative price of steel to iron has been reached, where, for the same size of ship, the cost is practically equal, and the shipowner has thus simply to decide whether his vessel shall be of steel or iron, and, seeing that he has with the former a considerable increase in the weight-carrying capacity, it certainly looks bad for the iron plate makers.[310]

Thus the puddling works suffered the same reversal on the market for ship's plate as they had ten years previously on the rails market. Now, however, there were no alternative opportunities, unlike when they turned from rails to plates for ships. The puddling process had passed its highpoint in 1882 in Great

308 Ibid., pp. 52f.
309 Table taken from McCloskey, p. 51.
310 *ICTR* 18 September 1885, p. 41.

Britain, and more and more works were having to adapt to steelmaking if they wanted to survive. At the start of 1883, the *ICTR*'s Scotland Report commented that:

the makers of finished iron in Scotland are keeping well abreast of the times in adapting their plant for the manufacture of steel, seeing that so large a quantity of that material is now required for shipbuilding and engineering purposes on the Clyde and elsewhere.[311]

Whereas there had never been many rails produced in Scotland, in the case of a few wrought-iron works on the north-east coast in Cleveland this meant their second changeover within a few years. The sale of iron plate had come to a standstill there too, and their attempts at stabilising prices by voluntarily combining to cut back production did not succeed.[312] In the long term the only way out left to them was to change over to the production of steel plates. In practical terms, the Siemens-Martin process was the only one to be considered, because getting their steel accepted by Lloyd's constituted an indispensable condition for its sale in Great Britain. Once it had been denied this general authorisation by Lloyd's, converter steel had been unable to get established because it was considered too brittle.[313]

Much time was spent puzzling why Bessemer steel and open-hearth steel, which were produced from the same pig iron and apparently showed the same chemical composition, still behaved differently. In vain had the Bessemer works tried, by referring to the identical chemical composition of their steel and of Siemens-Martin steel, to find a new market in the shipyards.[314] The latter showed that they were not impressed by any analyses, and quite simply refused to accept Bessemer steel. They were subsequently justified by experiments conducted in Sweden at the end of the eighties, which showed that the presence of small quantities of nitrogen in both acid and basic Bessemer steel made it brittle.[315] The basis for this was the refining process, during which prodigious quantities of nitrogen were projected through the steel in the converter with the air blast. In the process, a tiny part was released into the steel, rendering it more brittle than the otherwise identical open-hearth steel, which was protected from harmful nitrogen during the refining process by its slag covering.

As so often in the iron industry, the steel users' empirical discoveries were a

[311] *ICTR* 5 January 1883, p. 13.

[312] For greater detail see p. 234 below.

[313] Carr and Taplin, pp. 110f.

[314] Ibid., pp. 109f.

[315] H. Tholander, The presence of nitrogen in steel, in: *Iron* 29 November 1889, p. 461. Ibid., Über Stickstoffgehalt im Flusseisen und darauf begründete Vergleichungen zwischen Bessemer- und Herdflusseisen, in: *Stahl und Eisen* 9 (1889), pp. 115–21. Both are translations from *Jernkontorets Annaler* 42 (1888), no. 7.

Table 29. *Open-hearth steel production in Great Britain, by region, 1879–1890*

Year	Scotland (tons)	Wales (tons)	Sheffield (tons)	N.E. coast (tons)	N.W. coast (tons)	Rest (tons)	Total (tons)
1879	50,000	85,000	21,000	1,000	15,000	3,000	175,000
1880	84,500	116,000	23,500	3,200	19,500	4,300	251,000
1881	166,200	102,000	34,000	5,500	24,000	6,300	338,000
1882	213,000	129,500	42,000	6,000	33,500	12,000	436,000
1883	222,000	136,000	40,000	10,000	12,000	35,500	455,500
1884	213,887	150,894	43,440	16,400	10,029	40,600	475,250
1885	241,074	172,861	47,799	75,504	25,974	20,706	583,918
1886	244,900	194,500	39,500	124,100	22,750	68,400	694,150
1887	334,314	225,520	59,292	248,287	67,204	46,487	981,104
1888	442,936	274,650	81,692	352,710	74,938	65,816	1,292,742
1889	440,065	242,618	121,747	437,100	116,612	71,027	1,429,169
1890	485,164	282,170	134,864	469,938	128,079	63,985	1,564,200

few years ahead of the research. Every new material produced by the steelmakers had to wait until it was approved by its users, before greater investment in the new technology was justified. The final sphere of application for a new material often differed from its first manufacturers' original concept. This had until now been the case with all three steelmaking processes. Acid Bessemer steel had initially been thought of as an alternative to crucible steel, and basic Bessemer and open-hearth steel as replacements for acid Bessemer steel in meeting the needs of the railways. The majority in the branch had learnt from this and waited at first for the steel users' reactions; so too with open-hearth steel.

Nevertheless, the speed with which they reacted to the new sales opportunities varied considerably, as the break-down by region in Table 29[316] shows. Whereas Scotland, within a few years following the first success of open-hearth steel as a shipbuilding material, had a large steel plate industry whose most important works had all started up by 1880 at the latest, the north-east coast only came on the market with a considerable volume of production in the second half of the decade.

Towards the end of the decade, the two regions which had specialised in shipbuilding materials, Scotland and the north-east coast had achieved the largest production. The north-west coast joined this market much later on.

316 Table taken from *BITA* 1890, p. 37.

Wales, which was still in the lead at the beginning of the decade, clearly showed reduced growth rates, since it had scarcely any big shipyards and the tin-plate manufacturers' requirements could mostly be covered by converting now-redundant rails capacity.

Sheffield had always been a centre for high-quality steel, for which, however, there was only a limited market. By far the biggest open-hearth works here was Vickers, Sons & Co., one of the pioneers in the manufacture of crucible steel, which primarily made cast and wrought pieces.[317] Together with John Brown and Charles Cammell, who owned the second biggest open-hearth works, it specialised in heavy armoured plates, for which it needed the most uniform, mildest steel possible.[318]

This industry's growth in the eighties was thus borne primarily by Scotland and the north-east coast. An important difference between these two regions lay not only in the timing of their expansion, but also in the structure of their firms. Whereas the four 'big ones' in Scotland, Steel Company of Scotland, Beardmore, Colville and Mossend, had no blast furnaces and thus no pig-iron base of their own, in Cleveland the integrated steelworks dominated the open-hearth process. This difference also explains the different tempo of reaction in both regions. In Scotland, the puddling furnaces simply had to be replaced by open-hearth furnaces. There was no need to take account of pig-iron supplies, as both pig for puddling and the now requisite Bessemer pig could be obtained without difficulty on the Scottish pig-iron market. Because the local ore deposits were due to run out in the foreseeable future, the Scottish blast furnaces had by the second half of the seventies already had to change over to importing more and more ore, among which was a great deal of Spanish hematite.[319] It is, all the same, fairly remarkable that none of the large blast furnaces in Scotland took up steelmaking in the early eighties.

At the time they were far more concerned with extending their coal base, in order to achieve greater independence and more stable supplies, after the bitter disputes with the coalmasters in the seventies.[320] Where the blast furnace works were concerned, it seemed more profitable to resume vertical integration into coal than to join up with endprocessing concerns.[321] In view of the sharp price competition on the pig-iron market, it seemed initially to make more sense to

317 The fact that Vickers had taken on the direct Siemens process in a trial operation provided Krupp with an additional reason for pushing on with this project in his own works. If Vickers was going to rely on this process for its production of quality steel, it would qualify almost as a guarantee of success – which was, in this case, incorrect. HA-Krupp, FAH II B 313: expert opinion by A. Longsdon to A. Krupp on 21 October 1873.

318 Carr and Taplin, p. 85. Here, they recall the difficulties which John Brown and Charles Cammell experienced in making mild steel in the Bessemer converter.

319 Byres, p. 610.

320 Ibid., pp. 378f. Re disputes with the coal masters, see *ICTR* 19 September 1879, p. 273.

321 An extensive discussion of this phenomenon is to be found in Payne, *Colvilles*, pp. 47–55.

effect a possible reduction in costs than to invest further in endprocessing expensively produced pig iron. In the long term, this strategy proved successful in as much as, towards the end of the nineteenth century, Scottish ironworks were finally earning more with their coal than with their pig iron.[322] Given that the mighty ironmasters' interests were linked to the fundamental reorganisation of their raw-material base, the iron processors could operate fairly unimpaired.

Their independence from the raw materials coal and ore and from the conversion problems associated with them, as well as having a sufficient supply base through the local market, enabled the Scottish plate-makers to go over quickly to steelmaking. They had no ballast to jettison apart from their old puddling furnaces.

The importance of their independence is shown by the example of the two biggest wrought-iron works in Scotland, the Glasgow Iron Co. and the Monkland Iron and Coal Co., both of which had their own raw materials and blast furnaces for making pig iron, and thus remained tied to processing their pig in puddling furnaces.[323] Once the raw materials had been acquired, they went on determining the firms' development for a long time. The Glasgow Iron Co. found its way to the Siemens-Martin process only in 1894, after it had written off its not very successful attempts at processing its pig according to the Thomas process. In the following year they employed Sir James Riley, who had left the SCS, as their General Manager to extend their new steel production.[324]

The integrated steelworks in Cleveland were faced with similar problems too, since they also had to ensure that their own ores were turned to account, which they had hitherto been able to process into pig iron for their own puddling works. On account of its high phosphorus content, this ore was not initially suitable for the Siemens-Martin process – just as little as it had been for the acid Bessemer process. Attempts in the eighties to transpose Thomas's invention from the converter to the open-hearth furnace still had not produced economically satisfactory results.[325] Thus importing additional Spanish ore remained the only way of making a suitable kind of pig iron.

Unlike Scotland, the Cleveland plate manufacturers could not simply roll away their pig-iron problems, which remained inherent. Hence, adapting their plate manufacture to steel also entailed the loss of an outlet for their own ore

322 Ibid., p. 51.

323 Of a total 380 puddling furnaces in 1880, Glasgow Iron Co. owned 99 and Monkland Iron and Coal Co. owned 52. *Mineral Statistics* 1880, p. 99.

324 Payne, *Colvilles*, pp. 71f.

325 The dolomite in the converter turned out to be unsuitable as a material for building furnaces. Although de-phosphorisation did occur successfully, it entailed too many repairs for the operation ever to become profitable. See above p. 27.

mines. The cheap phosphoric ores near to the rich coal seams at Durham, which had made Cleveland's comet-like ascent in the sixties possible, were now becoming ever more of a drag because of being tied to the puddling process. Nor could this structural problem be solved by the production cutbacks implemented by the Cleveland Iron Masters' Association. Moreover, their efficacy was restricted in the early eighties by the long-drawn-out price war with the Scottish blast-furnace works. The latter were suffering, to the benefit of pig-iron consumers, under similar difficulties with regard to sales, but they nevertheless failed to get a cartel going.[326] It was only in the period between 1 October 1881 and 30 September 1882 that the Cleveland Iron Masters' Association and the Scottish Ironmasters concluded an agreement to cutback production together.[327] Subsequently, the Scots withdrew again and opened the next round in the price war.

The lack of cohesion among the Scottish pig-iron producers had its origins in Baird's excessively mighty position; it produced about a quarter of Scottish pig iron on its own, and had a strong financial backing in coal, coke and the chemical industry. Careful plans were laid for a price war, which would enable them to clean up the market vigorously when the smaller firms collapsed.[328] An indirect result of this policy has already been described; it was when Merry & Cunninghame and the Glasgow Iron Co. in Wishaw veered over to the production of basic Bessemer steel in the eighties.[329] For the integrated blast furnace and wrought-iron works in Cleveland this meant that only by end-processing their own pig iron as extensively as possible, could they be freed effectively from the keenly disputed pig-iron market. It is thus scarcely surprising that the integrated works in Cleveland kept to the production of iron plates for shipbuilding for as long as possible, in order to exploit their raw materials. Due to their extraordinarily low costs for pig iron, these works were able to delay the demise of their iron-plate production for a few years more, but when the price gap between iron and steel plates closed around 1886, all that was left to them was headlong flight – changing over to open-hearth steel.[330]

326 A cut-back in production was briefly implemented in common in 1879. BSC-London, CIA MSS: minutes 15 August 1879. Re the internal disagreements between the Scottish ironmasters, R. D. Corrins, *William Baird & Company. Coal and Iron Masters, 1830–1914*, diss. Strathclyde 1974, pp. 122–6.

327 BSC-London, CIA MSS: secretary's report for the year 1882. Corrins, pp. 122f.

328 Corrins, p. 123 and p. 125. Byres, p. 369.

329 See above p. 172.

330 The spread of open-hearth steel production in Cleveland coincided with the moment when the price relationship for steel and iron plates reached the critical figure of 1.2. See Table 28, taken from McCloskey.

(e) Consett and the changeover in Cleveland

Among the wrought-iron works on the north-east coast, which were already manufacturing plates for the shipyards, Consett provided the best preconditions for the changeover to open-hearth steel. It held a special position among the large steelworks of the region on account of its geographical situation alone. Unlike most of the others, it did not lie on the iron-ore fields near Middlesbrough, near the coast, but a bit further north and inland on the coal mines south of Newcastle. Consett was thus based on coal and not on iron ore. The firm's extraordinarily high profits – share dividends of over 20 per cent were almost the rule – rested, as Richardson and Bass have shown, to a good extent on this coal.[331] Consett had to buy the iron ore it needed to produce iron. The pig iron was then endprocessed to rails and plates in the firm's wrought-iron works and mills.

At the start of the seventies, when the American trade in iron rails was at its zenith, Consett was greatly embarrassed by its lack of an iron-ore base. The Middlesbrough works were processing as much of their iron as they could, in order to profit from the enormous demand for rails. Consett was no longer able to satisfy its ore requirements on the market, and there was a danger that its newly finished blast furnace could not be put into blast.[332] It looked for a while as if Consett had missed the boat, while others, such as Bolckow Vaughan, who had invested massively in iron and coal mines, were able to exploit all the advantages of Cleveland's situation. In 1872 Consett even had to procure pig iron from Bolckow Vaughan in order to be able to meet its engagements.[333] To keep the new blast furnace busy, it was altered to produce Bessemer pig, although this proved hard to sell at first. The iron ore came from the west coast and was truly expensive.[334] At the same time, however, Dowlais had already been contacted with a view to their eventual participation in the Spanish iron-ore mines, which led very quickly to founding Orconera, together with Dowlais, Krupp and Ybarra.[335]

331 H. W. Richardson and J. M. Bass, The Profitability of Consett Iron Company before 1914, in: *Business History* 7 (1965), p. 81. A. S. Wilson, *The Consett Iron Company, Limited: A Case Study in Victorian Business History*, M.litt. thesis, Durham 1973, pp. 101–45. This statement, as postulated by Richardson and Bass, is confirmed in a detailed manner at this point in this regrettably unpublished work.

332 BSC-Northern RRC, loc. no. 04687: minutes, board meeting 7 September 1872.

333 BSC-Northern RRC, loc. no. 04698: cost-analysis book, p. 73.

334 BSC-Northern RRC, loc. no. 04687: minutes, board meeting 10 March 1874. In the minutes taken in this period can be found current complaints about the high prices for ore, which were raised further still by the comparatively high railway tariffs set by the North Eastern Railway.

335 BSC-Northern RRC, loc. no. 04708: Consett Spanish Ore, minutes, board meeting 16 April 1872 and 9 July 1872. Since the statutes of the mother company did not permit holding shares in other enterprises, the Consett Spanish Ore Co. was founded for the Spanish ore trade by the shareholders of Consett Iron Company, Ltd. *De facto*, however, they were one and the same enterprise, and both branches were run by the same management.

Unlike Krupp and Dowlais, who intended to use the Spanish ores to ensure supplies to their Bessemer works, Consett needed the ore to keep its blast furnaces busy. At the time, not a thought had been given to building a Bessemer works. Decarburizing continued to be done by the puddling process, which required pig iron made of Cleveland iron ore. It was only in 1875, when it was apparent that the rails market had finally been lost, that the works manager was instructed to find out how much a steelworks would cost. Twenty months then elapsed before serious consideration was given to building a Bessemer works and an internal report to this effect was issued.[336] As explained above, the firm decided against building a Bessemer works, on account of the high capital input required, although the report had come out in favour of so doing. Instead, the rail rolling mills were altered for manufacturing plates.[337] The production of plates proved profitable. It earned between 8s. and 12s. per ton of finished product; this was equal to 7 to 10 per cent of the costs.[338] Consett was producing cheaply enough to be able to off-load large quantities of its plates on the Scottish shipyards, despite the disadvantages of freight expenses.[339] The firm's plates had thus already well penetrated the Clyde when steel first began to be used there.

Early in 1879 the demand for steel plates had risen buoyantly and it was at the time already apparent that they would supplant iron plates. Consett had to work on stock, in order to keep the mills employed and they initiated a short-term agreement among the Cleveland iron-plate makers.[340] In order to stabilise prices, they were to implement a mutual reduction in their plate production over three months. No success was reported from this venture. It is possible that the renewed bankruptcy of the Britannia Iron Works, which had been bought by Skerne Iron Co. and had been equipped for the production of plates, had the same effect, so that for the time being a formal contract would be postponed.[341] After its second bankruptcy, the unlucky Britannia works was

336 BSC-Northern RRC, loc. no. 04687: minutes, board meeting 2 March 1875, and board meeting 7 November 1876.

337 See above p. 126, BSC-Northern RRC, loc. no. 04687: minutes, board meeting 3 October 1876 and 6 March 1877.

338 BSC-Northern RRC, loc. no. D/Co 89: cost-analysis book, cost and profit steelplates, pp. 178, 198 and 226. In the Consett Co.'s records there existed two cost-analysis books for the general manager, kept parallel to one another. Whereas the one quoted here obviously provided the basis for price calculations and contained information about the profits for individual products, the one quoted in note 340 served for supervising internal costs. In it are to be found detailed half-yearly schedules of costs for all the enterprise's products. These schedules of costs corresponded broadly with Krupp's, which are set out above.

339 *ICTR* 17 January 1879, p. 72.

340 Wilson, p. 266. BSC-Northern RRC, loc. no. 04687: minutes, board meeting 1 April 1879, 3 May 1879 and 17 May 1879.

341 All the same, 120 puddling furnaces had to close in Cleveland. *ICTR* 16 May 1879, p. 529.

bought by Long and Co. and continued the production of plates, and for a brief time made rails for the American market as well.[342]

At Consett, the problem of an outlet provided the motive for resuming the discussions about building a steelworks. The general manager and an assistant were sent to Germany to obtain information about the relative advantages of the different steelmaking processes from Krupp, among others. They also visited the Steel Company of Scotland. They compiled the results of their enquiries, viewings and reflections in a new report about steel production, which was presented to the board on 6 September 1879.[343] In it they stressed that Consett could not be beaten in terms of price in its production of iron plate. Since the demand for steel plate had quietened in the meantime, no precipitate action was indicated. So long as iron plate could still be sold at all, Consett could keep to it. It was, however, foreseeable that this would not remain the case for long. The price gap between iron and steel plate would have closed within a few years and by then Consett would have had to go over to steel. However, at the time it was still uncertain which processes would catch on. In the event of the acid process (both Bessemer and Siemens-Martin) succeeding, there would be no grounds for concern because they owned low-phosphorus ore in Spain and had splendid coal. However, should the Thomas process prevail, for which Cleveland ore could be used, the question that would arise for Consett was whether it would not be better to abandon their works and build them up again near the others on the Middlesbrough iron-ore fields. In the first instance, however, they advised against this. Their report came to the astonishing conclusion that they supposed that the future would belong to the acid Bessemer process, as had been the case until now in Cleveland. The works' present situation was best suited for this process, since they would anyway have to work with imported ores. In order to be able to bridge the period until the final changeover to acid Bessemer steel, and to gain an introduction to the steel market for Consett, they suggested building two open-hearth furnaces. The steel from them could be immediately made into plates for shipbuilding in the existing mills, and then slowly introduced onto the market. Anyway, once a Bessemer works had been erected later to replace the puddling furnaces, these two open-hearth furnaces could then be used for processing their own steel scrap.

Looking back, it is remarkable how Consett, on the basis of false premises and much hesitation, came to the optimal strategic decision in the matter of steelmaking. This example makes it particularly clear how important are the different strategic terms employed in our investigation, which we discussed at

342 See above, pp. 176–7.

343 BSC-Northern RRC, loc. no. 04687: minutes, board meeting 8 August 1879 and report concerning steel production, 5 September 1879.

the beginning of the book. The content of their strategic decision, to build two open-hearth furnaces, proved pure gold, in the truest sense. The decision-making process, on the other hand, was characterised by gross mis-estimates. It seems rather curious that the Consett directors could still believe in 1879, when the disadvantages of Bessemer steel were already well known, that this was the process of the future and that Consett, whose future main market would lie with plates for shipbuilding, would soon have to go over to the Bessemer process.[344]

Furthermore, it is apparent from the report that Consett felt threatened by the Thomas process. It seemed as if the lack of an iron-ore base of their own in Cleveland would have an adverse effect on the firm for the second time in one decade. Its unusual site upon coal was apparently considered within the firm itself as not entirely unproblematic. All the same, it stayed put because sudden, bold decisions were not characteristic of Consett. The construction of a big steelworks was put off for the second time, although they were convinced that such an undertaking would be correct in principle. In this lay an essential difference between their policy and that of their neighbour, Bolckow Vaughan. In the latter, every new opportunity was put into action as soon as possible and if the project promised to become profitable, it was immediately given lavish funding. In this manner, huge supplies of raw materials had been amassed, and thus, too, the biggest basic Bessemer works in the world had been built within the shortest time. At the annual general meetings, this policy led to constant arguments with a few shareholders, who criticised the policy of forced expansion on the grounds that it would lead to over-capitalisation of the enterprise and would in this way reduce their dividends.[345]

They were supported in their criticisms by the nearby example of Consett, and subsequent developments in the eighties proved Consett right. Dithering had paid off. Consett's steelworks did not have to carry either the ballast of Cleveland iron ore or the bigger Bessemer plant, and drew splendid profits. Bolckow Vaughan ended by following Consett in the production of open-hearth steel, but only after huge sums had already been tied up in other quarters.[346]

The construction of the Consett steelworks proceeded extremely gradually. It was only in June 1883, three and a half years after the decision to build had been taken, that the two open-hearth furnaces were put into operation.[347] In

344 This opinion was explicitly represented in the report. Apart from this they stressed the possibility that it would then be possible to get back into the production of rails – which was not exactly an example of great perspicacity.

345 BSC-Northern RRC, loc. no. 04902: report of directors for the year ending 31 December 1880, p. 17; 31 December 1881, pp. 16–19; 31 December 1883, pp. 20f.

346 BSC-Northern RRC, loc. no. 04898: minutes book 7, board meeting 25 July 1884 and 24 April 1885.

347 BSC-Northern RRC, loc. no. 04687: minutes, board meeting 5 June 1883.

comparison, Beardmore in Scotland took only four months to build its open-hearth works and to fire its first charge.[348] The extraordinary shipbuilding boom at the beginning of the eighties[349] had enabled Consett to extend its iron-plate production by almost a half, as its profits rose.[350] It was only when it became apparent that the boom was subsiding at the same time as the proportion of steel plate was rising, that any emphasis was put on finishing the steelworks.

In November 1882 the dwindling demand for iron plates in Cleveland had led to renewed efforts, following the example of the Cleveland Iron Masters' Association, to reach an agreement about limiting production.[351] Finally, in March 1883 at the instigation of the workers' representative (!), the wrought-iron works stopped production on Mondays. However, this agreement was shaky from the start, since Dorman, Long & Co. with their two works, Britannia and West Marsh, did not join it. Nevertheless, the other works did stop production on Mondays until the beginning of May.[352] After this nothing more is heard about voluntary production cutbacks. For Consett, in any case, this development provided the latest motive for pushing its steel plans along.

After it had been established that open-hearth steel came up to expectations with regard both to commercial and technical demands, the board resolved as early as November 1883 to build a further four furnaces, which would be supplemented soon afterwards by two more.[353] Steel plates proved at once to be much more profitable for Consett than iron plates. The production costs for both sorts became increasingly similar, whereas steel was able to maintain its price lead. Table 30 provides a comparison of the two production processes at Consett.[354]

At the same time, the production of steel plate was not limited by lack of demand, but by the works' productive capacity. In order to employ the six additional open-hearth furnaces which had become operational in the course of spring 1885, a further blast furnace had then to be got ready and put into blast.[355] On the other hand, iron plates could only be sold in such quantities because the shipyards could not secure as much steel plate at they wanted.[356]

348 Payne, *Colvilles*, p. 46.

349 Pollard and Robertson, p. 26. *ICTR* 6 August 1880, p. 157.

350 BSC-Northern RRC, loc. no. D/Co 89: cost-analysis book, cost and profit ironplates, pp. 250, 275, 302 and 331. Subsequently, in 1881, the net profit per ton of iron plate exceeded £1 with £5 9s. costs per ton.

351 *ICTR* 17 November 1882, pp. 544f.

352 *ICTR* 4 May 1883, p. 514.

353 BSC-Northern RRC, loc. no. 04687: minutes, board meeting 6 November 1883 and 16 August 1884.

354 BSC-Northern RRC, loc. no. D/Co 89: cost-analysis book, cost and profits steel plates, cost and profits iron plates, pp. 358, 382 and 393.

355 BSC-Northern RRC, loc. no. 04687: minutes, board meeting 31 March 1885.

356 *ICTR* 17 April 1885, p. 513.

Table 30. *Cost of production, selling prices and profits in iron and steel plates at Consett, 1884–1886*

Year	Production		Costs per ton £ s.d.	Price ex works per ton £ s.d.	Profits per ton £ s.d.
1884	iron	68,777 t.	103/ 4.33	115/11.8	12/ 6.75
	steel	5,298 t.	133/ 1.85	153/11.53	20/ 9.68
1885	iron	58,606 t.	90/ 9.45	96/ 9.17	5/11.72
	steel	13,012 t.	105/ 4.28	128/ 4.48	23/ 0.20
1886	iron	42,195 t.	88/ 5.98	93/ 7.33	5/ 1.35
	steel	26,686 t.	91/11.49	121/ 3.88	29/ 4.39

The number of open-hearth furnaces was being increased in many places to satisfy the demand; the aim now was not to be the last one left with a supply of iron plates. So the Consett directors once again allowed a report on steelmaking to be put out by their General Manager.

This new report was laid before the directors on 16 February 1886.[357] In it, unlike the previous report, the question as to which steelmaking process would be most suitable for Consett was clearly answered in favour of the Siemens-Martin process: 'as regards the special process of steel making I am convinced from the recent supplies of steel plates for shipbuilding in various qualities of Bessemer hematite (acid Bessemer UW), Bessemer basic and Siemens hematite (acid open-hearth UW) material, we cannot do better than to duplicate the type of furnace we have already adopted'.[358] This decision had surely been made easier by the prohibition issued two months earlier by Lloyd's against using basic steel for shipbuilding purposes.[359] Using Cleveland ores thus no longer came into the discussion. Resuming production of rails was also quite unthinkable now, in view of the Railmakers' Association, quite apart from the trade's gloomy prospects.

In his report, the general manager concluded that the large profits which had so far been drawn from steelmaking justified increasing their existing capacities by more than double. He proposed building a further ten open-hearth furnaces and extending the mills. The Board of Directors agreed to these plans. The financial backing for them and for the acquisition of additional coal supplies was provided by increasing capital gradually over a period of years.

357 BSC-Northern RRC, loc. no. 04687: minutes, board meeting 16 February 1886.
358 Ibid.
359 The prohibition was announced on 17 August 1887, p. 177.

Table 31. *Cost of production, selling prices and profits in iron and steel plates at Consett, 1887–1890*

Year	Production		Costs per ton £ s.d.	Price ex works per ton £ s.d.	Profits per ton £ s.d.
1887	iron	34,015 t.	85/ 9.43	91/ 8.01	5/10.18
	steel	39,704 t.	77/ 8.17	108/ 8.79	31/ 0.62
1888	iron	25,921 t.	89/ 9.36	93/ 7.45	3/10.04
	steel	66,390 t.	83/11.39	112/ 5.14	28/ 5.75
1889	iron	20.771 t.	93/ 4.30	107/10.17	14/ 5.87
	steel	107,055 t.	84/ 0.90	118/ 9.88	34/ 8.89
1890	iron	20,432 t.	100/ 9.52	132/ 4.70	31/ 7.18
	steel	111,654 t.	89/ 8.22	132/ 1.93	62/ 5.71

This was conceived from the start so that the shares, which were allotted over a five-year period to previous shareholders, could be financed by current dividends.[360] Since the total of £39,600 payable yearly for them was set against dividends amounting to between £55,000 and £230,000, payment did not present any problems.[361] This was even more the case when the enlargements to the steelworks were already written off in 1889, whereas the report had cautiously reckoned on this happening only in 1894–6.[362]

These enlargements made Consett by far the biggest open-hearth works on the north-east coast, and by 1889 and 1890, with over 160,000 tons alone it was providing a third of the region's total production.[363] Its production of iron plates fell even further, as expected, and it was able, once the steelworks was finished, to restrict itself to particularly favourable transactions.[364] Consett's plate production for the years 1887 to 1890 is shown in Table 31.[365]

Demolition of the first puddling furnaces was started in 1887 to make space for the production of steel. They then turned most of their energy to

360 BSC-Northern RRC, loc. no. 04687: minutes, board meeting 6 April 1884.

361 BSC-Northern RRC, loc. no. 04689: directors' report for the year ending 30 June 1886, 30 June 1887, 30 June 1888, 30 June 1889 and 30 June 1890.

362 Ibid.: directors' report for the year ending 30 June 1889.

363 Production of open-hearth steel amounted on the north coast to 437,000 tons in 1889 and to 470,000 tons in 1890. *BITA* 1889, p. 39 and *BITA* 1890 p. 39. Consett's steel production amounted to 154,000 tons in 1889 and to 167,000 tons in 1890. BSC-Northern RRC, loc. no. 04698: cost-analysis book, steel ingots 1889 and 1890.

364 BSC-Northern RRC, loc. no. 04687: minutes, board meeting 6 March 1887.

365 Table according to BSC-Northern RRC, loc. no. D/Co 89: cost analysis book, cost and profits steel plates, cost and profits iron plates, pp. 406, 421, 432 and 452.

obtaining raw materials, so as to be able to make the necessary pig for producing steel.[366] This was all connected with the purchase of further coal supplies mentioned above, the expansion of steel production and the provision of additional low-phosphorus iron ore on the market, since the 100,000 tons contracted from Orconera had long ago not been enough and, even with supplementary deliveries, did not meet all their requirements.[367] By the beginning of the nineties, Consett was producing about as much steel as Krupp, although its quota from Orconera was only half as great.[368]

The construction of the open-hearth works at Bolckow Vaughan was completed under much less favourable circumstances because, unlike Consett, they had in most respects pursued almost the opposite policy until the early eighties. These differences were manifested primarily in their huge Bessemer steelworks which, according to their own figures, was easily capable of meeting the entire rails requirement for Great Britain. However, even after the dissolution of the IRMA, there was no question of fully employing this capacity. Opportunities for selling Bessemer steel in forms other than rails were urgently sought; limited success was achieved only in the production of steel sleepers, following the German example.[369] The sleepers were made according to a patent by James Riley, general manager of the Steel Company of Scotland. Riley advised his own firm against taking up the production of steel sleepers, on the grounds that it did not pay.[370] In any case, there was no mention of producing large quantities at Bolckow Vaughan either. The lamentable end to the attempts at introducing onto the market plate for shipbuilding made of basic steel has already been described.

For this reason it was decided in May 1884, shortly after the IRMA was wound up, to turn a rails mill into a plate mill.[371] However, since there was little chance of making either acid or basic Bessemer steel attractive to the shipyards, they decided no more than three months later 'owing to the strong prejudice of buyers in favour of steel made by Siemens-Martin process' to build two open-hearth furnaces.[372] While the directors' meetings were wholly dominated by their concern about the basic Bessemer steelworks and pig-iron sales, the open-hearth works was being extended until it had six furnaces with

366 Ibid., minutes, board meeting 5 March 1889 and 2 April 1889.

367 Consumption of Spanish ore amounted to 300,000 tons in 1890. BSC-Northern RRC, loc. no. 04698: cost-analysis book, pig iron 1890. In this connection it is clear that 'Spanish ores' were understood to include all low-phosphorus iron ores, i.e. also those imported from Elba and North Africa. See too Richardson and Bass, pp. 83f.

368 Consett produced 174,000 tons in the year ending 1891. See note 91. In the same period Krupp produced 157,000 tons of Bessemer steel. HA-Krupp, WA VII f 484: Produktion an Bessemerstahl.

369 BSC-Northern RRC, loc. no. 04898: minutes book 7, board meeting 22 October 1885.

370 BSC-Scottish RRC, loc. no. UA 77 box 14: minutes, board meeting 1 September 1886.

371 BSC-Northern RRC, loc. no. 04898: minutes book 7, board meeting 23 May 1884.

372 Ibid.: minutes book 7, board meeting 25 July 1884.

capacities of 15 tons in 1886. The furnaces were located in the Eston Bessemer works, the scrap of which they were working up. Their steel was processed into plates in two plate mills.[373] However, since they could process only phosphorus-free materials, Bessemer pig or steel scrap, they did not make a great contribution towards overcoming the firm's structural problem, to wit the use of phosphoric ores. After several bad years, the general meeting said in its report for 1888, that the pig iron had done badly as usual, 'but we had something worse than that, we had our steel trade'.[374] Only two departments in the firm were able to deliver satisfactory results: the Spanish iron-ore mines and, the latest branch, the salt-extraction venture using residual heat from the blast furnaces.

The purchase of new blast furnaces and the construction of the basic Bessemer works had not been able to set the exploitation of their own Cleveland ores on a secure basis and, viewed retrospectively, they only created further problems. Since it could not sell as much basic steel as it needed to, the firm was furthermore obliged to restrict its pig-iron production to within the limits set by the Cleveland Iron Masters' Association. Consett, on the other hand, although continuing as a member of the CIA, was exempted from their restrictions on production in 1885, because it was able to guarantee that it would endprocess its entire pig-iron production itself, insofar as it came under the terms of the agreement.[375] Consett had thereby reached, by conducting a policy of cautious expansion, the freedom of movement which Bolckow Vaughan had been straining after with its massive investment in the Thomas process. Bolckow Vaughan was able only slowly to escape the bottle-neck into which it had manoeuvred itself. Its converter steelworks were finally closed in 1911, after which it made only open-hearth steel, as had long been the case at Consett.[376]

Along the north-east coast, the slow changeover to the Siemens-Martin process on the part of the other iron-plate manufacturers was due in part to their taking their own blast-furnace works into consideration. The Weardale Iron Co., Palmer's Iron and Shipbuilding Co. and the West Hartlepool Iron Works had run their blast furnaces on Cleveland pig iron and were thus necessarily still interested in keeping their wrought-iron works employed. After some delay compared with Scotland, they too ended by following the market and

373 Ibid.: minutes book 7, board meeting 18 December 1885. Compare too *BITA* 1886, p. 46 and *BITA* 1887, p. 116.

374 BSC-Northern RRC, loc. no. 04898: minutes book 7, board meeting 23 May 1884 and 24 April 1885.

375 BSC-Northern RRC, loc. no. 04902: report of directors for the year ending 31 December 1888, pp. 17ff.

376 Re limiting production within the framework of the CIA at B, V & Co., BSC-London, CIA MSS: minutes book 3, meeting 7 January 1886, 8 November 1886, secretary's report for 1888. Re Consett's release of pig-iron production, ibid.: minutes book 3, meeting 2 February 1885.

erecting open-hearth furnaces.[377] Britannia, too, which had in the meantime been taken over by Dorman Long, changed over to steel in 1886/7 once the price gap between iron and steel plates had closed, immediately erecting the second biggest open-hearth works in the region after Consett.[378] Although being tied to their own pig iron meant that the changeover in the works on the north-east coast was delayed, compared with the independent processors in Scotland, it could not prevent it from happening. By the end of the eighties, both regions were producing about the same quantity of open-hearth steel and together almost two-thirds of the entire British production.

Since the biggest customers of plates for shipbuilding stood as before on the Clyde, this meant that the north-east also delivered to Scotland and competed with the established steelworks there.[379] In order to protect their oligopolistic position, the four big Scottish works (SCS, Colville, Beardmore and Mossend) combined in January 1885 to set minimum prices for their products. Although Beardmore refused formally to enter into the agreement, it did take part in the meetings of the 'Scotch Steel Manufacturers' and gave tacit support to price-fixing. For the most important article, ship's plate, a price ex works of £6 15s. free of dockyard was established.[380] However, given the tougher competition above all from Consett, this price could not be defended and over the year it had to be reduced by 7s.6d.[381] Consett could lower its prices without incurring much risk, because it had, since the beginning of 1886, been able to make its pig iron about 10s. a ton cheaper than the SCS had to pay for it in Scotland at the time.[382] This was why the SCS began making enquiries in Spain that very year about securing their own supplies as soon as possible – unsuccessfully, as it turned out.[383]

Price agreements between the Scottish and Cleveland works did not come into being. It was only in connection with planning a cut in wages that they discussed adopting common measures against the Steel Smelters' Union.[384] In

377 *Iron* 24 August 1888, p. 186; 16 December 1887, p. 552. Ibid. *BITA* 1887, p. 116.

378 *BITA* 1886, p. 46 and *BITA* 1887, p. 116. In 1887 they already had eight open-hearth furnaces working.

379 *BITA* 1886, p. 40. It says here that the works in Cleveland expanded their production of open-hearth steel to the detriment of Scottish producers.

380 BSC-Scottish RRC, loc. no. UA 77 box 14: minutes, board meeting 7 January 1885. Compare too Byres, p. 389. *ICTR* 6 March 1885, p. 305; 24 April 1885, p. 548.

381 BSC-Scottish RRC, loc. no. UA 77 box 14: minutes, board meeting 24 February 1886.

382 At this time, the SCS free works in Hallside reckoned with 47s. per ton for all the big Scottish pig iron producers: Coltness, Dalmellington, Summerlee and Baird. Ibid.: minutes, board meeting 20 January 1886. Consett on the other hand had costs of 36s. per ton in the first half of 1886. BSC-Northern RRC, loc. no. 4698: cost analysis book, pig iron 1886.

383 Their enquiries in Spain are reported in BSC-Scottish RRC, loc. no. UA 77 box 14: minutes, board meeting 15 September 1886, 20 October 1886 and 20 March 1888. The search was abandoned only in April 1888. Ibid.: minutes, board meeting 4 April 1888.

384 Ibid.: minutes, board meeting 16 April 1890, 9 July 1890 and 17 July 1890.

matters of price, on the other hand, the 'Scotch Steel Manufacturers', which survived as a loose organisation into the nineties, had to content themselves with the obvious protection afforded them by the cost of freight from Cleveland.

(f) The development of the open-hearth process in Germany

In Germany, high-quality open-hearth steel spread comparatively hesitantly during the eighties. Apart from Krupp and the Bochumer Verein, there were as before no significant open-hearth works. The country still lacked a large specific market such as the shipbuilding and, with reservations, the tin-plate manufacture in Great Britain. Until 1887 the entire German production of steel sheet, plate and tin plate amounted to not even 100,000 tons. This would have employed precisely one works of the magnitude of Consett or the Steel Company of Scotland.[385] It should be taken into account that these 100,000 tons included both acid and basic Bessemer steel. In this period, the Siemens-Martin process had not developed materially beyond its role as a lucrative scrap processor for the Bessemer works. The Swedish metallurgists who in 1885 had travelled around the German steelworks observed in their reports that 'the process [Siemens-Martin, U.W.] has been labelled in the works we visited as a consumer of the works' scrap'.[386] They supposed that the proportion of pig iron was only 10 per cent. In any case, they were denied access to Krupp's and the Bochumer Verein's works. Whether their large production was also based mainly on scrap, part of which would then in any case have had to be bought in, cannot be established with any certainty. That both works employed the designation *Martinwerk* does, however, allow one to conclude that they worked according to the Martin pig-iron and scrap process, although this tells us nothing about the input ratio of the two components. Nevertheless, the relative position of Krupp and the Bochumer Verein within the German open-hearth industry had clearly been weakened in the course of the eighties. By 1885 the two had a share of only just 40 per cent of the total production, although they had almost doubled their own production since 1879.[387]

The typical development was thus not represented by Krupp and the Bochumer Verein, but by the processing of home scrap in the remaining Bessemer works, which has been described by Daniellson and Wijkander. The independent open-hearth works which they visited, such as Poensgen, Wittener

385 Byres, p. 389. H. W. Macrosty, *The Trust Movement in British Industry*, London 1907, pp. 66–72.

386 *Statistisches Jahrbuch* 1907, p. 267. *Thomas- und Martinwerke*, p. 662.

387 The Bochumer Verein had at this time a production of around 60,000 tons that year. Däbritz, p. 228. Krupp produced about 34,000 tons during the same year. HA-Krupp, WA VII f 767: Produktionszahlen 1876–1887. For the total production see Table 26, p. 223.

Gußstahlfabrik, Annener Gußstahlwerk, Asbeck-Osthaus-Eiken & Cie., and Stahlwerk Grafenberg, also processed mainly scrap.[388] The endproducts of these works constituted, as indicated above, a rich palette. Without emphasising anything in particular, the two Swedes' works mentioned steel tyres, axles, complete train wheels, artillery steel, plates for shipbuilding, wire, armoured plates, tram rails, sleepers, boiler plate, gun barrels and propeller axles.[389] It may be presumed that a similar variety could at that time be found in England in the Sheffield district, where successful experiments had already been conducted using acid Bessemer steel for a wider range of applications, although no new mass market for it had been found.

In any case, the products listed above had all of them already been manufactured from Bessemer steel, with a fair degree of trouble and mostly of lower quality, and they corresponded thoroughly to Bessemer's original notions about the potential uses for his steel. The Bessemer process, however, had soon developed into a mass-production process, whereby the production of variable qualities of precisely stipulated compositions would have been a nuisance only. The Siemens-Martin process immediately sought to fill this gap, assisted by its better quality; the market for it, however, remained very restricted. It determined the Siemens-Martin process's fields of application in Germany during the eighties, when it could as yet not be supported on a developed market in scrap or on a pig-iron base of its own, which would have enabled it to compete in price terms with converter steel or wrought iron, as was already the case in Great Britain with the manufacture of plate for shipbuilding. Until then the preconditions compiled by R. M. Daelen applied to the installation of open-hearth furnaces.

> These are given, if the production is such a limited one that the Bessemer process does not appear profitable, but on account of the higher selling value of the manufactured object the higher running costs [of the open-hearth furnace, U.W.] are however covered. The open-hearth furnace is thus particularly suitable for furnishing smaller plant, where the smaller construction costs also come into consideration.[390]

Compared with the seventies, no important new perspectives had emerged. Unlike in Great Britain, there were not large markets to steal from the puddling process. The wrought-iron works had until 1889 been able more than to offset their losses with wire, tin plate and sleepers by increasing their sales of 'Handelseisen' (wrought iron of all kinds – fashioned iron, construction iron and iron girders). This situation altered only in 1890, when 'Handelseisen' from the steelworks first became cheaper than that from wrought-iron works (see Table 32).[391]

388 *Thomas- und Martinwerke*, pp. 662–7.
389 Ibid.
390 Daelen, *Fortschritte in der Darstellung*, p. 641.
391 Table according to *Statistisches Jahrbuch* 1907, pp. 264–7.

Table 32. *Production and prices for 'Handelseisen' according to steel processes in Germany, 1885–1895*

	Puddled iron		Steel (Bessemer, open-hearth)	
Year	Quantity (1,000 t.)	Average price/t.	Average price/t.	Quantity (1,000 t.)
1885	921 t.	197.92 Mk	114.04 Mk	57 t.
1886	841 t.	98.45 Mk	107.25 Mk	69 t.
1887	1,015 t.	103.15 Mk	111.61 Mk	112 t.
1888	1,036 t.	112.55 Mk	113.54 Mk	192 t.
1889	1,109 t.	122.99 Mk	123.49 Mk	281 t.
1890	1,027 t.	138.46 Mk	136.69 Mk	308 t.
1891	973 t.	123.54 Mk	118.78 Mk	362 t.
1892	887 t.	114.88 Mk	101.17 Mk	515 t.
1893	808 t.	112.25 Mk	95.83 Mk	695 t.
1894	821 t.	106.21 Mk	91.09 Mk	875 t.
1895	790 t.	104.30 Mk	90.99 Mk	1,021 t.

It was thus apparent that quality criteria played only a subordinate role in the case of 'Handelseisen'. However, this also meant that the open-hearth works could not exploit the special strength of their product, its superior quality, on what was by far the biggest market for German heavy industry. This meant that they continued to lag behind the basic Bessemer works, which were able to manufacture an equally mild and easily malleable steel, which, given the low quality demanded, could replace wrought-iron – and that at a price beneath that for open-hearth steel. Nor indeed would the introduction of the basic Siemens-Martin process since about 1887 have brought about any fundamental changes, as demonstrated by the sales pattern for basic Bessemer and open-hearth steel during the following decade (see Table 33).[392]

The buoyant rise in the production of open-hearth steel between 1893 and 1894 signified only that a relative decline in the preceding years had been made good.

The overwhelming importance which the basic Siemens-Martin process was to have for the German and British steel industries in the twentieth century only emerged after the turn of the century, when the gap between open-hearth steel and basic Bessemer steel in Germany was steadily narrowing. The heavy industry of the Ruhr was now investing mainly in new open-hearth works.[393]

392 Table according to Beck, *Geschichte*, p. 1057.

393 In Rhineland-Westphalia, open-hearth steel had already achieved a 48 per cent share in 1913. Feldenkirchen, *Eisen- und Stahlindustrie*, p. 81 and p. 143.

Table 33. *Production of basic Bessemer and open hearth steel in Germany 1888–1898 (1,000 tons)*

Year	Basic Bessemer steel	Open- hearth steel
1888	1138	409
1889	1306	465
1890	1493	388
1891	1780	476
1892	2013	491
1893	2345	537
1894	2342	1069
1895	2520	1189
1896	3005	1478
1897	3234	1404
1898	3607	1459

On the one hand, this expressed the higher quality requirements of the export market, with the purchaser insisting on open-hearth steel immediately and, on the other, the higher standards of the German metal-processing industry, which was doing increasingly less well with lower quality basic Bessemer steel. Despite massive propaganda on the one side and energetic research on the other, the basic Bessemer works did not in the long term succeed in preventing critical consumers from rejecting their products.

6 Efficiency and capacity for innovation

1 Entrepreneurial decline and steel: the British debate

The question about enterprise strategies and technical progress in the German and British steel industry must in the end refer to the assessment of the relative efficiency and capacity for innovation of these two industries, including the entrepreneurs who ran them. While the technical competence and managerial ability of the German steel industrialists was never in doubt and the discussion was primarily sparked off by the political consequences of their dominant position in the Reich, their British colleagues were subjected to occasionally destructive criticism and were made partly responsible for Great Britain's gradual loss of her hegemony around the turn of the century. The different structures of the German and British steel industries, whose development has been studied in this book, have led to the conclusion that, by the late nineteenth century, there was a successful German way based on the Thomas process and the use of poor, domestic iron ores. The origins of the British steel industrialists' tendential decline lay in their failure to follow this successful model, which is attributed solely to incompetence on the part of British entrepreneurs. This orthodox picture was subjected to a fundamental revision by McCloskey at the beginning of the 1970s, which gave a strong boost to the debate about 'entrepreneurial decline' in Great Britain. Before we undertake a comparative assessment of the efficiency and capacity for innovation of the German and British steel industrialists on the basis of our preceding research, we intend to make a critical analysis of the positions adopted by the two most important antipodes of this debate: Burn and McCloskey.

Burn is convinced that the iron ores from Lincolnshire and Northamptonshire, which were processed to basic Bessemer steel by Stewarts & Lloyds in Corby from 1934 onwards, could have been profitably exploited as early as the nineteenth century as well. At that time, basic steel could already have been made on that site, which would have been cheaper than acid Bessemer steel.[1] He then tests his hypothesis by comparing the manufacturing conditions for

[1] Burn, p. 182.

basic steel in Westphalia, the minette districts and Cleveland with conditions in Lincolnshire and Northamptonshire. He considers the minette from Luxemburg and Hayange (de Wendel, Lorraine) to have been the ideal iron ore for the Thomas process in the eighties and nineties, since it was very cheap to extract and could produce a sufficiently phosphoric pig without difficulty in a blast furnace. He makes no mention of other iron ores' supplements. According to him, Westphalia was less well situated, given its lack of suitable iron ores, and so imported Minette and pig iron from Lorraine and Luxemburg in the eighties. For this reason, Thomas pig iron was more expensive in Westphalia than in the minette districts.[2] According to him, Thomas pig iron in Cleveland was not quite as cheap as minette iron, but still considerably cheaper than Bessemer pig iron in the same plate.[3] Lincolnshire and Northamptonshire iron ores were very similar to minette, and just as easy and cheap to dig up. Since they lay closer to the coal, their pig iron would have come at least as cheap as the Lorraine and Luxemburg types.[4] By the beginning of the twentieth century, Lincolnshire pig iron could have been sold profitably at 30s. per ton. Indeed, by around 1890, this pig iron was already being processed in a basic open-hearth furnace by the Frodingham Iron and Steel Co.[5]

Compared with Germany, then, conditions in Great Britain were supposed to have been even more favourable for the manufacture of basic Bessemer steel. The only hitch was that pig iron from Lincolnshire contained too little phosphorus for the early application of the Thomas process. The addition of basic slag, however, could have raised its phosphorus content to nearly 1.8 per cent. In the late nineties, this amount of phosphorus did suffice in Westphalia, after the phosphorus content of the pig iron there had necessarily dropped for lack of further supplies of puddling slag. The Northamptonshire iron ores contained more phosphorus and, in Staffordshire in the eighties, they had already been used as one of the ingredients for making Thomas pig iron. Although their phosphorus content was meagre, in Burn's opinion, this did not present an obstacle as the example of Westphalia demonstrated.[6]

It must be said of his comparison, that he overestimates the importance of minette for the manufacture of pig iron in Rhineland-Westphalia. Brugmann, in an extensive paper about the manufacture of pig iron in Germany after 1880 presented to the Iron and Steel Institute, gave the following ratios for the iron ores which were fed to the blast furnaces to make Thomas pig iron after the reserves of puddling slag had dried up:[7]

2 Ibid., pp. 155f. 3 Ibid., pp. 167f. 4 Ibid., p. 168.

5 Ibid., p. 168, notes 3 and 4. 6 Ibid., pp. 168f.

7 W. Brügmann, The Progress and Manufacture of Pig Iron in Germany since 1880, in: *JISI* 1902 part 2, p. 20. The compositions of the blast-furnace materials in the Ruhr point even more clearly in this direction, in Feldenkirchen, *Eisen- und Stahlindustrie*, Tables 12–16. Feldenkirchen thinks that the works in the Ruhr would have smelted even more Swedish ore, had its export not been subjected to a quota system. Ibid., p. 63.

35–40% minette
35–40% Swedish iron ore
10% Nassau iron ore
Rest various iron ores.

Since minette contained at most 40 per cent iron, but Swedish ores 60 per cent, the latter were much more important. However, Burn's faulty estimates concerned only the competitive conditions within Germany and do not affect the question about the internal British cost comparison, which can be the only basis for deciding for or against the Thomas process there. It is in this context that Burn has mooted a price of 30s. per ton for pig iron in Lincolnshire at the beginning of the twentieth century, one which totally contradicts the known price quotations, which vacillated between 40s. and 50s. per ton.[8] Burn based his price on a passing reference in a discussion at the Iron and Steel Institute about American and British rolling mill practices. No explanation for the unusual prices was provided in the discussion, not even a note about the composition of the pig iron concerned.[9] Burn's assurance that this pig iron was already being successfully processed in the open-hearth furnace is not even alluded to in his source references. Since Burn did not make further use of this obscure price reference in his argument, in particular not producing it for purposes of comparison, it is not after all very important to his conclusions. They are based far more on a hypothetical cost comparison between acid Bessemer steel in the existing coastal locations and basic Bessemer steel in a hypothetical works in Lincolnshire. He thereby refers to conditions in Cleveland, where both processes were worked side by side.

He starts by questioning the data provided by Harbord in his handbook *Metallurgy of Steel*, according to which a price difference of 6s. to 8s. per ton between Bessemer and Thomas pig iron was necessary before basic steel became cheaper to make than acid Bessemer steel.[10] Harbord, who was himself experienced in the manufacture of basic steel in Staffordshire, considered this price difference, whereby the Thomas process was only just more cost effective, to be the normal difference between the two sorts of pig iron. According to Burn, the Thomas process was undervalued by this. His view was supported by the success of the North-Eastern Steel Co., which worked exclusively according to the Thomas process and was profitable until 1900.[11] Its subsequent problems could be explained by higher coal prices and German dumping exports. Since its start-up in 1884, the North-Eastern had conducted a profitable business in rails, competing successfully with acid Bessemer works. Burn's claims were confirmed by J. Smith, who had already conjectured

[8] McCloskey, p. 62, note 16.
[9] Contribution to the discussion by W. H. Bleckly in: *JISI* 1901 part 1, p. 120.
[10] Burn, pp. 170f. F. W. Harbor, *The Metallurgy of Steel*, London 1904, p. 207.
[11] Burn, p. 171.

in 1886 that costs for both acid and basic Bessemer rails from Cleveland were the same. However, the production costs for basic rails at the North-Eastern subsequently fell on account of (a) having Thomas slag to their credit from 1887 onwards; (b) the removal of their patent fee of 1s. per ton from 1894; (c) the introduction of the pig-iron mixer at the same time. Burn concludes from all this that the North-Eastern could manufacture basic Bessemer steel more cheaply than other works could make acid Bessemer steel, since the acid Bessemer works had not managed to profit from the same proportional reductions in costs. However, if basic steel was cheaper to make in Cleveland, where conditions were less favourable than in Lincolnshire, then, according to Burn, it is still completely incomprehensible why the Thomas process was not introduced in grand style in Lincolnshire. The difference in pig-iron prices in favour of the Thomas process must have been greater in Lincolnshire than Harbord admitted. Harbord had presumably retained, not the potential price difference for the Lincolnshire location, but the price difference between existing works known to him. These Midlands works, however, had to pay supplementary freight expenses for their iron ores, since they were not working in the optimal location for iron-ore supplies. The South Staffordshire Steel Co., for instance, had freight expenses of 3s.3d. to 3s.10d. per ton of iron ore, and required over two, probably three tons, for one ton of pig iron. Avoiding these freight expenses would thus have made their pig iron considerably cheaper to manufacture. A completely obvious opportunity, so Burn's conclusion ran, had not been taken up by the British steel industry.[12] This analysis of Burn's was of fundamental importance for the adherents of the pessimistic school in the debate about the British entrepreneur in the nineteenth century.

Burn was recognised, even more than Burnham and Hoskins,[13] as an authority on the question of available technologies and investment opportunities. 'It is in the area of technological development, where quantitative investigation flounders, that a good part of the explanation for the United Kingdom's loss of leadership will be found. The subject of innovation within this area is vast, but only Burn's study is truly adequate.'[14] Landes in particular, whose *Unbound Prometheus* was described as 'the high-water mark of the critical school'[15] based his damning judgement on Victorian entrepreneurs to a great extent on examples taken from Burn.

[12] Ibid., p. 182.

[13] In their resumé they stated: 'British ores were adaptable to the Thomas process, but due to prejudice they were not exploited', and 'what was required was the setting up by 1890 of Thomas plants in Lincolnshire or Frodingham'. T. H. Burnham and G. O. Hoskins, *Iron and Steel in Britain, 1870–1930*, London 1943, pp. 120 and 180.

[14] Thomas J. Orsagh, Progress in Iron and Steel: 1870–1913, in: *Comparative Studies in Society and History 3* (1960–1), p. 228.

[15] Peter L. Payne, *British Entrepreneurship in the Nineteenth Century*, London 1974, p. 46.

McCloskey and Sandberg criticised this slender basis for his far-reaching judgement.[16] McCloskey has also provided a revised picture of the British iron and steel industry in his dissertation, in which he concludes that: 'Late nineteenth century entrepreneurs in iron and steel did not fail. By any cogent measure of performance, in fact, they did very well indeed.'[17] If, however, the show-piece of the pessimistic school was dismantled by this, then the entire hypothesis of entrepreneurial failure as the origin of Britain's lesser growth rate, compared with the USA and Germany, would have to collapse. McCloskey's criticism of Burn thus assumes a central role in the debate about the tendential decline of the British economy and the responsibility of the entrepreneurial class for it.

McCloskey takes up two of Burn's reservations against the use of Lincolnshire iron ore, which the latter for his part, had rejected.[18] The first related to the quality of these ores. Burn had described them as 'rather lean, rather variable in quality, and not very phosphoric'. According to Burn, this meant that they were similar to minette, apart from their low phosphorus content and, for this reason, were suitable after all.[19] The second reservation concerned the unfavourable economic and geographical position of the iron ore, being far removed from the coast and the centres of the steel-processing industry. Burn's contention was: 'what they gained on the ore . . . they might lose on the trains.'[20] Nevertheless, in order to be able to exploit the advantage of the lowest possible steel prices in Great Britain, lavish investment would have been necessary, which would straightaway have involved the steel users. 'In the circumstances great capital was required for rapid advance, and the most potent influence would have been capital associated either with a big consuming interest or with an important existing producer of basic steel. For several reasons the British economy did not favour a development of this kind.'[21] According to Burn, these reasons were founded in the inability of the entrepreneurial class to realise opportunities as they presented themselves as quickly and thoroughly as their German and American competitors.

McCloskey then checked whether Burn's own reservations were after all sounder than he himself believed. The first test applied to freight expenses.[22] McCloskey calculated the freight expenses for conveying Lincolnshire pig iron to Cleveland and then compared the price of Lincolnshire pig iron from Cleveland with that of Cleveland pig iron ex works. His prior assumption was

16 Donald N. McCloskey and Lars G. Sandberg, From Damnation to Redemption: Judgements on the late Victorian Entrepreneur, in: *Explorations in Economic History* 9 (1971), p. 97.

17 McCloskey, p. 127.

18 McCloskey, p. 60.

19 Burn, p. 168.

20 Ibid., p. 181.

21 Ibid., p. 182.

22 McCloskey, pp. 61–5.

that:

If Lincolnshire pig iron sold for less than Cleveland pig iron of the same grade in the same market, continued exchanges of Cleveland pig iron would demonstrate the static irrationality of consumers in not buying iron at the lowest price and the static irrationality of sellers in not selling at the highest price. The relevant statistic for this test is the prevailing market price, inclusive of rents: if the price differs substantially there is substantial static irrationality.[23]

In default of other price quotations, he establishes by this test the price for foundry pig no. III.[24] The comparison reveals that foundry pig no. III from Lincolnshire would have been more expensive in Cleveland than foundry pig no. III from Cleveland itself. The price difference in favour of Lincolnshire was less than the freight expenses to Cleveland.[25] A comparison of the added up costs of raw materials for pig iron in Lincolnshire and Cleveland also produced the same result. The higher price of ore in Cleveland was partly offset by the lower freight expenses for coke. In no case, however, did the difference in Lincolnshire's favour match the freight expenses to Cleveland, even by this method of calculation.[26] Thus, according to McCloskey, Burn's reservations about the freight expenses were pertinent after all. In Lincolnshire, they would actually have lost in freight expenses what was to be gained by the ore.

In a second test, McCloskey compares the manufacturing costs on site, in order to exclude possible differences in rents.[27] 'Since the issue is one of an allegedly larger incentive to expand the output of pig iron using Lincolnshire ore relative to using Cleveland ore, the "dynamic" test should be undertaken excluding rent in both locales: a nationwide incentive to expand output in all regions may exist without differential incentive among regions.'[28] 'If the costs excluding rents of Lincolnshire pig iron were substantially less than those of Cleveland pig iron, Lincolnshire pig iron was earning a differential rent and its output should have been expanded relative to Cleveland pig iron.'[29] The result was: 'If pig iron made in Cleveland had required 3.3 tons of ore and 1.5 tons of coke per ton of pig iron, as it did at the time in Lincolnshire, it would have cost 1.4 shillings per ton less than Lincolnshire pig iron.'[30] The difference is so slight that it is not attributed any significance. 'What is important is that the

23 Ibid., p. 61.

24 Ibid., pp. 62f. It is the price noted in Lancashire for this pig iron according to the *ICTR*'s weekly quotations from 1890. McCloskey subtracted a further 9s. per ton from this price quotation for freight expenses to Lancashire. Ibid., p. 63 notes 17 and 18.

25 Ibid., p. 63.

26 Ibid., pp. 63ff.

27 Ibid., pp. 65ff.

28 Ibid., p. 65.

29 Ibid., p. 62.

30 Ibid., p. 66. Emphasis as in original text.

costs were virtually equal . . . nothing would have been gained by an increase in the relative output of Lincolnshire pig iron . . . The relative neglect of the Lincolnshire ores, in short, appears to have been a rational adjustment to the balance of locational advantage.'[31] On the basis of these results to his quantitative tests, McCloskey rejects Burn's condemnation of the British steel industrialists. Their behaviour had not been irrational at all, but had corresponded to a high degree to market demands and natural conditions.

McCloskey's criticism of Burn has two decisive weaknesses. The first lies in the unfounded choice of Cleveland as the processing site for the pig iron.[32] One can only speculate as to why McCloskey thought of conveying the pig iron over a distance of 250 km to Cleveland, and of only converting it to steel once it was there. It is possible that he started from the premise that this could only occur in existing steelworks. These however, stood much nearer to hand in Sheffield, where Brown, Bayley & Dixon had already made basic steel employing Lincolnshire pig iron.[33] Furthermore, the Sheffield district disposed of a large and traditional steel-processing industry, which was much more likely to have come into consideration as purchasers for the basic steel than the shipyards on the coast. In his deliberations, Burn had always started from the premise that pig iron should be converted to steel and where possible end-processed on the spot. He gives no indication of having thought of transporting the pig iron.

Burn was thinking in terms of an earlier precursor of the basic Bessemer works in Corby, not even a basic open-hearth works, which also existed in Frodingham from 1890 onwards. His deliberations about the phosphorus content of the pig iron would otherwise be completely incomprehensible, since 1.8 per cent phosphorus presented a critical limit only in the case of converter steelmaking.[34] This sort of pig iron would not have been suitable for open-hearth furnaces. It was precisely the special suitability of the pig iron for the Thomas process in a converter, which authorised Stewarts & Lloyds to build their Corby steelworks in the 1930s. Their decision was influenced by calculations made by Brassert & Co. in London, whereby steelmaking according to the Thomas process in Corby would come much cheaper than with any other steelmaking process.[35] In Corby, then, the main product was tubes, for which the basic Bessemer steel was especially suited, on account of its soft-

31 Ibid., p. 67.

32 'The price of Lincolnshire pig iron at the point of consumption, taken to be Cleveland'. Ibid., p. 60.

33 C. B. Holland and A. Cooper, On the Manufacture of Bessemer Steel and Ingot Iron from Phosphoric Pig, in: *JISI* 1880 part 1, p. 80.

34 Burn, p. 172, refers expressly to 'Thomas-pig', which should have been made in Lincolnshire.

35 Carr and Taplin, p. 533. F. Lilge, Das Thomas Stahlwerk Corby (England), in *Mitteilungen aus den Forschungsanstalten des GHH-Konzerns* 4 (1936), pp. 147f.

ness and good weldability.[36] The endprocessing thus also took place on the spot.

Given the importance of the basic Bessemer works in Corby for the discussion about the competitivity of the British steel industry in the thirties,[37] it is not surprising that in Burn's opus, which appeared in 1940, it played such a role. For Charles Loch Mowat, Corby was 'an epic of pioneering', which elicited great hopes of surmounting the depression.[38] The Corby works rose up on green fields, and was employing 4,000 people within the shortest possible time. In this, it corresponded fairly precisely to the notions which Burn, as much as Burnham and Hoskins, had entertained about the missed opportunities of the 1890s. One will hardly be mistaken in supposing that their criticism of the entrepreneurs of the late nineteenth century was stimulated by the works at Corby. McCloskey's calculations about the freight expenses for conveying pig iron from Lincolnshire to Cleveland are irrelevant to these notions and wholly contradict the ironmasters' practice of having their pig iron and steel production in the same place, on grounds of cost. With the processes imputed by McCloskey, still many more regions would presumably have 'lost on the trains what they gained on the ore'.

The second inadequate aspect of McCloskey's argument is that he uses prices for foundry pig no. III, a pig iron employed for making castings but never for basic steel, for which purpose it was totally unsuited.[39] For this reason, McCloskey's calculations revealed only that the market for foundry pig iron was balanced equally between Cleveland and Lincolnshire, and that there was no justification for expanding production of foundry pig iron in Lincolnshire. Burn had certainly never disputed this. His lengthy, cautious discussion of the pig-iron question thus turned precisely around establishing the probable manufacturing costs of Thomas pig iron as opposed to other sorts of pig. Unlike McCloskey, he was not content to take the next best price quotations for any pig iron, and with good reason. The core of his comparative survey of potential basic steel production in Lincolnshire and the existing production in Cleveland, was his conviction that the manufacture of basic steel in Lincolnshire would not have involved the same problems as in Cleveland and, consequently, that the oncost, compared with the usual pig irons such as foundry pig III, would have been less in Lincolnshire.[40] His is not at all an eccentric assumption, since ordinary pig iron from Lincolnshire did not contain even half as much sulphur as, for instance, that made by Bolckow Vaughan or

36 Lilge, p. 147.
37 Carr and Taplin, pp. 532ff.
38 Charles Loch Mowat, *Britain between the Wars 1918–1940*, London 1964, p. 444.
39 McCloskey's reasons for using the price quotations for 'foundry pig No. III' were that he could find no others. McCloskey, pp. 62f.
40 Burn, pp. 171f.

Bell in Cleveland.[41] This was why Burn stressed that the relative prices for the various sorts of pig iron existing at the time, were not adequate 'to show the potential strength of East Midlands for steelmaking'.[42] In fact, no Thomas pig iron was being produced at that time in Lincolnshire, although its manufacturing costs could, on their own, have served as a rule for a comparison with other steelmaking centres and other processes. In this, according to Burn, lay the failure of the entrepreneurial class, but also the difficulty of estimating the competitivity of a basic Bessemer works in Corby or Frodingham. McCloskey's criticism cannot touch him, because the preconditions for both quantitative tests are not in keeping with their object.

Any examination of Burn's hypothesis must start with the possible suitability of East Midlands ores for making a Thomas pig iron, and attempt, as he does, to estimate the costs of such a product. This very important figure must be found before it can be compared with other ores. Burn's way thus appears, as before, to be the only viable one, although he incurs of course all the perils of a counter-factual assumption. One cannot, however, do justice in any other way to Burn's argument, which had become so very important to the discussion about entrepreneurial decline in Great Britain. Obviously, this also means that we must once again deal briefly with a few metallurgical problems, whereby we benefit from our present deeper knowledge of the subject.

The obvious starting point for examining the suitability of these ores for the Thomas process is their use in Stewarts & Lloyds's basic Bessemer works in Corby from 1934 onwards. They had until then been reckoned unsuitable for making Thomas pig. The most important thing wrong about Midland pig iron for the Thomas process was its high silicon content, which gave rise to high running costs, due to the rapid destruction of the converter's fireproof lining which is affected, and to the disproportionately large quantities of slag produced by it. In Scotland the Thomas process had on these grounds to be suspended yet again, because it was too expensive compared with the open-hearth process, in spite of using old slag from puddling for the pig iron.[43] It was precisely for this reason that East Midlands ores were mainly processed to foundry pig, which McCloskey had used as the basis for his comparison. With foundry pig, a silicon content markedly above 0.5 per cent, representing the upper limit for an economical Thomas process, was thoroughly desirable.

The problems presented by processing these ores to a viable Thomas pig iron

41 Frodingham 'white iron' (Lincolnshire) contained in 1885, according to an analysis published in *Engineering*: 0.12 per cent sulphur with 0.629 per cent silicon. *Engineering* 21 August 1885, p. 177. White pig iron from Bell contained in 1880 0.38–0.49 per cent sulphur with 0.58–0.70 per cent silicon. Bell, p. 416. In 1880, white pig iron from Bolckow, Vaughan & Co. had 0.25 per cent sulphur with 0.5 per cent silicon. *Zeitschrift des berg- und hüttenmännischen Vereins von Steiermark und Kärnten* 12 (1880), p. 493.

42 Burn, p. 172.

43 Payne, *Colvilles*, p. 71.

were outlined in Corby in 1936 as follows:

It is well known that when a blast-furnace, used for the smelting of ores of this type [Northamptonshire ores, U.W.], is burdened in accordance with common practice in Great Britain when making basic iron, the slags formed are high in alumina and lime, and of a very refractory character. As a result, the normal blast-furnace practice with these ores has been to work with a high temperature, necessary to melt the slag formed from the mixture of ore and limestone used, in accordance with the principles of blast-furnace burdening commonly adapted. This has only been possible at the expense of a high coke consumption, and the high hearth temperatures have provided conditions which, while suitable for the manufacture of foundry iron, have made the manufacture of basic iron of low silicon content very difficult. It has, in fact, been commonly accepted that such ores are unsuitable for the manufacture of basic iron, except in limited proportion in admixture with other ores.[44]

Part of these problems could be overcome by preparing the ores and altering the construction of the blast furnaces at the beginning of the 1930s. The pig iron's unusually high silicon content, which was deleterious for steelmaking, remained nevertheless almost unchanged. The attempt to manufacture a Thomas pig iron always ended the same way; the blast furnace 'operates in cycles, alternately hot and cold, the production fluctuates, and the quality of the iron is very variable'.[45]

The means of overcoming this difficulty had been very well known since 1880; adding ores with a strong manganese content, in other words, the way the Hörder Verein made Thomas pig. But, as already in the case of the Cleveland works, 'the use of these added ores, in some cases imported from abroad, has almost invariably resulted in increased costs of production, due to the freight charges involved in the assembly of the ores required'.[46] The added expense of importing ores to make pig iron could be avoided in Corby by a newly developed method of removing the slag in the blast furnace. This new method, which made it possible for the first time to make a satisfactory Thomas pig iron out of East Midlands ores, was based on American research from the years between 1915 and 1931.[47] By applying it under otherwise identical conditions, the daily rate of production of Thomas pig iron in Corby was increased by 21 per cent, with a simultaneous reduction by about 15 per cent in the consumption of coke.[48] However, the uniform working of the blast furnaces was viewed as the 'most important feature' of the new method of slag removal,

[44] T. P. Colcough, The Constitution of Blast-Furnace Slags in Relation to the Manufacture of Pig Iron, in: *JISI* 1936 part 2, p. 548.

[45] Colcough, p. 559.

[46] Ibid., p. 560.

[47] This method is extensively described in Colcough, pp. 560–70. In this connection, the preliminary scientific work by Rankin, Wright and McCaffery is expressly acknowledged as being a decisive precondition for it.

[48] Ibid., p. 572.

which led to the desired 'uniformity of composition of the iron'.[49] It was this lack of consistency, carried through to the steel, which had given British basic Bessemer steel products their bad reputation before the First World War.

However, Thomas pig iron from Corby still had one disadvantage. With the new blast furnace process, the sulphur content of the pig iron could not be taken into consideration, which meant that it was too high for steelmaking and had first of all to be reduced by adding manganese. It was thus not possible, in Corby either, to make any basic steel entirely without the addition of manganese, although it was thought that this way of removing sulphur was cheaper than adding more manganese in the blast furnace.[50]

The scientific premises for the manufacture of basic steel in Corby were thus the most recent and were not available in the nineteenth century. Conditions for the production of a Thomas pig iron in the late nineteenth century were basically as bad in the East Midlands as in Cleveland. No satisfactory Thomas pig iron could be made in either place without the additional cost of imported ores containing manganese. This was the reason why Middlesbrough, before the First World War, was by far the most important British port for importing manganese ores.[51] For this, however, the East Midlands were in a less favourable position, since they would have required additional rail transport from the coast.

Burn obviously thought that Cleveland pig iron was the only one requiring the addition of manganiferous ores. He thus mentioned the high sulphur and silicon content of Cleveland pig iron, which necessitated the import of 'relatively expensive ores' for making Thomas pig iron, while simultaneously producing the impression that this had not been necessary for East Midlands ores.[52] In their case, we read only that 'some other ores would possibly have had to be charged'.[53] These ores, however, as the references from Corby quoted above show, were the same 'relatively expensive' ores as those which also had to be imported in Cleveland. Before the new method for removing slag was applied in the blast furnace, Corby pig iron also contained too much silicon and its sulphur content was very high, even with the new method and in spite of

[49] Ibid., p. 573.

[50] Ibid., p. 572 and Colcough's contribution to the discussion which followed on his lecture, p. 575.

[51] Pothmann, p. 269. At the same time, Middlesbrough was also becoming the most important British port for importing Swedish iron ores, which, as in the Ruhr, were supplanting domestic ores more and more. This did not say much for the advantages of British ores in making phosphoric pig iron. The biggest single buyer there in the decade before the First World War was the North-Eastern with over a million tons. Martin Fritz, *Järnmalmsproduktion och Järnmalmsmarknad 1883–1913. De svenska exportföretagens produktionsutveckling, avsättningsinrikting och skeppningsförhållanden*, Göteborg 1967, Table 198.

[52] Burn, p. 173, esp. p. 174 note 1.

[53] Ibid., p. 169.

burdening the blast furnace with manganiferous charges.[54] In the course of the discussion about the Corby method at the Iron and Steel Institute, this high sulphur content of the pig iron was regretted above anything else, since it could be removed only by adding expensive manganiferous charges in the mixer.[55] It was claimed there that it might possibly be cheaper to increase the charge of imported manganese ores in the blast furnace, after all.

Burns's supposition, that it was easier and thus cheaper to manufacture Thomas pig iron in the East Midlands than in Cleveland, was not confirmed by experiences in Corby. The cost of local ores and blast furnace coke amounted to about the same in both places; McCloskey's 'dynamic' criticism is in this respect relevant. The cheaper Midlands ore was offset by the cheaper Cleveland coke.[56] It must, all the same, have been cheaper to import manganese ores in Cleveland, given that the works lay directly on the coast and did not require extra transport by rail. When the reintroduction of the Thomas process into Great Britain was being discussed at the beginning of the thirties, the Cleveland site was rejected at the time only because all the free ore supplies there were running out, and all the lucrative deposits were unlikely to change owners. On the other hand, the suitability of these ores compared to those of the East Midlands was not in doubt.[57]

Thus, when taking into account the essential additional charges which would have been needed for manufacturing Thomas pig iron in the East Midlands, there is hardly any reason why the cost of production for Thomas pig iron in the late nineteenth century should be estimated less there than in Cleveland. The references which Burn provides here are not sufficient. At the time, all the reservations against Thomas pig iron from Cleveland applied to pig iron from Lincolnshire and Northamptonshire as well. It would have been difficult to establish 'a new low level of steel prices for England' there.

The fact that the Thomas process was not expected to cost less in the East Midlands, however, does not mean that it would have been foolish to build basic Bessemer works there. Even with slightly higher costs of production for basic steel, there would have been the possibility, as in Cleveland, of producing steel for the domestic market, especially the nearby industries around Birmingham, since the freight costs would have provided a degree of

54 In spite of a manganese content of 1.4 per cent it still contained 0.13 per cent sulphur. In the discussion following Colcough's lecture, it was mentioned that with this average value traces of over 0.2 per cent sulphur had to emerge, which was very worrying for the manufacture of Thomas steel. Colcough, *Discussion*, p. 575.

55 Ibid., pp. 575, 577 and 580. One participant in the discussion considered basic steel with 0.13 per cent sulphur, as it was made in Corby, to be unusable. His works would not accept this material.

56 McCloskey, pp. 65ff.

57 F. W. Harbord, The Basic Bessemer Process. Some Considerations of its Possibilities in England, in: *JISI* 1931 part I, pp. 185f.

protection. According to McCloskey's figures, this protection would have amounted to 9s. per ton over against the basic Bessemer works in Cleveland. Harbord had suggested a production of this kind for the inland market in 1931, when he made a speech in favour of reintroducing the Thomas process in Great Britain.[58] Burn's argument also hinted that since the nineties basic steel was cheaper to make in Cleveland than acid steel, and by 1886, the manufacturing costs for steel rails were already the same for both processes.[59] This meant that a profitable production of basic steel could have been built up in the East Midlands from the mid-eighties, even without a cost advantage over Cleveland. Unlike in Cleveland, the richest iron-ore deposits had not all been allocated and the works would not have had to buy all its raw materials on the market, as the North-Eastern had to.

The viability of such a works would have depended on its sales opportunities on the internal market. Given costs almost on a par with Cleveland, the extra rail freight to the shipping ports meant that exporting it would only have been thinkable under extraordinarily good economic conditions. Its situation would have been similar to that of the Sheffield steelworks. For this reason, the production of rails suggested by Burn for the purposes of comparison would not have been considered, given that this market was anyway oversubscribed. Apart from which, the claim that by 1886 basic and acid rails were equally expensive to produce in Cleveland, is extremely dubious. Burn's source for this was a guess, made by J. T. Smith on a public occasion.[60] At the time, however, he was president of the rails cartel, which was fighting for survival in 1886. Certainly, no other information could be expected from him in this role. On the other hand. Arthur Cooper's evidence, which has already been quoted, ran along quite different lines. He was head of the North-Eastern Co., which Burn pointed to as a model of a profitable enterprise. In 1894 he explained why the Thomas process had spread so little in Great Britain by pointing out that the railway companies could still buy Bessemer rails at the same price or cheaper than Thomas rails. His works had a guaranteed outlet with the North-Eastern Railway, since it profited from transporting raw materials to Cleveland.[61]

> It is . . . against the interests of most of the big railway companies to buy such rails, as are manufactured from a works lying alongside another railway and especially those which are made by a process in competition with the other steelmaking processes, whose works carry out their traffic on these lines. The traffic conditions have much influence on the English railway managements and, for these reasons, the Great-

[58] Ibid., p. 187.

[59] Burn, p. 171.

[60] Ibid.

[61] Here, according to Thyssen-Archiv, RSW, 640-00 D b) 4: Der Thomasstahl in seiner Bedeutung als Schienenmaterial, Entwurf nach den Beschlüssen der Commissionssitzung am 28 Mai 1895, p. 7.

Western Railway buys the bulk of its requirements from manufacturers in South Wales; the Great-Northern, Midland and Great-Eastern go to Sheffield, whereas the London and North-Western meet their requirements in South Wales, or from the West coast and the North-Eastern from manufacturers in the Middlesbrough district.

At the time Cooper was delivering this estimate, which can hardly be considered an advertisement for his firm, all three of the cost-cutting operations promulgated by Burn had already been effected by the North-Eastern: the slag credit, the introduction of the pig-iron mixer and the removal of the patent fee.[62] In spite of all this, the North-Eastern could at best just keep its prices level. The other rails customers named by Cooper were, apart from the North-Eastern Railway, either Scottish railways, whose network joined up with the North-Eastern network, and to which Cleveland could send freight at advantageous rates, over other British works, or railways in the colonies, whose orders were dealt with through the colonial service. There was no question of doing business on the hotly disputed 'neutral' markets.

This meant that production in the East Midlands would have been limited to mild steel, as urged by Harbord in 1931 and realised by Stewarts & Lloyds in 1934, in order to exploit the strengths of the Thomas process in the case of this product.[63] This sort of production already existed in the neighbourhood, in Alfred Hickman's South Staffordshire Steel Co. There they were already using Northamptonshire ores for making Thomas pig iron.[64] Hickman's basic steel production was the direct precursor of the works in Corby. Stewarts & Lloyds had bought Hickman's steelworks in 1920, as well as acquiring the iron-ore deposits in Northamptonshire, which later formed the basis of the new steel works.[65] Nevertheless Hickman had still been using the Northamptonshire ores merely as additional charges and his pig iron was blasted mainly from puddling slag.[66] As the puddling process gradually disappeared, however, this slag became increasingly scarce and expensive. So Hickman enlarged his works with basic open-hearth furnaces, which did not rely on a high phosphorus content in their pig iron.[67]

Burn names Hickman's works as an example of how Northamptonshire pig iron could successfully be processed to basic steel according to the old method. According to him, the works in Bilston were simply operating on the wrong site. The ores had to be conveyed by the railway, which rendered them unnecessarily expensive at the blast furnace. In any event, this criticism of the

62 Almond, *Making Steel*, pp. 211 and 224. Burn, p. 171.

63 Harbord, Basic Bessemer, p. 185.

64 Burn, pp. 169 and 172, attributes great importance to this example. Compare too Harbord, *Thomas-Gilchrist*, p. 86.

65 Burn, p. 437 note 2.

66 Harbord, *Thomas-Gilchrist*, p. 86.

67 Ibid., p. 87. When processing puddling slag first began, the Hickman Thomas pig still contained 3 per cent phosphorus.

location applied only when ores were mainly being charged. So long as Hickman was working with up to 80 per cent slag, he had lower freight costs in Bilston, which lay close to the puddling works. However, Hickman does not appear to have been held back by the freight costs ensuing from his greater use of Midland ores. At the beginning of the twentieth century, when phosphoric additions were already very scarce, trials were conducted in Bilston for refining a pig iron with a high silicon content in the basic converter. These trials were abandoned once again when their results did not come up to expectations.[68]

Similarly, the two basic Bessemer works in Cleveland, which shared the same problems, also tried to refine a pig iron with a higher silicon content. Bolckow Vaughan developed the two-slag process proposed by O. Massenez until it was ready for production, and ran its basic Bessemer works with it from July 1905.[69] The advantage of this process was that ordinary siliceous Cleveland pig iron could be used without the addition of manganic or phosphoric charges. This pig iron was first blown in the converter until its silicon was completely oxidised. Then all the siliceous slag was run off, after which the usual Thomas process was carried through. This meant that they could do with less phosphorus, since the iron in the converter was already extremely hot as a result of burning off the silicon. Basically, this entailed a simplification of the unsuccessful 'transfer-systems' from the early eighties, which were now conducted in one single converter. A. W. Richards claimed that he had thereby achieved considerable savings, since with this pig iron he could avoid adding manganese and phosphorus, which had been 'a very expensive matter'[70] that he was finally in a position to be able to manufacture steel of uniform quality, so that the reservations against the English basic Bessemer steel became untenable. He based this claim on the pig iron's lower phosphorus content, which meant that the after-blow ended sooner.[71] However, this dubious explanation did not catch on with his colleagues at the North-Eastern Steel Co. It was rather directed towards a credulous clientele, which had hitherto rejected basic steel on account of its lesser quality.

The process was also introduced as a trial at the North Eastern, but proved unprofitable there, because it took up too much time. In addition, the steel did not reach the quality usually achieved when pig iron with a higher manganese content was used, which was why they abandoned it. Cooper explains Bolckow Vaughan's different assessment of its profitability by pointing out that Bolckow Vaughan 'had a very much more extensive

68 Thomas Turner 1907 in a discussion at the Iron and Steel Institute, in: *JISI* 1907 part 1, p. 109.

69 A. W. Richards, Manufacture of Steel from High-Silicon Phosphoric Pig Iron by the Basic Bessemer Process, in: *JISI* 1907 part 1, pp. 104–8.

70 In the discussion following his lecture, ibid., p. 112.

71 Ibid., p. 106f.

converter plant than the North Eastern Steel Co., and in consequence of this the appreciable delay in the process was more serious at the latter works, and reduced the output materially and thus increased the cost'.[72] For Bolckow Vaughan, too, this latter attempt to refine Cleveland pig iron without additional charges in the basic converter did not admit any prospects for the future. The steelworks in Eston was finally closed in 1911. The North-Eastern managed to survive the war and gave up the Thomas process in 1919.[73]

The problems about the quality of Thomas steel could not be solved. Being rejected as a material for shipbuilding and the ensuing prohibition by Lloyd's had proved particularly damaging, and had thoroughly ruined the reputation of basic steel in Great Britain. 'The mention of basic-Bessemer, it was said, was like a red rag to a bull to the ship-builders of the North-East Coast' Burn tells us.[74] Even tube and plate manufacturers rejected basic steel, although its properties by far came closest to their requirements.

In his criticism of Burn, McCloskey fell far below the level of discussion which Burn had striven to achieve, since he did not take note at all of the technical problems linked to the introduction of the Thomas process in the East Midlands and did not enter at all into the kernel of Burn's argument. The sort of pig iron which lies at the basis of his quantitative tests was completely unsuitable for steelmaking in the basic Bessemer converter; furthermore, he was alone in suggesting transporting pig iron to Cleveland, a notion which was not only totally devoid of any kind of business sense, but which was also never propagated in any way by Burn. McCloskey's results are thus irrelevant to the discussion about the introduction of the Thomas process in the East Midlands.

Burn built his argument on a counter-factual reconstruction of supposed manufacturing conditions and costs for the Thomas process in the East Midlands. He included the production of Thomas steel in Cleveland to provide a comparative measure within Britain. Such proceedings, which must make do without direct quantitative comparisons of real quantities, can lead to credible conclusions only when a fairly large number of clear indicators point in one direction. This, however, is not the case, as shown in particular by the discussion in the Iron and Steel Institute about the blast-furnace process in Corby. This discussion tended rather more closely towards the opposite conclusion, that the scientific premises for the production of a competitive Thomas pig from Midland ores were still lacking before the First World War. Burn knew about the main contributions to this discussion;[75] he nevertheless did not make any use of its findings and relied instead on a series of

72 Contribution to the discussion by A. Cooper, ibid., pp. 109f.

73 Almond, *Making Steel*, pp. 217 and 219.

74 Burn, p. 175.

75 All the lectures and discussions held in the Iron and Steel Institute about the Thomas process quoted here are also specified by Burn.

suppositions and claims, expressed in connection with other events. In one remark only does he concede that the metallurgical knowledge of the time was not yet adequate. In his opinion, this obstacle could have been overcome simply by 'groping in the dark', even without the metallurgical research conducted over the next twenty years by Rankin, Wright, McCaffery etc. up to Clough.[76] In reaching this conclusion, Burn stepped beyond the reasonable bounds of a counter-factual investigation into the realms of mere wishful thinking.

It was certainly no failure on the part of British steel industrialists that their technicians did not discover by chance in the 1880s that which American and British research scientists were to discover in their laboratories a few decades later. Correspondingly, Burn's comparison between a hypothetical basic steel production in the Midlands and the existing one in Cleveland, having furthermore overestimated its competitivity, started out from false premises. Indeed, the British basic Bessemer works can scarcely be accused of being insufficiently keen to innovate, given the long-drawn-out trials and developmental work in Cleveland and Staffordshire. The fact that they were not granted a lucky chance hit, as the Hörder Verein was right at the start of its efforts, cannot be ascribed to obduracy or insufficient knowledge. In the 1930s, when they finally, with American help, did find a suitable smelting process for their domestic ores, it was immediately exported to Germany where it served for processing ores of a similar composition around Salzgitter and in southern Germany, which the German steelworks, like the British ones in the case of Lincolnshire ores, had previously been unable to exploit.[77] In this way, the second vital innovation for using phosphoric ores came to Germany from Great Britain. One cannot consequently speak in terms of the British steel industry's scientific or technological backwardness in connection with the Thomas process. The same thing applies to the British entrepreneurship's readiness to innovate prior to the First World War.

Taking the level of knowledge at the time into account, one must assess the prospects for a successful basic Bessemer works in the East Midlands in the late nineteenth century very sceptically. Under those conditions, it would already have required a certain foolhardiness on the part of an entrepreneur to have invested his capital in a works whose production had no prospects of being superior, in terms of quality or price, to that of the two basic steelworks

[76] Burn, p. 169 note 1.

[77] In Germany, Paschke and Petz built on these research findings and also acquired patents for the so-called 'acid smelting' process, which was applied for the first time in Corby. The process established the metallurgical basis for the exploitation of the Salzgitter ores and the construction of the Hermann Göring Werke by the American Hermann Brassert, who had also erected the works in Corby. Further details can be found in Matthias Riedel, *Eisen und Kohle für das Dritte Reich. Paul Pleigers Stellung in der NS-Wirtschaft*, Göttingen 1973, pp. 134–7.

in Cleveland, but which, on account of the poor transport conditions, would have entailed even higher production costs. What was more, the steel industry offered an alternative investment opportunity, one which promised very safe profits – building an open-hearth steelworks. This path was also followed by the British basic Bessemer steelworks, apart from the North Eastern, from the nineties at the latest. According to McCloskey, the British open-hearth works were fitted out with the very latest plant.[78] Open-hearth steel fulfilled all the requirements for shipbuilding and bridge construction, and was unreservedly approved not only by Lloyd's but also by the Admiralty. This seal of approval contributed towards securing the esteem of all those branches of industry and firms which could not carry out tests of their own. Since its price still lay not inconsiderably above that of both acid and basic Bessemer steel, but its quality was not questioned, it made a rapid breakthrough in the course of the eighties. Whereas wrought iron was mainly replaced by basic Bessemer steel in Germany, in Great Britain this was done by the qualitatively superior open-hearth steel. In the two countries, the different raw materials on which their manufactures were based had allowed them to develop along divergent lines.

They drew closer together only when more demands began to be made on the quality of steel in the German market and, as was already the case in Great Britain and most of the export markets, open-hearth steel was increasingly insisted on. After the turn of the century, then, as previously in Great Britain, the Siemens-Martin process was the preferential choice for new plant. Basic Bessemer capacities were extended further only in minette regions, where the ore could not be processed economically to open-hearth steel, in spite of extensive trials. All the research efforts in the existing basic Bessemer works were now being directed to achieving 'Siemens-Martin equality', which dominated the ongoing discussion among the German foundry workers in the thirties.[79] In other words, they tried with their existing basic Bessemer plant, to produce a steel which reached the higher quality of open-hearth steel, thereby ensuring that their plant would in future be worked to capacity. Nevertheless, the bulk of production was lost to open-hearth steel.

The differences in the quality of basic Bessemer and open-hearth steel also create a few problems for comparing the competitivity of the German and British steel industries prior to the First World War. Thus Allen came to the conclusion, based on the average price of selected steel products, that after the turn of the century the German steel industry was able to reduce its manufacturing costs below the level of British works, whereas in the eighties they

[78] McCloskey, pp. 71f.

[79] Karl Ernst Mayer and Helmut Knüppel, Entwicklungslinien des Basischen Windfrischverfahrens in Deutschland, in: *Stahl und Eisen* 74 (1954), pp. 1271–4.

had still stood just above it.[80] It was thus obvious that the British steel industry did not manage to reduce its manufacturing costs like its German competitors. Allen sees the reason for this in the British works' lower productivity and higher costs for raw materials. The thesis about entrepreneurial decline is thereby confirmed, contradicting McCloskey. In any case, he fails in his collected data to distinguish between different quality steels. This means that he has necessarily to conclude that after the eighties, when the Thomas process was becoming established, German steel was on average cheaper than British steel. Once again, this emphasises the preponderance of the qualitatively superior and therefore more expensive open-hearth steel in Great Britain over the simpler and cheaper basic Bessemer steel in Germany. Allen compares the price and factor inputs of different products, which were also valued differently on the market. His conclusions are thereby derived from false premises, which means that they remain irrelevant. The same applies to Webb's productivity comparison on the basis of German export prices and freight expenses to Great Britain.[81]

Indeed, it was precisely this which marked the development of the German and British steel industries from the eighties, in that, on the basis of different structures of demand and on account of different iron-ore supplies, the average qualities – and thereby the costs – of their steel products developed separately in the first instance. Correspondingly, before the First World War only the simplest basic steel qualities were exported from Germany to Great Britain – those which could not be manufactured there, or only with the greatest metallurgical problems.[82] This was why these export successes cannot without further ado count as an indicator of higher productivity. On the other hand, as the value of their steel products rose, the imports from Germany onto the British market became increasingly insignificant. These findings directly contradict the productivity thesis, since much more work and capital were embodied in such products than in the simple steel ingots and girders which

80 Robert C. Allen, International Competition in Iron and Steel, 1850–1913, in: *Journal of Economic History* 39 (1979), p. 912.

81 Steven B. Webb, Tariffs, Cartels, Technology and Growth in the German Steel Industry, 1879 to 1913, in: *Journal of Economic History* 40 (1980), pp. 322f. Distinctions drawn according to the shape of the produced and exported steel, as undertaken in trade statistics, do not help here. Plates and girders etc. could in fact be manufactured just as much from basic Bessemer as from open-hearth steel. This difference was then also expressed in the price, and Webb and Allen have found this difference in the production and export statistics. Their calculations would be relevant only if they had compared, for instance, German open-hearth plates with British open-hearth plates, and not undifferentiated plates. Unfortunately the official statistics do not permit this differentiated comparison.

82 The conspicuous German steel exports to Great Britain shortly before the war consisted quite overwhelmingly of semi-finished products, which were end-rolled there. They must have been considerably cheaper than British open-hearth steel, if they were supposed to replace it. *JISI* 1931 part 1, p. 205; Statistik des Deutschen Reichs. Auswärtiger Handel, 271 A 819139, Teil IX, p. 56.

came from Germany. Since, as Allen established, all the signs indicate that the raw materials in both countries were equally productive,[83] this can only mean that the productivity of labour and capital in the British steel industry, for all products superior to the simple basic Bessemer quality, can hardly have been worse than that in German works. This, however, still does not take account of the fact that the German works often offered their wares on the British market at dumping prices, below the average yield for these products.

Thus, that which at first glance gives the impression that the German steel industry was on average more productive, turns out to be the result of British steel production being undershot in terms of quality by products manufactured in German basic Bessemer works. In the case of products of equal quality and value, meaning primarily open-hearth steel, on the other hand, German industry did not enjoy spectacularly successful exports to the British market. In the case of shipbuilding materials, the most important branch of the British steel industry, the picture is even reversed. In their case, the big German shipyards could, at the turn of the century, only be induced by contractual means to use more expensive German steel instead of English steel.[84]

2 Precarious protectionist measures versus entrepreneurial autonomy

In spite of such trade restrictions, which were certainly not limited to Germany or to only one product, British heavy industry was able to protect its share of the export trade until 1909, as Saul has shown.[85] Only afterwards did it clearly lose ground against a strengthened German competition entrenched behind its protective tariffs. As we have seen, this applied to those sorts of steel which it could not manufacture for lack of the right ore basis. It has been made plain by Temin and Payne, who builds on Temin's research, that by the late nineteenth century it was already impossible to defend the industry's share of world production.[86] The main features of Temin's argument are that, since the eighties, the British share of the export trade would have had to increase in order to keep up with the faster growth of internal consumption in the USA and in Germany. The American and German markets were protected by tariff walls,

[83] Allen, pp. 919f.

[84] Marina Cattaruzza, *Arbeiter und Unternehmer auf den Werften des Kaisserreichs*, Stuttgart 1988, p. 35.

[85] S. B. Saul, The Export Economy, in: *Yorkshire Bulletin of Economic and Social Research* 17 (1965), p. 17.

[86] Peter Temin, The Relative Decline of the British Steel Industry, 1880–1913, in H. Roskovsky (ed.), *Industrialisation in Two Systems: Essays in honor of Alexander Gerschenkron*, New York 1966, pp. 148–51. Peter L. Payne, *Iron and Steel Manufactures*, in: Derek H. Aldcroft (ed.), *The Development of British Industry and Foreign Competition 1875–1914*, London 1968, pp. 97f.

which excluded the British steel industry from these fast-expanding markets. Since steel sales in the rest of the world were not growing to the same extent, the relative decline of the British share in the world production of steel was therefore unavoidable, even when their share of exports remained constant. It did not lie in the power of the British steel industry to maintain their leading position.

Temin's argument is convincing, although, as Payne pointed out, his statistical basis is relatively sparse. Assessments by contemporaries of the relative backwardness of British sales quotas are thereby confirmed, which had long been relegated to the background in the ongoing economic historical discussion. An extract from a memorandum by James Bryce, the then Under Secretary for Foreign Affairs, reads like a preview of Temin and Payne's argument. He wrote in 1886 that

> Under the changed condition of the world, with telegraphs and lines of steamers everywhere, with some large markets closed by protective tariffs, with native dealers supplanting the old system under which British mercantile houses did business through their branches abroad, the competition to which our commerce is exposed is far more severe than at any previous time. We must face this, perceiving that it was impossible under these changed conditions to retain the sort of monopoly which we practically enjoyed in many parts of the globe, and comforting ourselves with the knowledge that we are still far ahead of any other people.[87]

Since these protective tariffs had been introduced with the express intent of deflecting cheaper British competition, further reductions in the price of British steel would only have led to increases in the relevant tariff levels. The British steelworks' sales opportunities would scarcely have been improved by this. The tariff, after all, protected not only the German steel industry's inner market, but also it made it easier for it to extend its share of exports at dumping prices. The dual character of the tariffs, which made them enormously effective in assigning shares of the world market, presupposed the formation of the national steel industry into a cartel, and on this basis prevented the German steel industry, when it was being built up in the seventies and eighties, from having to conform to the international structure of costs. On the other hand, this did enable it, unlike the British steel industry, to stabilise low manufacturing costs by means of more intensive plant utilisation which, even in times of economic recession, was then effected with the help of dumping exports at the expense of consumers inside the Reich.

As we have seen, the German steel industry's lower prices on the export markets had since the seventies no longer meant that they also had lower costs

[87] Extract from a Memorandum by Mr James Bryce respecting the question of Diplomatic and Consular Assistance to British Trade Abroad, Foreign Office, 17 July 1886, here from D. C. M. Platt, pp. 414f.

than their British competitors. It is therefore not required to explain their export success since the last third of the nineteenth century, as do Burn and many others of his persuasion, by assuming, on the grounds of their more advanced technology, that costs in Germany were lower. The strength of German exports lay far more in their large internal market, hedged about by tariffs, which by the beginning of the twentieth century was already absorbing more steel than the British. Since the seventies its over-average growth and high level of prices had established the basis for its export success.[88]

Should one wish to discuss any failings on the part of British steel industrialists, then they are to be seen in their hesitant attitude in the seventies and eighties towards the protectionist policies of their competitors and their respective governments. In retrospect, they could be accused of having failed to use their collective market power in time and to best effect. In this, and not in the technological sphere, they lagged behind their German colleagues.

Unwillingness to give up their entrepreneurial autonomy in favour of the collective enforcement of their market power within the framework of a national cartel was peculiar to the British steel industry in our research period. Nor can it be explained by any lack of opportunity for such a proceeding under the prevailing political conditions in Great Britain. After all, both the railway companies and the Colonial Service had shown by means of concrete instances that they were willing to give home works the edge even when their prices were underbid by foreign competition. A better response than this unspoken favouritism was not to be expected without relevant public proceedings by the steelworks. The German works had consequently exploited this opportunity in the VDESI. The BITA did not even make the attempt, although on the Continent it was very much expected, above all when Great Britain had a Conservative government.[89]

When the British steel industry also finally went over to the practice of dumping exports, it had already lost its dominant position on the world market and could no longer prevent the German and American competition, which had in the meantime grown as strong but was better organised, from gaining more ground. It would still have been possible to do so without further ado on the basis of the British and colonial markets in the seventies and eighties, when the cartels had come into being in Germany as the 'children of necessity',[90] and

88 In the survey conducted by the Tariff Commission in 1904 J. S. Jeans, Secretary to the BITA, referred to the fact that German heavy industry had in the meantime won a great advantage in relation to the introduction of new processes and plant, in that its internal market absorbed a total of 5 million tons a year instead of only 3 to 3.5 million tons as in Great Britain. Report of the Tariff Commission para. 1024, here from Payne, *Iron*, p. 97, note 3.

89 See above p. 134 and GHH-Archiv, 300 108/13: Bericht für den Aufsichtsrat, August 1885.

90 Blaich, p. 54, disputes the notion that they were crisis cartels and bases his opinion on the long lives which these associations enjoyed, through all the economic fluctuations of business cycles. We must maintain in contradiction to this that the reasons which led to the rigid

this eventuality had also been feared. Whether this sort of heightened international competition, in which Landes has seen the single most important factor for the new imperialism,[91] would also have been politically desirable, was indeed another question. Such deliberations may nevertheless scarcely be imputed to those representing the interests of enterprises in the nineteenth century. What was certainly more important was that the same urgency about forming a cartel was not felt in Great Britain at the time. I. L. Bell made the following comment about this to the Royal Commission:

I have, however, been informed that the average dividends of about fifty German iron companies was only about 2 per cent in 1879, and about 5¼ per cent in 1884, which, on such undertakings, is not a high rate of interest. Generally then, you would say the German and Belgian companies are not prosperous? They may be moderately profitable.[92]

If that, in the view of British steel industrialists, was the result of protective tariffs and cartels, then it was certainly no grounds for imitation and makes Bell's great scepticism about the usefulness of trade restrictions easy to understand.

On the other hand, a principal revulsion against market agreements can certainly not be attributed to British steel industrialists in general and to Bell in particular. Bell was, with his iron works, a member of the Cleveland Iron Masters' Association and was involved in their quantity and price accords. Apart from which he was a director of the Barrow Steel Co., which led the British group of the IRMA from there. However, these market accords were only occasionally effective and served in the main for containing and smoothing economic slumps and not, as was the case with the German cartels, for fighting against a persistent inferiority. The IRMA and CIA adapted to a recessive economic situation by means of concerted restrictions on production and they did not try to outmanoeuvre it by stepping up dumping exports. Even in years of severe slump, as in 1877–9 and 1884–6, the BITA's reports proclaimed all the more their confidence in the competitivity of British heavy industry and in the further capacity for absorption above all of their colonial

cartelisation of the industry were structural rather than short-term fluctuations. What is more, the truly non-obligatory railmakers' association of the sixties and early seventies with its very 'English' and encouragingly loose agreements is no longer comparable with the new railmakers' association, formed in 1876 with the support of the Disconto Gesellschaft. Apart from their name (*Schienenvereinigung*) and a few members, these two associations had scarcely anything in common. The new railmakers' association was a very strictly led band, whose members belonged to it under threat of going down, whereas its precursor had presented more of a gentlemen's agreement, without having a really obligatory character. In any case, this difference only becomes obvious when based on a technical inquiry into enterprise history, presented here.

91 Landes, p. 228.

92 R.C. Depression, evidence Bell, p. 122.

markets. The fact that they had long overestimated the demand from the colonies and less developed countries in South America and Asia and had simultaneously underestimated the developments on the Continent may possibly have constituted grounds for the disparity between the often remarkably encouraging optimism in the British steel industry and its comparatively minimal rates of growth and strong cyclical sales.

For decades, British heavy industry had dominated the world market with such sovereign power that differentiating between market behaviour in the internal market and in exports, which was of existential importance for building up the German heavy industry, was considered expendable. It did not appear necessary to the majority of enterprises to renounce their entrepreneurial autonomy in favour of a comprehensive system for safeguarding the market. Their occasionally worsening profit situation was merely conjunctural and not – as in Germany – also structural in nature. Accordingly, the numerous associations restricted themselves to – mostly unbinding – price quotations for regional groups and – in the most extreme cases – to collective crisis management for a limited period. They had no motive for permanently relegating their entrepreneurial freedom of decision on the market to a superior organisation. Only under this condition, however, would it in the last analysis have been possible for them to defend their own share of the export markets.

In any case, when discussing the German and British steel industries' shares of the market, one should not overlook the fact that expansion *per se* is not what a firm and, consequently, an industry is aiming at, but a high level of the most stable possible returns. This, as the monopoly theory teaches us, is not best achieved by endless expansion, since this gives rise to overproduction and pressure on selling prices.

In the seventies and eighties, the British steel industry's returns were obviously better than the Germans' – and this the British heavy industrialists were conscious of, as I. L. Bell's statement shows. One might therefore ask whether these good returns could not have been improved still further by acting in an expansive manner on the world market while attempting to block the German steelworks' intrusion onto the same. Only then would it have made economic sense to expand further. Counter-factual questions about historical development cannot be answered with certainty, but in this case there is a series of important indicators which suggest the conclusion that such expansive behaviour on the part of the British steel industry would have promised little success from the economic point of view, and for this reason was duly avoided.

In view of falling prices, rising customs tariffs and markets controlled by cartels, there can be no question of a supply bottle-neck in the case of steel during the whole of our research period. A surplus supply on the part of British steelworks would thus definitely have led to further price reductions on the unregulated export markets, which would also have established the conditions for suppressing the German works there. Since a notable surplus supply could only be aimed at by investing in new, additional works – and not merely by operating the existing ones more intensively as during the wave of rationalisation in the seventies and early eighties – the average costs of production would at best have remained constant – and with simultaneously falling prices, the returns would have fallen in any case.

In Germany, on the other hand, the initial situation was completely different. Here it was not a matter of stabilising good returns and, instead of flooding the market to the detriment of prices, of preferring to invest in foreign branches. After the devastating effects of the wave of rationalisation in the seventies and under the more difficult terms of higher costs for raw materials, any return at all was initially desirable. Here, the alternative lay between either mass bankruptcy, as desired by Krupp and initially by the Disconto-Gesellschaft too, in order to match production capacities to domestic needs, which would then indeed have permitted the survivors to obtain a satisfactory return, and between collectively rescuing the capital funds which had been invested in the steel industry. The latter course was adopted, whereby free competition was widely excluded.

The British steel industry had never been reduced for long to this quandary which, as Bell correctly remarked, had led to a long period of lower, if any, dividends and whose only advantage was to differentiate them from the otherwise threatening wave of bankruptcies. Thanks to its lower raw-materials costs, it had not been forced in the seventies to adopt such intensive rationalisation measures as the German works had, and had correspondingly suffered less from their effect on capacity. In 1883/4, nevertheless, as soon as a similar situation to that in Germany ten years previously arose in Great Britain, steps were taken there too towards forming a cartel, which, for lack of any permanently superior competition, could assume only a temporary character.

The British steel industry was subjected to much less pressure to adapt. Each individual firm's room for decision-making was (thanks to their unusually good situation *vis-à-vis* raw materials and consequently costs compared with their German rivals), with the exception of the IRMA's brief existence, always large enough to allow them to stabilise their anyway fairly high returns without partly delegating their decision-making autonomy to superior organisations. One expression of this was apparently that the British steel industry did not grow any further once it found a profitable outlet on the free market, thereby avoiding their German competitors' not very lucrative immobilism during the seventies and eighties.

One essential effect of the system of administered prices and quantities in Germany was the possibility of consolidating sales. On the basis of cartel and tariff the German steel industry had been the first, unlike the British and at the time the Belgian steel industry as well, to promote a collective anti-cyclical sales policy in the seventies, whereby it was able to avoid dismantling its capacity, as would otherwise have been necessary. The success of this mutual action for rescuing superfluous production capacities during economic slumps – quite apart from the self-inflicted effects of rationalisation – had had vital consequences for the assessment of risk in later investment decisions, which ended by being completely disassociated from sales expectations. Feldenkirchen refers, for the beginning of the twentieth century, to examples from Hoesch and the Rheinische Stahlwerke, according to which an increase in the relevant cartel quota provided the sole motive for purchasing new works, and no thought at all was given to whether the newly acquired capacities could be fully exploited.[93] At the same time, the two big basic Bessemer works in Cleveland were driven out of business, or were at least reduced to subsistence levels, by the systematic dumping of German exports. That this was not the result of inadequate capacity for innovation on their part has been demonstrated by the foregoing comparative inquiry into the German and British steel industries to the beginning of the nineties, when the technical structures of the two industries were already firmly established.

In that our investigation has excluded technical superiority on the part of the German steel industry as one cause for their expansion at the expense of the British, it refers all the more clearly to the industry's collective behaviour and to the national economic policy, especially to customs policies, as dominant factors in this process. Although modern steel technology with the Bessemer, Thomas and Siemens-Martin processes was the basis for this shift in the industrial balance in Europe, it was, however, not the determinant quantity. It was not the way steelmaking processes developed technically which created Germany's growing strength on the world markets, but the industry's tight organisation, formed in the 1870s with the support of the banks and the state, which enabled it to suspend the market as a regulatory mechanism, in the majority of cases indeed forcing them to do so given that they were otherwise threatened with bankruptcy. The new steel technology did play a role here, in as much as the very reduced divisibility of plant, compared with the old puddling process, together with the considerable cost degression linked to the level of capacity utilisation, made it all the harder to adapt to dwindling sales.

Their fear of going down together had kept the German steel industrialists' coalition of interests together in our research period and very certainly beyond it too. One may speculate about the social and psychological effects of this constellation on entrepreneur and manager. It will, however, have certainly made a difference whether an industry reacted to its shrinking internal market by stepping up expansion on the world markets at the expense of its internal consumers, or by occasionally throttling production in concert, as in Great

93 Feldenkirchen, *Kapitalbeschaffung*, p. 43 notes 17 and 18.

Britain. The pattern of reaction evinced by the collective behaviour of the German steel industry was coined at a time of persistent subordination. Indeed, they did not lose this coinage even when they had long surpassed the quantity produced by the British competition. Their newly-won leading position in Europe and the stability of their industry were not the results of entrepreneurial autonomy and were consequently always felt to be precarious.

Sources and bibliography

1 UNPUBLISHED SOURCES

(A) GREAT BRITAIN

1 National archives

Glamorgan County Record Office, Cardiff
The Dowlais Iron Company Papers
Gwent County Record Office, Cwmbran
The Blaenavon Iron Company Papers
National Library of Wales, Aberystwyth
The Cyfarthfa Papers (Crawshay Dynasty)
The Tredegar Iron Works Papers
City of Wakefield Metropolitan District Archives
Goodchild loan MSS (Joseph Abbott MSS/Nantyglo & Blaina)

2 Company archives

British Steel Corporation Headquarters, London
The Cleveland Iron Masters' Association Records
British Steel Corporation East-Midlands Regional Records Centre, Irthlingborough
The Samuel Fox & Company Papers
British Steel Corporation Northern Regional Records Centre, Stockton (now Middlesbrough)
The Bell Brothers Iron Company Papers
The Consett Iron Company Papers
The Workington Iron Company Papers
British Steel Corporation Scottish Regional Records Centre, Glasgow
The Steel Company of Scotland Papers

(B) GERMANY

1 National archives

Bundesarchiv, Koblenz
Bestand R 13 I, Verein Deutscher Eisen- und Stahlindustrieller

2 Company archives

Historisches Archiv der Gutehoffnungshütte Aktienverein, Oberhausen
Archivmaterial der Hüttengewerkschaft und Handlung Jacobi, Haniel und Huyssen
Archivmaterial der GHH Oberhausen

Friedrich Krupp Hüttenwerke, Werk Bochum, Werkarchiv
Bestand Bochumer Verein
Bestand Gesellschaft für Stahlindustrie, Bochum (formerly Daelen, Schreiber und Co.)
Historisches Archiv der Fried. Krupp GmbH, Villa Hügel, Essen
Bestand Fried. Krupp
Mannesmann-Archiv, Düsseldorf
Bestand Phönix AG für Bergbau und Hüttenbetrieb
Archiv der Thyssen Industrie AG, Essen
Bestand Rheinische Stahlwerke

(C) FRANCE

Archives Nationales Paris
Série C: Assemblée Nationale et Chambre des Députés, Ia, projets de lois, 1871–1885, no. 3223

2 PUBLISHED SOURCES

(A) GREAT BRITAIN

Board of Trade, Returns of Iron and Steel Imports and Exports, in: Parliamentary Papers, Sessional Papers
British Iron Trade Association, *Annual Reports*
Mineral Statistics, ed. R. Hunt, in: Geological Survey until 1881, after which in: Parliamentary Papers, Sessional Papers
Royal Commission on the Depression in Trade and Industry, Minutes of Evidence and Reports, in Parliamentary Papers 1886
Select Committee on Companies Acts of 1862 and 1867, in: Parliamentary Papers 1877

(B) GERMANY

Protokolle über die Vernehmung der Sachverständigen durch die Eisen-Enquête-Commission, Berlin 1878
Statistisches Jahrbuch für das Deutsche Reich, publ. Kaiserliches Statistisches Amt, Berlin 1881
Statistisches Jahrbuch für das Deutsche Reich, publ. Kaiserliches Statistisches Amt, Berlin 1907
Die Produktion der Bergwerke, Salinen und Hütten im Deutschen Reich und in Luxemburg, in: *Vierteljahreshefte zur Statistik des Deutschen Reiches* until 1876, after which in: *Monatshefte zur Statistik des Deutschen Reiches*
Produktion der Hütten in dem preussischen Staate, in: *Zeitschrift für das Berg-, Hütten- und Salinenwesen in dem preussischen Staate, Statistischer Ergänzungsband*

3 CONTEMPORARY PERIODICALS

Annales des mines
Berg- und hüttenmännische Zeitung

Economist
Engineering
Iron
Iron and Coal Trades Review
Journal of the Iron and Steel Institute
Revue universelle des mines, de la métallurgie, des travaux publics, des sciences et des arts appliqués à l'industrie
Stahl und Eisen
Zeitschrift für das Berg-, Hütten- und Salinenwesen in dem Preussischen Staate
Zeitschrift des berg- und hüttenmännischen Vereins für Steiermark und Kärnten
Zeitschrift des Vereines deutscher Ingenieure

4 SELECT BIBLIOGRAPHY

The simultaneous pursuit of technical, economical and historical lines of inquiry has meant that to document a comprehensive bibliography in each specific field would take up a disproportionate amount of space. Instead, the following summary has been kept as short as possible. It cannot and does not claim to be complete.

Allen, Robert C., International Competition in Iron and Steel, 1850–1913, in: *Journal of Economic History* 39 (1979), pp. 911–37

Almond, J. K., Making Steel from Phosphoric Iron in the Converter – The Basic-Bessemer Process in Cleveland in the Years from 1879; in: C. A. Hempstead (ed.), *Cleveland Iron and Steel, Background and 19th Century History* (The British Steel Corporation) 1979, pp. 177–230

Steel Production in North England Before 1880, in: ibid., pp. 157–76

Ansoff, H. Igor, *Corporate Strategy*, New York 1965

Babbage, Charles, *On the Economy of Machinery and Manufacture*, London 1832

Beck, Ludwig, *Die Geschichte des Eisens in technischer und kulturgeschichtlicher Beziehung*, vol. 5, Das 19. Jahrhundert von 1860 an bis zum Schluß, Braunschweig 1903

Zum fünfzigjährigen Jubiläum des Regenerativofens, in: *Stahl und Eisen* 26 (1906), pp. 1421–7

Bell, Isaac Lowthian, *Principles of the Manufacture of Iron and Steel*, London 1884

Bessemer, Henry, *An Autobiography*, London 1905

Billy, Edouard de, Note sur l'invention du procédé Bessemer pour la fabrication de l'acier, in: *Annales des mines* 14 (1868), pp. 17–46

Birch, Alan, *The Economic History of the British Iron and Steel Industry 1784–1879*, London 1967

Blaich, Fritz, *Kartell- und Monopolpolitik im Kaiserlichen Deutschland. Das Problem der Marktmacht im deutschen Reichstag zwischen 1879 und 1914*, Düsseldorf 1973

Böhme, Helmut, *Deutschlands Weg zur Großmacht. Studien zum Verhältnis von Wirtschaft und Gesellschaft während der Reichsgründungszeit 1848–1881*, 2nd edn, Köln 1972

Boelcke, Willi A., *Krupp und die Hohenzollern in Dokumenten. Krupp-Korrespondenz mit Kaisern, Kabinettschefs und Ministern 1850–1918*, Frankfurt am Main 1970

Bower, James L., *Managing the Resource Allocation Process*, Boston 1970

Brandi, Hermann, Entwicklung der Thomasstahlerzeugung in Europa und die bauliche Ausgestaltung der Thomaswerke, in *Stahl und Eisen* 74 (1954), pp. 1262–7

Brandt, Gerhard, Bernard Kündig, Zissis Papadimetriou, Jutta Thomae, *Sozio-ökonomische Aspekte des Einsatzes von Computersystemen und ihre Auswirkungen auf die Organisation der Arbeit und der Arbeitsplatzstruktur, Bundesministerium für Forschung und Technologie*; Forschungsbericht DV77–04, Frankfurt am Main 1977

Brauns, Hermann, Die nordamerikanische und die deutsche Flußstahlerzeugung, in: *Ztschr. d. VDI* 21 (1877), cols. 91–4

Brügmann, W., The Progress and Manufacture of Pig Iron in Germany since 1880, in: *JISI* 1902 part 2, pp. 10–45.

Die Burbacher Hütte 1856–1906, Saarbrücken 1906

Burn, Duncan L., *The Economic History of Steelmaking, 1867–1939*, 2nd edn Cambridge 1961 (1st edn, 1940)

Burnham, T. H. and G. O. Hoskins, *Iron and Steel in Britain, 1870–1930*, London 1943

Byres, T. J., The Scottish Economy during the 'Great Depression', 1873–1896, with special reference to the Heavy Industries of the South-West, B.Litt. thesis, Glasgow 1962

Carnegie, David and Sidney C. Gladwyn, *Liquid Steel – its Manufacture and Cost*, New York 1913

Caron, François, Le Rôle des compagnies des chemins de fer en France dans l'introduction et la diffusion du procédé Bessemer, in: *L'Acquisition des techniques par les pays non initiateurs*, Pont-à-Mousson 1970, pp. 561–75

Carr, J. C. and W. Taplin, *History of the British Steel Industry*, Oxford 1962

Chandler, Alfred D., *Strategy and Structure. Chapters in the History of the Industrial Enterprise*, Cambridge, Mass. 1962

The Visible Hand. The Managerial Revolution in American Business, Cambridge, Mass. 1977

Checkland, S. G., *The Mines of Tharsis – Roman, French and British Enterprise in Spain*, London 1967

Clapham, John H., *An Economic History of Modern Britain*, vol. 2, Cambridge 1932

Colcough, T. P., The Constitution of Blast-Furnace Slags in Relation to the Manufacture of Pig Iron in: *JISI* 1936 part 2, pp. 547–86

Commissions-Bericht über den derzeitigen Stand der Entphosphorung des Eisens im Bessemer Converter nach Thomas'-Gilchrist's patentiertem Verfahren, in: *Ztschr. des berg- und hüttenmännischen Vereins für Steiermark und Kärnten* 12 (1880), pp. 217–60

Cossons, Neil, *The BP Book of Industrial Archaeology*, Newton Abbot 1978

Däbritz, Walther, *Bochumer Verein für Bergbau und Gußstahlfabrikation in Bochum, neun Jahrzehnte seiner Geschichte im Rahmen der Wirtschaft des Ruhrbezirks*, Düsseldorf 1934

Daelen, R. M., Über neuere Bessemerwerke, in: *Ztschr. d. VDI* 29 (1885), pp. 544–55 and 1016–18

Fortschritte in der Darstellung von Flußeisen und Flußstahl durch das Herdschmelzverfahren, in: *Ztschr. d. VDI* 28 (1884), pp. 641–3 and 925–7

Über die Fortschritte in der maschinellen Einrichtung der Bessemer-Stahlwerke, in: *Ztschr. d. VDI* 25 (1881), cols. 318–20 and 388–90

Zusammenstellung verschiedener Äußerungen über den Herdofen und das Herdofenschmelzen, in: *Stahl und Eisen* 12 (1892), pp. 12–17

Darby, J. H., The Manufacture of Basic Open-Hearth Steel, in: *JISI* 1889 part 1, pp. 78–83

Dermietzel, Otto, *Statistische Untersuchungen über die Kapitalrente der größeren deutschen Aktiengesellschaften von 1876–1902*, Diss. Göttingen 1906

Dickmann, Herbert, Die Einführung der Thomasverfahrens in Deutschland im Jahre 1879, in: *Stahl und Eisen* 74 (1954), pp. 1257–62

Dürre, Ernst Friedrich, Notizen über das Bessemerwerk zu Seraing (Aktiengesellschaft John Cockerill) mit besonderer Berücksichtigung einer späteren Verwendung des fabricierten Stahls, in: *ZBHSW* 18 (1870), pp. 262–73

Reisenotizen, betreffend die Anlage und den Betrieb der neu erbauten Bessemerhütte auf dem v. Arnim'schen Werke Königin Marienhütte zu Cainsdorf bei Zwickau, in: *ZBHSW* 15 (1867), pp. 256–62

Über die relativen Vorzüge des Cupoloofen- und des Flammofenbetriebes für das Einschmelzen der Bessemerchargen; mit besonderer Beziehung auf die Bessemerwerke von Rheinland-Westphalen, sowie die Cupoloofenprofile der Neuzeit, in: *Berg- und hüttenmännische Zeitung* 28 (1869), in instalments

Elsner, Kurt, Wachstums- und Konjunkturtheorie, in: Werner Ehrlicher et al. (eds.), *Kompendium der Volkswirtschaftslehre*, vol. 1, 4th edn, Göttingen 1975, pp. 246–96

Erickson, Charlotte, *British Industrialists: Steel and Hosiery 1850–1950*, Cambridge 1959

Fairbairn, William, *Iron – its History, Properties, & Processes of Manufacture*, 3rd edn, Edinburgh 1869

Faulkner, Henry Underwood, *American Economic History*, 8th edn, New York 1960

Feldenkirchen, Wilfried, Banken und Schwerindustrie im Ruhrgebiet. Zur Entwicklung ihrer Beziehungen 1873–1914, in: *Bankhistorisches Archiv* 1979, pp. 26–52

Die Eisen- und Stahlindustrie des Ruhrgebiets 1879–1914. Wachstum, Finanzierung und Struktur ihrer Großunternehmen, Wiesbaden 1982

Kapitalbeschaffung in der Eisen- und Stahlindustrie des Ruhrgebiets 1879–1914, in: *Ztschr. für Unternehmensgeschichte* 24 (1979), pp. 39–81

Die wirtschaftliche Rivalität zwischen Deutschland und England im 19. Jahrhundert, in: *Ztschr. für Unternehmensgeschichte* 25 (1980), pp. 77–107

Flinn, Michael W., British Overseas Investment in Iron Ore Mining 1870–1914, M.A. thesis, Manchester 1952

British Steel and Spanish Ore: 1871–1914, in: *Economic History Review*, 2nd ser., 8 (1955–6), pp. 84–90

Florence, Philip Sargant, *The Logic of British and American Industry*, rev. edn, London 1961

Ford, A. G., Overseas Lending and Internal Fluctuations 1870–1914, in: A. R. Hall (ed.), *The Export of Capital from Britain 1879–1914*, London 1968, pp. 84–102

Fritz, Martin, *Järnmalmsproduktion och Järnmalmsmarknad 1883–1913. De svenska exportföretagens produktionsutveckling, avsättningsinriktning och skeppningsförhallanden*, Göteborg 1967

Gale, W. K. V., The Bessemer Steelmaking Process, in: *Transactions of the Newcomen Society* 46 (1973/4), pp. 17–26

Gemeinfaßliche Darstellung des Eisenhüttenwesens, Verein Deutscher Eisenhüttenleute (Eds.), 14th edn, Düsseldorf 1937

Gonzalez Portilla, Manuel, *La formacion de la sociedad capitalista en el pais vasco (1876–1913)*, vol. 1, San Sebastian 1981

Gouvy, A., Die Flußeisenerzeugung auf basischem Herde in Resicza, in: *Stahl und Eisen* 9 (1889), pp. 396–403

Gruner, Louis, *The Manufacture of Steel*, translated from the French by Lenox Smith, A.M.E.M. with an appendix on The Bessemer Process in the United States by the translator, New York 1872

Gutenberg, Erich, *Grundlagen der Betriebswirtschaftslehre*, vol. 2: Der Absatz, 10th edn, Berlin 1967

Hackney, William, The Manufacture of Steel, in: *Excerpt Minutes of Proceedings of the Institution of Civil Engineers*, vol. 42, Session 1874–75 – Part IV

Harbord, F. W., The Basic Bessemer Process. Some Considerations of its Possibilities in England, in: *JISI* 1931 part 1, pp. 183-214

The Metallurgy of Steel, vol. 1: *Metallurgy*, London 1904

The Thomas-Gilchrist Basic Process, 1879–1937, in: *JISI* 1937 part 1, pp. 77–97

Hardach, Karl W., *Die Bedeutung wirtschaftlicher Faktoren bei der Wiedereinführung der Eisen- und Getreidezölle in Deutschland, 1879*, Berlin 1967

Harrison, J. K., The Development of a Distinctive 'Cleveland' Blast Furnace Practice 1866–1875, in: C. A. Hempstead (ed.), *Cleveland Iron and Steel, Background and 19th Century History* (The British Steel Corporation) 1979, pp. 81-115

The Production of Malleable Iron in North East England and the Rise and Collapse of the Puddling Process in the Cleveland District, in: ibid., pp. 117–55

Hartmann, Carl, *Der practische Puddel- und Walzmeister, oder Anleitung zum Verpuddeln des Roheisens mit Steinkohlen, Braunkohlen, Torf, Holz und brennbaren Gasen, sowie zur weiteren Verarbeitung des Puddeleisens und Puddelstahls zu Stabeisen aller Art, zu Eisenbahnschienen, Spurkranzeisen, zu Blech und Draht*, 2nd edn, Weimar 1861

Hatzfeld, Lutz, *Die Handelsgesellschaft Albert Poensgen Mauel-Düsseldorf. Studien zum Aufstieg der deutschen Stahlrohrindustrie 1850–72*, Cologne 1964

Hasslacher, Anton, *Der Werdegang der Rheinischen Stahlwerke*, Essen 1936

Healey, B. D., Manufacture of Cast Steel, in: *JISI* 1872 part 2, pp. 179ff

Heuss, Ernst, *Allgemeine Markttheorie*, Tübingen 1965

Heymann, Hans Gideon, *Die gemischten Werke im deutschen Großeisengewerbe, ein Beitrag zur Frage der Konzentration der Industrie*, Stuttgart 1904

Heyvaert, Hubert, *Stratégie et innovation dans l'entreprise*, Louvain 1973

Hoerder Bergwerks- und Hüttenverein, 50 Jahre seines Bestehens als Aktiengesellschaft, 1852–1902, Aachen 1902

Hogan, William T., *Economic History of the Iron and Steel Industry in the United States*, vol. 1, Lexington, Mass. 1971

Holland, C. B., On the Manufacture of Bessemer Steel and Steel Rails at the Works of Messrs Brown, Bayley and Dixon Limited, in: *Iron* 13 April 1878, pp. 456–7

and A. Cooper, On the Manufacture of Bessemer Steel and Ingot Iron from Phosphoric Pig, in: *JISI* 1880 part 1, pp. 79–88

Holley, Alexander Lyman, Bessemer Machinery, in: *Engineering*, 22 November 1872 in instalments

Setting Bessemer Converter Bottoms, in: *JISI* 1874 part 1, pp. 368–72

Holtfrerich, Carl-Ludwig, *Quantitative Wirtschaftsgeschichte des Ruhrkohlenbergbaus im 19. Jahrhundert*, Dortmund 1973

Howe, Henry Marian, *The Metallurgy of Steel*, 2nd edn, vol. 1, New York 1891

Hupfeld, Wilhelm, Die Ausdehnung der Bessemer- und Martinstahl-Fabrikation in Deutschland, in: *Ztschr. d. berg- und hüttenmännischen Vereins für Steiermark und Kärnten* 6 (1874), pp. 9–13

Der Einfluß der Roheisen-Entphosphorung auf die Weiterentwicklung der Bessemer-Industrie, in: ibid. 11 (1879), pp. 249–57

Eisen und Stahl im Jahre 1874, in: ibid. 7 (1875), in installments

Hyde, Charles K., *Technological Change and the British Iron Industry 1700–1870*, Princeton 1977

Illies, Hermann, Das Bessemerwerk der Königshütte, in: *Stahl und Eisen* 33 (1913), pp. 225–34

Jeans, J. S., *The Iron Trade of Great Britain*, London 1906

Steel: Its History, Manufacture, Properties and Uses, London 1880

Jeidels, Otto, *Das Verhältnis der deutschen Großbanken zur Industrie mit besonderer Berücksichtigung der Eisenindustrie*, Leipzig 1905

Jordan, Samson, *Notes sur la fabrication de l'acier Bessemer aux Etats-Unis d'après MM. Holley, Smith, etc.*, Paris 1873

Métallurgie du fer et de l'acier. Etudes pratiques et complètes sur les divers perfectionnements apportés jusqu'à ce jour dans la fabrication de ces deux métaux, Paris 1872

Kendall, J. D., The Iron Ores of Spain, in: *Transactions of the Federated Institution of Mining Engineers*, 3 (1891–2), pp. 604–16

Köstler, Hans-Jörg, Einführung und Beginn der Stahlerzeugung nach dem Bessemerverfahren in Österreich, in: *Berg- und hüttenmännische Monatshefte, vereinigt mit Montan-Rundschau*, 122 (1977), pp. 194–206

Krupp 1812–1912. Zum 100jährigen Bestehen der Firma Krupp und Gußstahlfabrik zu Essen/Ruhr, Essen 1912

Kunze, Walther, *Der Aufbau des Phoenix-Konzerns*, Diss. Frankfurt am Main 1926

Lambi, Ivo Nikolai, *Free Trade and Protection in Germany, 1868–79*, Wiesbaden 1963

Lancaster, J. Y. and D. R. Wattleworth, *The Iron and Steel Industry of West Cumberland. A Historical Survey* (BSC Teesside Division) 1977

Landes, David S., *Der entfesselte Prometheus. Technologischer Wandel und industrielle Entwicklung in Westeuropa von 1750 bis zur Gegenwart*, Köln 1973

Learned, Edmund P., C. Roland Christensen, Kenneth R. Andrews, William D. Guth, *Business Policy. Texts and Cases*, Homewood, Ill. 1965

Leese, Die Erhöhung des Gütertarifs der deutschen Eisenbahnen im Jahre 1874, in: *Jahrbuch für Gesetzgebung, Verwaltung und Volkswirtschaft 1893*, pp. 199-215

Liefmann, Robert, *Die Unternehmerverbände (Konventionen, Kartelle). Ihr Wesen und ihre Bedeutung*, Freiburg 1897

Lilge, F., Das Thomas-Stahlwerk Corby (England), in: *Mitteilungen aus den Forschungsanstalten des GHH-Konzerns* 4 (1936), pp. 147–56

Lilienberg, N., Meddelanden från utlandet, in: *Jernkontorets Annaler, Ny serie, Tidskrift för Svenska Bergshandteringen* 35 (1880), pp. 257–73

Löffelholz, Josef, *Repetitorium der Betriebswirtschaftslehre*, 5th edn, Wiesbaden 1975

Lord, W. M., The Development of the Bessemer Process in Lancashire, 1856–1900, in: *Transactions of the Newcomen Society* 25 (1945–7), pp. 163–80

Macar, J. de, Note sur les aciéries allemandes et belges en 1882, in: *Revue universelle des mines, de la métallurgie, des travaux publics, des sciences et des arts appliqués à l'industrie, 2ème sér.*, 12 (1882), pp. 143–78

McCloskey, Donald N., *Economic Maturity and Entrepreneurial Decline: British Iron and Steel, 1870–1913*, Cambridge, Mass. 1973

and L. G. Sandberg, From Damnation to Redemption: Judgements on the late Victorian Entrepreneur, in: *Explorations in Economic History* 9 (1971), pp. 89–108

Macrosty, H. W., *The Trust Movement in British Industry*, London 1907

Mai, Joachim, *Das deutsche Kapital in Rußland 1850–1894*, Berlin 1970

Martell, B., The Present Position occupied by Basic Steel as a Material for Shipbuilding, in: *Iron* 19 August 1887, pp. 176–7

Maurer, Eduard, Empirie, Produktionszahlen und Wissenschaft bei der Stahlherstellung, in: *Abhandlungen der Deutschen Akademie der Wissenschaften zu Berlin*, Klasse für Mathematik, Physik und Technik, No. 3 1961

Mayer, Karl Ernst and Helmut Knüppel, Entwicklungslinien des basischen Windfrischverfahrens in Deutschland, in: *Stahl und Eisen* 74 (1954), pp. 1267–75

Mensch, Gerhard, *Das technologische Patt. Innovationen überwinden die Depression*, Frankfurt am Main 1975

Milkereit, Gertrud, Das Projekt der Moselkanalisierung, ein Problem der westdeutschen Eisen- und Stahlindustrie, in: *Beiträge zur Geschichte der Moselkanalisierung*, Cologne 1967

Mönnich, Horst, *Aufbruch ins Revier – Aufbruch nach Europa. Hoesch 1871–1971*, München 1971

Mottek, Hans, *Die Ursachen der preußischen Eisenbahn-Verstaatlichung des Jahres 1879 und die Vorbedingungen ihres Erfolges*, Diss. Berlin 1950

Mowat, Charles Loch, *Britain between the Wars 1918–1940*, 2nd edn, London 1964

Müller, Friedrich C. G., Untersuchungen über den deutschen Bessemerprocess, in: *Ztschr. d. VDI* 22 (1878), pp. 385–404 and 453–70

Müller, H., *Georgs-Marien-Bergwerks- und Hütten-Verein, Osnabrück. Die Geschichte des Vereins*, Osnabrück 1896

Müller, Hermann, *Die Übererzeugung im Saarländer Hüttengewerbe von 1856 bis 1913*, Jena 1935

Müller, J. Heinz, Produktionstheorie, in: Werner Ehrlicher et al. (eds.), *Kompendium der Volkswirtschaftslehre*, vol. 1, 5th edn, Göttingen 1975, pp. 57-113

Orsagh, Thomas J., Progress in Iron and Steel: 1870–1913, in: *Comparative Studies in Society and History* 3 (1960–1), pp. 216–30

Osann, Bernhard, Lehrbuch der Eisenhüttenkunde, verfaßt für den Unterricht, den Betrieb und das Entwerfen von Eisenhüttenanlagen, vol. 2: *Erzeugung und Eigenschaften des schmiedbaren Eisens*, 2nd edn, Leipzig 1926

Payne, Peter L., *British Entrepreneurship in the Nineteenth Century*, London 1974

Colvilles and the Scottish Steel Industry, Oxford 1979

Iron and Steel Manufactures, in: Derek H. Aldcroft (ed.), *The Development of British Industry and Foreign Competition, 1875–1914*, London 1968, pp. 71–99

Pelatan, L., Eisenhüttenmännisches aus England, in: *Berg- und hüttenmännische Zeitung* 37 (1878), in instalments

Penrose, Edith T., *The Theory of the Growth of the Firm*, Oxford 1959

Platt, D. C. M., *Finance, Trade, and Politics in British Foreign Policy 1815–1914*, Oxford 1968

Pollard, Sidney and Paul Robertson, *The British Shipbuilding Industry, 1870–1914*, Cambridge, Mass. 1979

Pollock, D., *Modern Shipbuilding and the Men Engaged in it*, London 1884

Porter, Jeffrey Harvey, Industrial Conciliation and Arbitration 1860–1914, Ph.D. thesis, Leeds 1968

Pothmann, Wilhelm, *Zur Frage der Eisen- und Manganerzversorgung der deutschen Industrie*, Jena 1920

Rabius, Wilhelm, *Der Aachener Hütten-Aktien-Verein in Rote Erde, 1856–1913: die Entstehung und Entwicklung eines rheinischen Hüttenwerkes*, Jena 1906

Reuleaux, Franz, *Briefe aus Philadelphia*, Braunschweig 1877

Richards, A. W., Manufacture of Steel from High-Silicon Phosphoric Pig Iron by the Basic Bessemer Process, in: *JISI* 1907 part 1, pp. 104–8

Richardson, H. W. and J. M. Bass, The Profitability of Consett Iron Company before 1914, in: *Business History* 7 (1965), pp. 71–93

Riedel, Matthias, *Eisen und Kohle für das Dritte Reich. Paul Pleigers Stellung in der NS-Wirtschaft*, Göttingen 1973

Riley, James, On recent improvements in the method of the manufacture of open hearth steel, in: *JISI* 1884 part 2, pp. 443–52

Rosenberg, Nathan, Factors affecting the diffusion of technology, in: *JISI*, Perspectives on Technology, Cambridge 1976, pp. 189–210

Sandberg, C. P., De la spécification et de la réception des rails en Europe, in: *Revue universelle des mines, de la métallurgie, des travaux publics, des sciences et des arts appliqués à l'industrie*, 2ème sér. 10 (1880), pp. 505–75

Saul, S. B., The Export Economy, in: *Yorkshire Bulletin of Economic and Social Research* 17 (1965)

Studies in British Overseas Trade 1870–1914, Liverpool 1960

Schmalenbach, Eugen, *Selbstkostenrechnung und Preispolitik*, 3rd edn, Leipzig 1926

Schmidt, F., *Kalkulation und Preispolitik*, Berlin 1930

Scholl, Lars U., Im Schlepptau Großbritanniens. Abhängigkeit und Befreiung des deutschen Schiffbaus von britischem Know-how im 19. Jahrhundert, in: *Technikgeschichte* 50 (1983), pp. 213–23

Schürmann, Eberhard, Der Metallurge Henry Bessemer, in: *Stahl und Eisen* 76 (1956), pp. 1013–20

Schumpeter, Joseph A. *Geschichte der ökonomischen Analyse*, Göttingen 1965

Konjunkturzyklen, Göttingen 1961

Selznick, Paul, *Leadership in Administration*, Evanston 1957

Sering, Max, *Geschichte der preußisch-deutschen Eisenzölle von 1818 bis zur Gegenwart*, Leipzig 1882

Smith, Lennox, Der Bessemerprozeß in den Vereinigten Staaten, in: *Berg- und hüttenmännische Zeitung* 31 (1872), in instalments

Sonnemann, Rolf, *Die Auswirkungen des Schutzzolls auf die Monopolisierung der deutschen Eisen- und Stahlindustrie, 1879–1892*, Berlin 1960

Spiethoff, Arthur, *Die wirtschaftlichen Wechsellagen. Aufschwung, Krise, Stockung*, Tübingen 1955

Stillich, Oskar, *Die Eisen- und Stahlindustrie*, Berlin 1904

Styffe, Knut, Über Bessemern und Gußstahlfabrikation, in: *Berg- und hüttenmännische Zeitung* 27 (1868), pp. 169–70

Temin, Peter, *Iron and Steel in Nineteenth-century America*, Cambridge, Mass. 1964

The Relative Decline of the British Steel Industry, 1880–1913, in: Henry Roskovsky (ed.), *Industrialization in Two Systems: Essays in honor of Alexander Gerschenkron*, New York 1966, pp. 140-55.

Tholander, H., Über Stickstoffgehalt im Flußeisen und darauf begründete Vergleichungen zwischen Bessemer- und Herdflußeisen, in: *Stahl und Eisen* 9 (1889), pp. 115–21

Thomas, Sidney Gilchrist and Percy C. Gilchrist, The Manufacture of Steel and Ingot Iron from Phosphoric Pig-Iron, in: *Iron* 12 May 1882, pp. 372–5 and 26 May 1882, pp. 410-11

Thomas- und Martinwerke, in: *Stahl und Eisen* 6 (1886), in instalments

Treue, Wilhelm, *Die Geschichte der Ilseder Hütte*, Peine 1960

Troitzsch, Ulrich, Die Einführung des Bessemer-Verfahrens in Preußen – ein Innovationsprozeß aus den 60er Jahren des 19. Jahrhunderts, in: Frank R. Pfetsch (ed.), *Innovationsforschung als multidisziplinäre Aufgabe*, Göttingen 1975, pp. 209–40

Technische Rationalisierungsmaßnahmen in Eisenhüttenwesen während der Gründerkrise 1873–1879 als Forschungsproblem, in: *Hamburger Jahrbuch für Wirtschafts- und Gesellschaftspolitik* 24 (1979), pp. 283–96

Tunner, Peter, Amerikanische Verbesserungen am Bessemer-Converter, in: *Ztschr. d. berg- und hüttenmännischen Vereins für Steiermark und Kärnten* 7 (1875), pp. 233–6

Ulrich, Aust and Jänisch, Die Darstellung und Weitere Verarbeitung von Bessemerstahl in England. Bericht über eine im Jahre 1867 ausgeführte Instructionsreise, in: *ZBHSW* 16 (1868), pp. 1–29 and 131–42

Ulrich, Wiebmer and Dressler, Reisenotizen über den englischen Eisenhüttenbetrieb, in: *ZBHSW* 14 (1866), pp. 295–339

Wackernagel, W., *Berichte über die Verhandlungen des 17. Kongresses deutscher Volkswirte in Bremen 25, 26 and 28 September 1876*, Berlin 1876

Wagenblass, Horst, *Der Eisenbahnbau und das Wachstum der deutschen Eisen- und Maschinenindustrie 1835 bis 1860. Ein Beitrag zur Geschichte der Industrialisierung Deutschlands*, Stuttgart 1973

Wagon, Eduard, *Die finanzielle Entwicklung deutscher Aktiengesellschaften, 1870–1900*, Jena 1903

Walker, B., Description of the latest improvements in appliances for the manufacture of Bessemer Steel, in: *JISI* 1874 part 1, pp. 374–83

Warren, Kenneth, *The British Iron & Steel Sheet Industry since 1840. An Economic Geography*, London 1970

Webb, Steven B., Tariffs, Cartels, Technology and Growth in the German Steel Industry, 1879 to 1914, in: *Journal of Economic History* 40 (1980), pp. 309–39

Wedding, Hermann, Die Fortschritte des deutschen Eisenhüttenwesens seit 1876, mit besonderer Berücksichtigung des basischen Verfahrens, in: *Stahl und Eisen* 19 (1890), pp. 927–47

Die Resultate des Bessemer'schen Processes für die Darstellung von Stahl und Aussichten derselben für die rheinische und westfälische Eisen- resp. Stahlindustrie, in: *ZBHSW* 11 (1863), pp. 232–70

Versuche zur Entphosphorung des Roheisens in Königshütte, in: ibid. 14 (1866), pp. 155–9 and 272

Die Darstellung des schmiedbaren Eisens in praktischer und theoretischer Beziehung. Erster Ergänzungsband. Der Basische Bessemer- oder Thomas-Process, Braunschweig 1884

Wehler, Hans-Ulrich, *Bismarck und der Imperialismus*, 3rd edn, Cologne 1972

Wengenroth, Ulrich, Die Entwicklung der Kartellbewegung bis 1914, in: Hans Pohl (ed.), *Kartelle und Kartellgesetzgebung in Praxis und Rechtsprechung vom 19. Jahrhundert bis zur Gegenwart*, Stuttgart 1985, pp. 15–27

Raw material ventures: multinational activities in acquiring and processing of iron ore before World War I, European University Institute, Colloquium Papers 158/84, Florence 1984

Technologietransfer als multilateraler Austauschprozeß. Die Entstehung der modernen Stahlwerkskonzeption im späten 19. Jahrhundert, in: *Technikgeschichte* 50 (1983), pp. 224–37

Williams, Edward, The Manufacture of Rails, in: *Engineering*, 1 October 1869, p. 225

Wilson, A. S., The Consett Iron Company, Limited: A Case Study in Victorian Business History, M.Litt. thesis, Durham 1973

Wood, C., Statistics respecting the production and depreciation of rails, and notes on the application of wrought iron and steel to permanent ways, with a description of a new kind of railway sleeper and clipchair, in: *JISI* 1878 part 1, pp. 74–82

Woot, Philippe de, *Stratégie et management*, Paris 1970

Zweig, Emil, *Die russische Handelspolitik seit 1877*, Leipzig 1906

Index